# OPERATIONS RESEARCH INTRODUCTION AND APPLICATIONS

*By*

**Dr.Vandana Bagla**

(MSc (Maths), M.Phil.(Maths), MBA (HR), Ph. D. (O.R.))

Assistant Professor, Department of Applied Sciences

Maharaja Agrasen Institute of Technology, Rohini Sec-22, Delhi

*By*

**Mr. Naveen Solanki**

(B. Tech (MAE), M.E. (Thermal Engr.),Ph. D. (P))

Assistant Professor, Department of Mechanical and Automatic

Maharaja Agrasen Institute of Technology, Rohini Sec-22, Delhi

**STANDARD BOOK HOUSE**

*unit of* : **RAJSONS PUBLICATIONS PVT. LTD.**

1705-A, Nai Sarak, PB. No. 1074, Delhi-110006 Ph.: +91-(011)-23265506
Show Room: 4262/3, First Lane, G-Floor, Gali Punjabian, Ansari Road, Darya Ganj, New Delhi-110002 Ph.: +91-(011) 43751128 Tel Fax : +91-(011)43551185, Fax: +91-(011)-23250212

**E-mail: sbh10@hotmail.com www.standardbookhouse.in**

*Published by:*
**RAJINDER KUMAR JAIN**
Standard Book House
*Unit of:* Rajsons Publications Pvt. Ltd.
1705-A, Nai Sarak, Delhi - 110006
Post Box: 1074
Ph.: +91-(011)-23265506 Fax: +91-(011)-23250212

*Show room:*
4262/3, First Lane, G-Floor, Gali Punjabian
Ansari Road, Darya Ganj
New Delhi-110002
Ph.: +91-(011)-43751128, +91-(011)-43551185
E-mail: sbh10@ hotmail.com
www.standardbookhouse.in

**First Published : 2017**



*Price :* ₹ **320.00**

**ISBN: 978-81-89401-56-6**

*Typeset by:*
C.S.M.S. Computers, Delhi.

*Printed by:*
R.K. Print Media Company

# About the Authors

***Dr. Vandana Bagla***, is presently working in Maharaja Agrasen Institute of Technology (Affiliated to GGSIPU), Rohini, Delhi. After graduating in Computer Science from Delhi University, she completed her Masters in Mathematics and in Philosophy from Maharishi Dayanand Saraswati University, Ajmer and MBA in Human Resource Management from Bundelkhand University, Jhansi and thereafter Ph.D in Operations Research from Teerthanker Mahaveer University, U.P. She has got a number of publications in renowned journals / books and also she has launched a website named theengineeringmaths.com for ready reference of the students. Her areas of interest are Operations Research, Numerical Analysis and Engineering Mathematics.

***Mr. Naveen Solanki***, is Assistant Professor of Mechanical and Automation engineering Department at Maharaja Agrasen Institute of Technology affiliated to Guru Gobind Singh Indraprastha University, Dwarka, New Delhi. He obtained his Bachelor's in Mechanical and Automation Engineering from Guru Gobind Singh Indraprastha University, New Delhi and completed his Master's Degree from the Delhi College of Engineering, New Delhi. He is pursuing his Doctorate from Delhi Technological University (formerly Delhi College of Engineering), New Delhi. He has 8 years of teaching experience in the field of engineering along with 1 Yearof experience in Industry. He has taught various subjects related to mechanical engineering which includes Engineering Drawing (Engineering Graphics), Engineering Mechanics, Thermal Science,Heat Transfer, Refrigeration and Air-Conditioning, Renewable Energy Resources and Operation Research at Undergraduate course. He has published many papers in renowned journals and conferences and few of them are under review. He has also guided undergraduate students on various projects in the field of Refrigeration and Air-Conditioning.

# *Preface*

This book titled "Operations Research: Introduction and Applications"provides undergraduate and graduate students with basicconcepts, techniques and applications of linear programming and related topics. With this first edition,we have tried to meet the expectations ofthe students by describing methodologies used in operationsresearch effectively from the introductory level.With a strong emphasis on conceptual knowledge, the book providesworking methodologies along with illustrations and examples. Suitable for individual and group learning, it bestows numerous worked out examples and questions inquired in the preceding years.Practicing engineers and managers will find it pragmatic in industry related application problems. Level of the book has been kept moderately elementary and plain sailed to provide its' readers with lucidity and perceptibility. It is hoped that this book will be advantageous to the tutees and prove to be serviceable.

Dr. Vandana Bagla

Mr. Naveen Solanki

DEDICATED

To

**My loving father Sh. Jagan Nath Dhingra**

***Dr. Vandana Bagla***

**"I dedicate this book to my family for imparting skills that have shaped my career and professional life"**

***Mr. Naveen Solanki***

# Contents


**Chapter 1 Introduction to Operation Research** **1–6**

1.1 Introduction 1

1.2 Meaning and Definition of O.R. 1

1.3 History 2

1.4 Characteristics of Operation Research 3

1.5 Scope of Operation Research 4

1.6 Limitations of Operation Research 5

1.7 Why do We Study Operation Research 6

**Chapter 2 Linear Programming (Formulation and Graphical Method)** **7–31**

2.1 Introduction 7

2.2 Mathematical Formulation of LPP 7

2.3 Graphical methods for solving linear programming problems (2 variables) 14

2.3.1 Corner point method 16

*Previous Years Questions* 23

*Multiple Choice Questions* 29

*Summary* 31

**Chapter 3 Linear Programming (Simplex Method)** **32–64**

3.1 Introduction 32

3.2 Important Definitions and Results 32

3.2.1 Slack, Surplus and Artificial Variables 32

3.2.2 Basic Solution 33

3.2.3 Basic Feasible Solution 33

3.2.4 Non-Degenerate Basic Feasible Solution 33

3.3 Canonical and Standard Forms of LPP 35

3.4 Simplex Method (Algorithm) 35

3.4.1 Multiple Optima (Alternative Optimal Solution) 39

3.4.2 Unbounded Solution 41

3.5 Simplex Algorithm (Inqualities of < , > or = type) 42

3.5.1 Big M Method or Charne's Penalty Method 42

3.5.2 Two-phase Method 45

3.6 Degeneracy In Linear Programming Problems 49

3.6.1 Steps to Resolve Tie in Entering Variables 49

3.6.2 Steps to Resolve Tie in Leaving Variables 49

3.6.2.1 Perturbation Technique 49

*Previous Years Questions* 52

*Multiple Choice Questions* 61

*Summary* 62

**Chapter 4 Duality in Linear Programming 65–84**

4.1 Introduction 65

4.2 Procedure to Convert Primal into Dual LPP 65

4.3 Results on Duality (Dual-Primal Relationship) 69

4.4 Economic Interpretation of Duality 73

4.5 Advantages of Duality 73

4.6 Dual Simplex Method 74

*Previous Years Questions* 77

*Multiple Choice Questions* 83

*Summary* 84

**Chapter 5 Transportation Problems 85–94**

5.1 Introduction 85

5.2 Initial Basic Feasible Solution 86

5.2.1 North-West Corner Rule 87

5.2.1 North-West Corner Rule 87

5.2.2 Matrix-Minima or Least-Cost Method 90

5.2.3 Row-Minima Method 91

5.2.4 Column-Minima Method 92

5.2.5 Vogel Approximation Method (VAM) or Unit Cost Penalty Method 92

5.3 Modi (Modified Distribution) Method— Optimal Solution 94

5.4 Degeneracy in Transportation Problems 100

5.4.1 Resolution of Degeneracy During Initial Stage 100

5.4.2 Resolution of Degeneracy During Solution Stage 102

5.5 Unbalanced Transportation Problem 102

5.6 Maximization Transportation Problem 104

*Previous Years Questions* 106

*Multiple choice questions* 122

*Summary* 124

**Chapter 6 Assignment Problems** **126–164**

6.1 Introduction 126

6.2 Solution of Assignment Problem 127

6.2.1 Enumeration Method 127

6.2.2 Simplex Method 127

6.2.3 Transprotation Method 127

6.2.4 Hungarian Method 127

6.3 Miscellaneous Assignment Problems 141

6.3.1 Travelling Salesman Problem or Routing Problem 141

6.3.2 Scheduling Problem 145

*Previous Years Questions* 148

*Multiple choice questions* 163

*Summary* 163

**Chapter 7 Queueing Theory** **165–203**

7.1 Introduction 165

7.2 Queueing cost structure 165

7.3 Queueing System 166

7.3.1 Input Sources 166

7.3.2 Queueing Process 167

7.3.3 Queue Discipline 167

7.3.4 Service Mechanism 167

7.3.5 Customer Behaviour 168

7.4 State of the System 168

7.5 Kendall's Notation for Representing Queueing Models 169

7.5.1 Symbols and notations 169

7.6 Classification of queueing models Distribution of Arrival and Service Time 170

7.6.1 Solution of model–I [ (M/M/1):(μ/FCFS) ] (Birth and death process) 170

7.6.2 Model II. (M/M/2) : (μ/FCFC) Multiservice model (C=2) 185

*Previous Years Questions* 191

*Multiple choice questions* 199

*Summary* 201

**Chapter 8 Sequencing Problems** **204–225**

8.1 Introduction 204

8.2 Types of Sequencing Problems 205

8.2.1 Sequencing of n Jobs on two Machines 205

8.2.2 Sequencing of n Jobs on m Machines 210

8.2.3 Sequencing of 2 jobs on m machines (Graphical Method) 215

*Previous Years Questions* 218

*Multiple choice questions* 224

*Summary* 225

**Chapter 9 Network Analysis : PERT and CPM 226–257**

9.1 Introduction 226

9.2 Rules for Network Construction 227

9.3 Common Errors in Network Construction 227

9.4 Critical path method (C.P.M.) 230

9.4.1 Floats (Slack) 235

9.5 PERT (Project Evaluation And Review Technique) 237

9.6 Crashing of Netowrk (Time-cost Relationship) 245

*Previous Years Questions* 248

*Multiple Choice Questions* 255

*Summary* 256

**Chapter 10 Game Theory 258–293**

10.1 Introduction 258

10.2 Two Person Zero Sum Game 258

10.2.1 Strategy 258

10.2.2 Minimax and Maximin Principle 259

10.2.3 Summary of Systematic Methods for Solving Rectangular Games 261

10.3 Games without saddle point 262

10.3.1 Solution of 2 × 2 Game Without Saddle Point 262

10.3.2 Dominance Rules 264

10.3.3 Graphical Methods for 2 × n or m × 2 Games 267

10.3.4 Method of subgames for 2×n or m×2 game not reducible by dominance rules 271

10.3.5 Algebraic Method for (3×3) or higher order matrix game. 275

10.3.6 Matrix Method (Arithmetic method) or Method of Oddment for (3×3) Game 276

10.3.7 L.P. method for solving m×n matrix game 279

*Previous Years Questions* 283

*Multiple Choice Questions* 290
*Summary* 291

**Appendix A C++ Programming (Operation Research)** **294–333**

Formulation of LPP 294
Simplex Table 296
Simplex method 299
Primal Into Dual 306
Transportation Problem (North-West) 309
Assignment Problem 314
Sequencing Problem 321
Queuing Theory 326
PERT and CPM (Network Analysis) 327
Game Theory 331

**Appendix B Normal Distribution Tables** **334**

*Index* 337

CHAPTER 1

# Introduction to Operation Research

## 1.1 INTRODUCTION

Operation Research (O.R.) is an analytical method of problem solving and decision making that helps in the management of profit and non-profit organizations. The word 'Operation' may be defined as planned activity involving many people performing different actions to some problems, and the word 'Research' means an attempt to find the root of problem in a systematic manner. In other words O.R. is a disciplien dealing with the application of advanced analytical methods to find better solution of any problem.

MC Closky and Trefthen of U.K. were the first to use the term 'O.R.' in 1940, to solve the complex problems of military.

The tools of O.R. are not from any single discipline but formed with contribution of mathematics, statistics, economics, engineering, psychology etc. This newer discipline of knowledge, i.e., O.R., thus formed is a combination of above different disciplines. Nowadays O.R. has become a latest professional discipline which is getting so much popularity. It involves application of scientific methods for selection of resources, and decision making to the opportioning of scare resources.

## 1.2 MEANING AND DEFINITION OF O.R.

O.R. is a science which delas with problem formulation, solution and some appropriate decision making and O.R. may assist decision makers in almost any management function.

It is very difficult to define 'O.R.' as its foundaries and content are not marked clearly till now. Many writers have given different significance to the meaning of O.R. as it is not possible to limit the range of O.R. in few sentences. Some useful definitions are given below which shows the importance with the development of subject :

"*(O.R.) is the art of giving bad answers to problems which otherwise have more defective answers*". — **T.L. Saaty**

"*(O.R.) is the scientific method of providing executive departments with a quantitative basis for decisions under their control*".

— **Morse and Kimball (1946)**

"*(O.R.) is the systematic method-oriented study of the basic structure, characteristics functions and relationships of an organisation to provide the executive with a sound, scientific and quantitative basis for decision making*".

— **E.L. Arnoff and M.J. Netzorg**

"*(O.R.) is an aid for the executive in making his decisions by providing him with the needed quantitative information based on the scientific method of analysis*". — **C. Kittel**

"*(O.R.) is essentially a collection of mathematical techniques and tools which in conjuction with system approach, are applied to solve practical decision problems of an economic or engineering nature*".

— **Daellenbach and George**

"*(O.R.) is a scientific knowledge through interdisciplinary team effort for the purpose of determining the best utilization of limited resources*".

— **H.A. Taha**

"*(O.R.) is scientific approach to problem solving for executive management*".

— **H.M. Wager**

"*(O.R.) is conerned with scientifically deciding how to best design and operate man-machine system usually requiring the allocation of scare resources*".

— **Operation Research Society America**

## 1.3 HISTORY

There is no clear data that marks the birth of Operation Research, it is generally accepted that Operation Research came into existance in World War II and has expanded tremendously in the last few decades. The British Scientists setup the first field installations of Radar system during the war for detecting the aircrafts. Their analysis increased the effectiveness of British defence. In 1941

Operational Research Section (O.R.S.) was established in coastal command which was to carry out some of the most wellknown Operation Research work in World War II. Thus Operation Research as a seprate field of specialization was born.

The effectiveness of operation reserach in military spread interest in it to other governmental departments and industry. In the U.S.A. the National Research Council formed a committe on Operation Research in 1951, and the first book on the subject "Method of Operations Research", by Morse and Kimball, was published. In 1952 the Operations Research Society of America came into being.

In India, First Operation Research unit was established at the Regional Research laboratory at Hyderabad in 1949. Ans another group was setup in Defence Science laboratory. In India Statistical Institute, Kolkatta, an Operation Research unit was established. In 1953, for the applications of Operation Research methods in National Planning and Survey. A journal of Operation Research in the form of OPSEARCH was started in 1963.

During the year 1950, Operation Research achieved recognition as a subject for the universities. Since then the subject has gained more and more importance for the students of Management, Public Administration, Engineering, Mathematics, Economic and Commerce. The Operation Research techniques began to be applied more widely to problems in business industry and society. Today Operation Research is used by virtually every business and government throughout the world for decision-making at all levels.

## 1.4 CHARACTERISTICS OF OPERATION RESEARCH

(a) Operation Comprises of Inter-disciplinary team approach, to determine the solution of any problem.

(b) Operation Research uses scientific techniques to determine the optimum solution of any problem.

(c) Operation Research is a continuous process. It is not restricted to the application of Model to one problem, rather new problems can be created in other sector and decision can be implemented there also.

(d) Operation Research attempts to find optimal solution to the problem under consideration.

(e) Operation Research optimise the total output by maximizing the profit and minimizing the loss.

(f) Operationg Research Comprises of inter-disciplinary team approach, to determine the solution of any problem.

(g) Operation Research cannot give perfect answer to the problem, thus it improve only the quality of solution.

(h) Operation Research focuses on enhancing the creative ability of the decision maker.

(i) Operation Research reveal new problems for study methods.

(j) Operation Research is an decision making or problem solving science.

## 1.5 SCOPE OF OPERATION RESEARCH

Operation Research has very wide range of scope. It has entered successfully in different areas of research. Operation Research is useful in following important fields.

1. **In Agriculture.** With the sudden explosion of population every country is facing problem of optimum distribution of water for irrigation and land for croping. Thus a lot of work can be done to determine the best policies under the given restrictions.

2. **In Finance.** Due to economic crisis it has become very important for every government to effectively plan for the economic progress of the country. Operation Research techniques can be effectively applied to determine the profit plan, Replacement policies and maximize per capita income with least resources.

3. **In Industry.** Operation Research is very helpful to the industry director in deciding the optimal dispersion of several resources like men, machines, materials, money and time etc. to reach at optimum decision.

4. **In Marketing.** Operation Research techniques help the marketing Administrator to decide :

- Where to allocate the product for sale in order to minimise the transportation cost.
- Minimum per unit sale price.
- Best advertising media with respect to cost and time.
- How, when and what to purchase at minimum cost.
- The size of the stock to cater future need.

5. **In Personal Management.** Operation Research techniques can be used by a personal manager :

- to appoint efficient person at minimum salary.
- to determine the retirement age.

6. **In Production Management.** Operation Research techniques can be used by a personal manager :

- to determine the number of item to be produced.
- to design and locate the production site.

7. **In Research and Development.** Operation Research is also useful to research and development manager :

- to determine area of research.
- to prepare budget for research.
- for selection of individual projects, size of R & D. facilities.
- to determine cost and time.

## 1.6 LIMITATIONS OF OPERATION RESEARCH

Operation Research is very powerful tool of getting the best out of limited resources. It is so much useful but problem occur in its application. Some of the limitations are :

1. **Depend on electronic media.** Operation Research techinques are best to find out an optimal solution of any problem which takes into account all possible factors. The computations involved voluminous calculations for expressing and establishing relationship among large variables.

2. **Quantifies problem.** Due to large number of intangible factors, all problems cannot be qualified properly. In Operation Research techniques solution can be provided when all the elements related to problem can be qualtified. But all relevant variables do not lead to quantification. Thus, factors that can't be quantified find no place in Operation Research.

3. **Implementation by the manager.** Operation Research is a specialists job, that requires mathematicians or statisticians, who can formulate the models and recommend for implementation. But manager fails to understand the compler working of Operation Research. Thus, this creates a 'gap' between the two.

4. **Cost factor.** In Operation Research models, the basic data are subjected to frequent changes raising its cost factor. In Operation Research models need to be updated frequently as, a fairly good solution at present may be perfect but not efficient after sometimes.

5. **Implementations.** In Operation Research technique the implementation of decisions is very exquisitely fine task. It must take into account the complexities of human behaviour.

## 1.7 WHY DO WE STUDY OPERATION RESEARCH

- We have the convenience to solve real-world problems in areas such as mathematics, engineering, healthcare services, manufacturing industries, finance field, military, pharmaceuticals, law enforcement, consultancy, marketing, resource management and even entertainment business and numerous other fields.
- We don't have to adhere to a dominant worldview. O.R. has no restricted mode of professional practice, so we never have to be assorted into a specific technique that never changes.
- We become better strategist and develop better insight into looking at problems, creating models and setting up analysis that helps us to make better personal and professional decisions.
- We can have fun at work. The versatile techniques of O.R. are all proven with hundreds of successful case studies. Each application area is evolving and so we have consistent opportunities of learning new things.

CHAPTER 2

# Linear Programming (Formulation and Graphical Method)

## 2.1 INTRODUCTION

Linear Programming (LP) deals with the optimization (maximization or minimization) of a linear function subject to linear constraints. Typically a linear programming problem (lpp) consists of an objective function to be optimized, subject to linear equality or inequality constraints.

Linear programming had its early usage for military applications but now it is being employed widely for all types of managerial problems that arise in trade, industry and day-to-day decision making scenarios. In this chapter we shall discuss mathematical formulation of these problems. We also analyse the graphical methods for solving the linear programming problems which involve only two variables.

## 2.2 MATHEMATICAL FORMULATION OF LPP

**Step I :** Identify the decision variables and denote them by $x_1$, $x_2$, $x_3$, .........$x_n$.

**Step II :** Identify the objective function and express it as a linear function of decision variables.

**Step III :** Determine the set of constraints and express them as linear inequations or equations as per given conditions.

**Step IV :** Apply the non-negetivety restrictions on decision variables as per requirements.

Thus the Linear Programming Problem (LPP) in its standard form is usually expressed as

Maximize (Minimize) $Z = C_1x_1 + C_2 x_2 + ..... + C_nx_n$

Subject to

$$a_{11}x_1 + a_{12}x_2 + ......+ a_{1n}x_n \le (=, \ge) b_1$$
$$a_{21}x_1 + a_{22}x_2 + ......+ a_{2n}x_n \le (=, \ge) b_2$$
$$\vdots$$
$$a_{m1}x_1 + a_{m2}x_2 + ......+ a_{mn}x_n \le (=, \ge) b_m$$
$$x_1, x_2, ......, x_n \ge 0$$

Here $x_1, x_2, ....., x_n$ are called decision variables. The linear function $Z = C_1 x_1 + C_2 x_2 + ......+ C_nx_n$ which is to be optimized is called the objective function.

The inequations or equations which describe the conditions under which the optimization is to be accomplished are called constraints.

**Example 2.1** *A firm manufactures two types of products A and B and sells them at a profit of ₹5 on type A and ₹6 on type B. Each product is processed on two machines $M_1$ and $M_2$. Type A requires two minutes of processing time on $M_1$ and four minutes on $M_2$; type B requires three minutes on $M_1$ and two minutes on $M_2$. The machine $M_1$ is available for not more than 8 hours 30 minutes while machine $M_2$ is available for 10 hours during any working day. Formulate the problem as LPP.*

***Solution*** Let $x_1$ – number of units of product *A*

$x_2$ – number of units of product *B*

| | *Product A* ($x_1$) units | *Product B* ($x_2$) units | |
|---|---|---|---|
| **Objective Function : Profit** | ₹ 5 | ₹ 6 | **Objective : Maximization** |
| **Constraints** | | | **Availability (min)** |
| Machine $M_1$ | 2 min | 3 min | 510 |
| Machine $M_2$ | 4 min | 2 min | 600 |

**Objective Function**

$$\text{Maximize } (z) = 5x_1 + 6x_2$$

**Constraints**

$$2x_1 + 3x_2 \le 510 \text{ (Machine } M_1)$$
$$4x_1 + 2x_2 \le 600 \text{ (Machine } M_2)$$

Also since the number of units to be produced are non negative,

$$x_1 \geq 1; x_2 \geq 0$$

∴ LPP model is

$$\text{Maximize } (z) = 5x_1 + 6x_2$$

Subject to

$$2x_1 + 3x_2 \leq 510$$
$$2x_1 + x_2 \leq 300$$
$$x_1, x_2 \geq 0$$

**Example 2.2** *A fruit seller invests at most ₹ 1000 in apple and grapes. A box of apples costs ₹ 70 and a box of grapes costs ₹ 80. He can sell a box of apples at a aprofit of ₹10 per box and that of grapes at a profit of ₹15 per box. He has a space to store at most 12 boxes and as per market demand, he wants to purchase apples and grapes boxes in the ratio 1 : 2 respectively. Assuming that he can sell all the items that he purchases, how many boxes of apples and grapes should he buy, so as its maximize the profit. Formulate the problem as LPP.*

***Solution*** Let $x_1$ – The number of boxes of apples

$x_2$ – The number of boxes of grapes

| | *Apple boxes* ($x_1$) | *Grape boxes* ($x_2$) | |
|---|---|---|---|
| **Objective Function : Profit** | ₹ 10 | ₹ 15 | **Objective : Maximization** |
| **Constraints** | | | **Availability** |
| Cost | ₹ 70 | ₹ 80 | ₹ 1,000 |
| Boxes | 1 | 2 | 12 boxes |

**Objective Function**

$$\text{Maximize } (z) = 10x_1 + 15x_2$$

**Constraints**

$$70x_1 + 80x_2 \leq 1000 \text{ (Cost)}$$
$$x_1 + 2x_2 \leq 12 \text{ (Boxes)}$$

Also since number of boxes to be purchased are non-negative

$$x_1 \geq 0, \ x_2 \geq 0$$

∴ LPP model is

$$\text{Maximize } (z) = 10x_1 + 15x_2$$

Subject to

$$7x_1 + 8x_2 \leq 100$$
$$x_1 + 2x_2 \leq 12$$
$$x_1, x_2 \geq 0$$

**Example 2.3** *A firm produces three types of toys, say $T_1$, $T_2$ and $T_3$ each with cost of production ₹200, ₹ 400 and ₹ 300 respectively manufacturing of these require three machines A, B and C. $T_1$ requires 3 hours on machine A, 2 hours on machine B and 1 hour on machine C. Each piece of $T_2$ requires 4 hours on machine A, 3 hours on machine B and 3 hours on machine C, while a piece of $T_3$ needs 5 hours on machine A, 2 hours on machine B and 2 hours on machine C. The total time available per week on machine A is 80 hours, on machine B is 70 hours and on machine C is 60 hours. Formulate the problem as LPP so as to minimize the cost.*

***Solution*** Let

$x_1$ – Number of pieces of $T_1$ per week
$x_2$ – Number of pieces of $T_2$ per week
$x_3$ – Number of pieces of $T_3$ per week

| | $T_1$ $(x_1)$ | $T_2$ $(x_2)$ | $T_3$ $(x_3)$ | |
|---|---|---|---|---|
| **Objective function** | | | | **Objective** |
| Cost | ₹ 200 | ₹ 400 | ₹ 300 | Minimization |
| **Constraints** | | | | **Available time (hour)** |
| Machine *A* | 3 hr | 4 hr | 5 hr | 80 |
| Machine *B* | 2 hr | 3 hr | 2 hr | 70 |
| Machine C | 1 hr | 3 hr | 2 hr | 60 |

**Objective Function**

$$\text{Minimize } (z) = 200\, x_1 + 400\, x_2 + 300\, x_3$$

**Constraints**

$$3x_1 + 4x_2 + 5x_3 \leq 80 \text{ (Machine } A)$$
$$2x_1 + 3x_2 + 2x_3 \leq 70 \text{ (Machine } B)$$
$$x_1 + 3x_2 + 2x_3 \leq 60 \text{ (Machine } C)$$

Also since number of pieces produced can not have negative values,

$$x_1 \geq 0, x_2 \geq 0, x_3 \geq 0$$

$\therefore$ LPP model is

$$\text{Minimize } (z) = 200\, x_1 + 400\, x_2 + 300x_3$$

Subject to

$$3x_1 + 4x_2 + 5x_3 \leq 80$$
$$2x_1 + 3x_2 + 2x_3 \leq 70$$
$$x_1 + 3x_2 + 2x_3 \leq 60$$
$$x_1, x_2, x_3 \geq 0$$

**Example 2.4** *A firm manufactures two type of dolls A and B. The respective profits on each type of doll are ₹ 40 and ₹ 30. Each doll of type A requires twice as much time as a doll of type B and the firm could make 1000 dolls per day of type B. The supply nof plastic is sufficient for only 800 dolls per day (both A and B combined). Doll A requires a fancy dress and only 500 dresses per day are available wheras 600 dresses per day are available for doll B. Formulate the problem as lpp to miximize the profit.*

***Solution*** Let $x_1$ – Number of dolls per day of type $A$

$x_2$ – Number of dolls per day of type $B$

| | *Type A* ($x_1$) | *Type B* ($x_2$) | |
|---|---|---|---|
| **Objective Function : Profit** | ₹ 40 | ₹ 30 | **Objective : Maximization** |
| **Constraints** | | | **Availability** |
| Time | $2t$ | $t$ | 1000 $t$ |
| | (If $t$ in the time required to produce a doll of type $B$) | | |
| Plastic | 1 | 1 | 800 |
| Fancy Dress for $A$ | 1 | – | 500 |
| Fancy Dress for $B$ | – | 1 | 600 |

**Objective Function**

$$\text{Maximize } (z) = 40x_1 + 30x_2$$

**Constraints**

$$2tx_1 + tx_2 \leq 1000\,t \text{ (Time)}$$
$$x_1 + x_2 \leq 800 \text{ (Plastic)}$$
$$x_1 \leq 500 \text{ (Fancy dress for type } A)$$
$$x_2 \leq 600 \text{ (Fancy dress for type } B)$$

Also since number of dolls to be produced of type $A$ and $B$ are non-negative,

$$x_1 \geq 0,\ x_2 \geq 0$$

∴ LPP model is

$$\text{Maximize } (z) = 40x_1 + 30x_2$$

Subject to

$$2x_1 + x_2 \leq 1000$$
$$x_1 + x_2 \leq 800$$
$$x_1 \leq 500$$
$$x_2 \leq 600$$
$$x_1, x_2 \geq 0$$

**Example 2.5** *A famer has 1000 acre farm. He can sell all tomatoes, potatoes or radish, he raises. The price he can obtain is ₹ 10 per kg of tomatoes and ₹ 5 per kg each for potatoes, and radish. The average yield per acre is 100 kg of tomatoes and radish and 200 kg of potatoes. Fertilizer is available at 75 paise per kg and amount required per acre is 200 kg each for tomatoes and radish and 100 kg for potatoes. Labour required for sowing, cultivating and harvesting per acre is 7 man-days for tomatoes, 6 man-days for potatoes and 5 man-days for radish. A total of 500 man-days of labour are available at ₹ 50 per man-day. Formulate this problem as LPP to maximize the famer's total profit.*

***Solution*** Let

$x_1$ – Acres of land to be allotted for tomatoes

$x_2$ – Acres of land to be allotted for potatoes

$x_3$ – Acres of land to be allotted for radish

| | *Tomatoes* ($x_1$ *acres*) | *Potatoes* ($x_2$ acres) | *Radish* ($x_3$ acres) | |
|---|---|---|---|---|
| **Objective function**<br>**Profit** | | | | **Objective**<br>**Maximization** |
| Selling Price (per kg)<br>Yield (per acre)<br>Fertilizes (per acre) | ₹ 10<br>100 kg<br>200 kg | ₹ 5<br>200 kg<br>100 kg | ₹ 5<br>100 kg<br>200 kg | Fertilizer cost<br>(₹ 75 per kg) |
| Labour (man-days) | 7 | 6 | 5 | Labour cost<br>(₹ 50 per man-day) |
| **Constraints** | | | | **Availability** |
| Land | 1 | 1 | 1 | 1000 acres |
| Labour (per acre) | 7 man-days | 6 man-days | 5 man-days | 500 man-day |

**Objective Function**

$$\begin{aligned}\text{Profit} &= \text{Selling price} - \text{expenditure on fertilizer} \\ &\quad - \text{expenditure on labour} \\ &= [10\,(100\,x_1) + 5\,(200\,x_2) + 5\,(100\,x_3)] \\ &\quad - \left[\frac{3}{4}(200x_1) + \frac{3}{4}(100x_2) + \frac{3}{4}(200x_3)\right] \\ &\quad - [50\,(7x_1) + 50\,(6x_2) + 50\,(5x_3)] \\ &= 500\,x_1 + 625\,x_2 + 100\,x_3\end{aligned}$$

**Objective function is**

$$\text{Maximize } (z) = 500\,x_1 + 625\,x_2 + 100\,x_3$$

**Constraints**

$$x_1 + x_2 + x_3 \leq 1000 \text{ (Land)}$$

$$7x_1 + 6x_2 + 5x_3 \leq 500 \text{ (Man-days)}$$

Also since area of land allotted can not be negative,

$$x_1 \geq 0, x_2 \geq 0, x_3 \geq 0$$

$\therefore$ LPP model is

$$\text{Maximize } (z) = 500\,x_1 + 625\,x_2 + 100\,x_3$$

Subject to

$$\begin{aligned}x_1 + x_2 + x_3 &\leq 1000 \\ 7x_1 + 6x_2 + 5x_3 &\leq 500 \\ x_1, x_2, x_3 &\geq 0\end{aligned}$$

**Example 2.6** *A factory owner wishes to hire extra workers during the Deepawali season due to extra demand. A man is paid ₹60 per day and a woman is paid ₹50 per day. Because of limited space and budgetary conditions, the number of temporary workers can not exceed 20. As per past experience a man can prepare 250 candles and 100 dias per day while a woman can prepare 200 candles and 130 dias per day. According to market demand the volume of extra candles and dias may not be less than 2000 and 1000 respectively per day. Formulate the problem as LPP is as to how many men and women would he hire to pay them minimum amount.*

***Solution*** Let $x_1$ - Number of men to be hired

$x_2$ - Number of women to be hired

| | *Men* ($x_1$) | *Women* ($x_2$) | |
|---|---|---|---|
| **Objective Function : Cost** | | | **Objective : Minimize** |
| Cost | ₹ 60 | ₹ 50 | |
| **Constraints** | | | **Rerquirement** |
| Workers | 1 | 1 | 20 |
| Candles | 250 | 200 | 2000 |
| Dias | 100 | 130 | 1000 |

**Objective Function**

$$\text{Minimize } (z) = 60x_1 + 50x_2$$

**Constrants**

$$x_1 + x_2 \leq 20 \text{ (Maximum 20 workers)}$$

$$250x_1 + 200x_2 \geq 2000 \text{ (Minimum 2000 candles)}$$

$$100x_1 + 130x_2 \geq 1000 \text{ (Minimum 1000 dias)}$$

Also since number of men and women can not be negative,

$\therefore$ $$x_1 \geq 0, x_2 \geq 0$$

$\therefore$ LPP model is

$$\text{Minimize } (z) = 60x_1 + 50x_2$$

Subject to

$$x_1 + x_2 \leq 20$$

$$250x_1 + 200x_2 \geq 2000$$

$$100x_1 + 130x_2 \geq 1000$$

$$x_1, x_2 \geq 0$$

## 2.3 GRAPHICAL METHODS FOR SOLVING LINEAR PROGRAMMING PROBLEMS (2 VARIABLES)

In the previous section, a mathematical model for a Linear Programming Problem (LPP) was incepted. Before coming to graphical solutions of these problems, let us have a look on some definitions, which will be frequently used further.

*Feasible solution* : A solution of the Linear Programming Problem which also satisfies its non-negatively restrictions, is called its feasible solution .

*Optical Soltuion* : A feasible solution which optimizes (maximizes, or minimizes) the objective function of the Linear Programming Problem (LPP) is called an optimal solution.

**Convex Set :** A convex set is a set of points such that, given any two points

$P,Q$ in that set, the line $PQ$ joining them lies entirely within that set. In Fig. 2.1 some examples are shown of convex and non-convex regions.

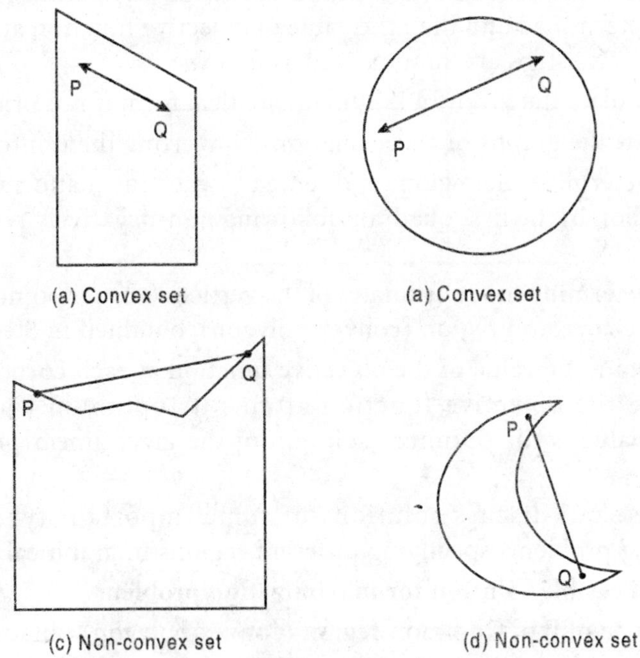


(a) Convex set (a) Convex set

(c) Non-convex set (d) Non-convex set

**Fig. 2.1**

The statements of the following theorem, are relevant while obtaining the optimal solution of a linear programming problem.

THEOREM **1.1.** *The collection of all feasible solution to a linear programming problem constitute a convex set whose extreme points correspond to the basic feasible solutions.*

THEOREM **1.2** *An optimum solution to a linear programming problem if it exists, occurs at one of the corner points of the feasible region.*

THEOREM **1.3** *There are finite number of basic feasible solutions with in the feasible solution space.*

THEOREM **1.4** *If the optimal solution occurs at more than one extreme point, i.e. an alternate optimal solution exists, then the value of the objective function will be the same for all convex combinations of these extreme points.*

Graphical methods can be used for solving linear programming problems having two variables only. The following methods are used to solve linear programming problems graphically:

(i) Corner-Point Method

(ii) ISO-Profit or ISO-Cost Method.

### 2.3.1 CORNER POINT METHOD

In this method, the coordinates of all corner (extreme) points of the feasible region are determined and then the value of objective function at these points is computed. The steps are summarized as follows:

**Step I :** Formulate the given LPP in mathematical form if not originally.

**Step II :** Draw the graphs of all inequations converting them into equations.

**Step III :** Determine the region represented by each inequation and mark the common region in the first quadrant following non-negatively restrictions on variables.

**Step IV :** Determine the coordinates of the vertices (corner points or extreme points) of the common region (convex polygon), obtained in Step III.

**Step V :** Obtain the value of the objective function at each corner point. The point where the objective function attains its optimum (maximum or minimum) value is the optimum solution of the given linear programming problem.

Now we shall discuss solutions of some important types of linear programming problems specifying different regions in graphical plane.

**Type I : Feasible solution for maximization problem**

In this type of lpp, Common region (convex polygon) obtained is closed and bounded.

**Example 2.7** *A jeweller produces two types of ornaments A and B, that require gold and silver. Each unit of type A requires 3 gm of silver and 1 gm of gold while that of type B requires 1 gm of silver and 2 gm of gold. The jeweller can obtain 9 gm of silver and 8 gm of gold per day. If each unit of the type A brings a profit of ₹ 40 and that of type B ₹ 50, find the number of units of each type that the jeweller should produce per day to maximize the profit. What is the maximum profit ?*

***Solution*** Let $x_1$ – Number of ornaments of type *A*

$x_2$ – Number of ornaments of type *B*

| | *Type A* ($x_1$ units) | *Type B* ($x_2$ units) | |
|---|---|---|---|
| **Objective Function : Profit** | ₹ 40 | ₹ 50 | **Objective : Maximization** |
| **Constraints** | | | **Availability (gm)** |
| Silver (gm) | 3 | 1 | 9 |
| Gold (gm) | 1 | 2 | 8 |

**Objective Function**

$$\text{Maximize } (z) = 40x_1 + 50x_2$$

**Constraints**

$$3x_1 + x_2 \leq 9 \text{ (Silver availability)}$$
$$x_1 + 2x_2 \leq 8 \text{ (Gold availability)}$$

Also since number of units of type *A* and type *B* can not be negative,

$$x_1 \geq 0, x_2 \geq 0$$

*Formulating the LPP in mathematical form.*

$$\text{Maximum } (z) = 40x_1 + 50x_2$$

**Subject to**

$$3x_1 + x_2 \leq 9$$
$$x_1 + 2x_2 \leq 8$$
$$x_1, x_2 \geq 0$$

Plotting the lines $3x_1 + x_2 = 9$ and $x_1 + 2x_2 = 8$ and marking the common region in first quadrant taking into account non-negativity restrictions.

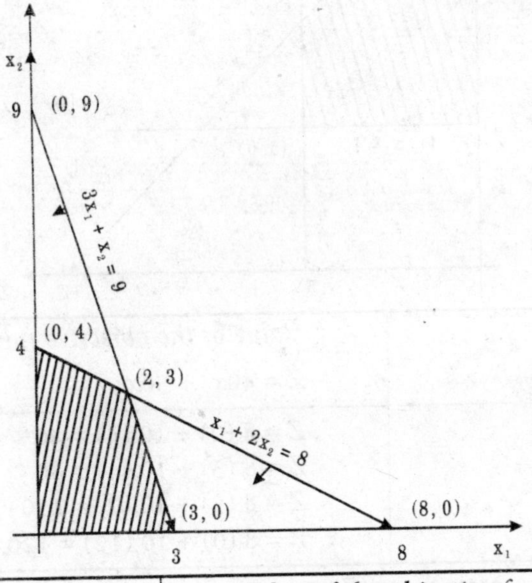


| Point $(x_1, x_2)$ | Value of the objective function $Z = 40x_1 + 50x_2$ |
|---|---|
| (3, 0) | $Z = 40\,(3) + 50\,(0) = 120$ |
| (2, 3) | $Z = 40\,(2) + 50\,(3) = 230$ |
| (0, 4) | $Z = 40\,(0) + 50\,(4) = 200$ |

Clearly $Z$ is maximum at $x_1 = 2$, $x_2 = 3$ and maximum value of $Z$ is ₹ 230.

∴ The company should produce 2 units of *A* and 3 units of *B* to obtain a maximum profit of ₹ 230.

**Example 2.8** *Solve the LPP*

$$\text{Maximize } (z) = 8x_1 + 10x_2$$

*Subject to*

$$x_1 + x_2 \le 12$$
$$x_1 \le 5$$
$$x_2 \ge 4$$
$$x_1, x_2 \ge 0$$

***Solution*** Plotting the lines $x_1 + x_2 = 12$, $x_1 = 5$ and $x_2 = 4$ in a two demensional plane and shading the common region.

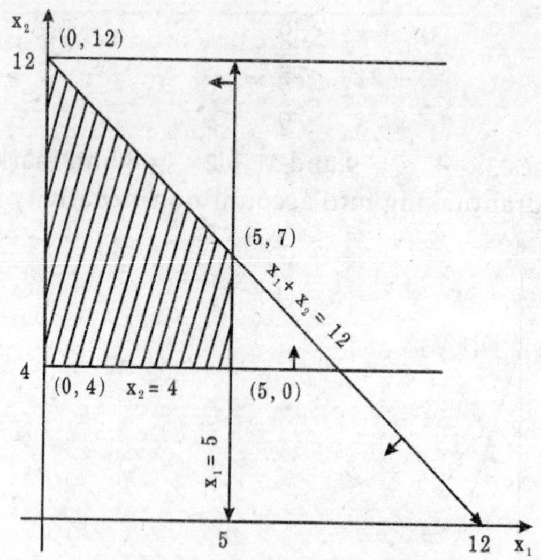


| Point $(x_1, x_2)$ | Value of the objective function $Z = 40x_1 + 50x_2$ |
|---|---|
| (0, 4) | $Z = 8\ (0) + 10\ (4) = 40$ |
| (5, 4) | $Z = 8\ (5) + 10\ (4) = 80$ |
| (5, 7) | $Z = 8\ (5) + 10\ (7) = 110$ |
| (0, 12) | $Z = 8\ (0) + 10\ (12) = 120$ |

$\therefore$ Maximum $(Z) = 120$

at $x_1 = 0$

$x_2 = 12$

**Type II : Feasible solution for minimization problem**

In this type of lpp. Common region obtained after plotting constraint inequation may be bounded or unbounded.

**Example 2.9** *A chemical company produces two compounds, A and B. The following table gives the units of ingredients C and D per kg of compound A and B as well as minimum requirements of C and D and costs per kg of A and B.*

*Find the quantities of A and B which would give a supply of C and D at a minimum cost.*

| | Compound | | Minimum Requirement |
|---|---|---|---|
| | A | B | |
| *Ingredient C* | 1 | 2 | 10 |
| *Ingredient D* | 3 | 1 | 15 |
| *Cost (₹) per kg* | 4 | 6 | |

**Solution** As per table, mathematical formulation of LPP is

$$\text{Maximize } (z) = 4x_1 + 6x_2$$

*Subject to*

$$x_1 + 2x_2 \geq 10$$
$$3x_1 + x_2 \geq 15$$
$$x_1, x_2 \geq 0$$

Plotting the lines $x_1 + 2x_2 = 10$ and $3x_1 + x_2 = 15$ and marking the common region, taking into account non-negativety restrictions.

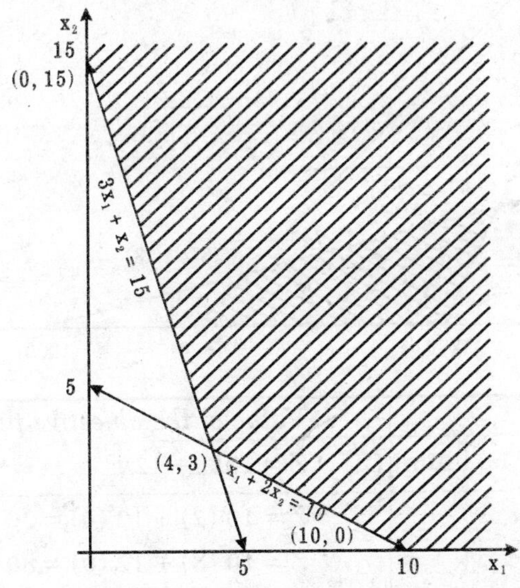


| *Point* $(x_1, x_2)$ | *Value of the objective function* $Z = 4x_1 + 6x_2$ |
|---|---|
| (10, 0) | $Z = 4\,(10) + 6\,(0) = 40$ |
| (4, 3) | $Z = 4\,(4) + 6\,(3) = 34$ |
| (0, 15) | $Z = 4\,(0) + 6\,(15) = 90$ |

$\therefore$ $Z$ is minimum at $x_1 = 4, x_2 = 3$

Hence the company should produce 4 kg of compound *A* and 3 kg of compound *B* which would give a minimum cost of ₹ 34.

**Example 2.10** *Solve the LPP*

$$\text{Minimize } (z) = 10x_1 + 12x_2$$

*Subject to*

$$2x_1 + x_2 \geq 4$$
$$x_1 + 4x_2 \leq 12$$
$$x_1 + x_2 \leq 8$$
$$x_1, x_2 \geq 0$$

***Solution*** Plotting the lines $2x_1 + x_2 = 4$, $x_1 + 4x_2 = 12$, $x_1 + x_2 = 8$ and identify the common region taking into account non-negativity restrictions.

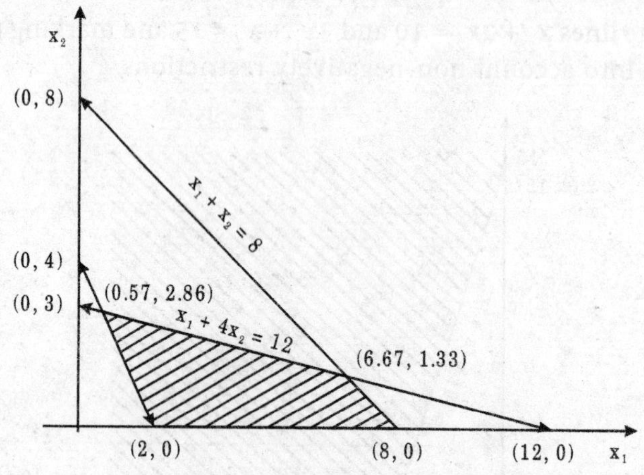


| *Point* $(x_1, x_2)$ | *Value of the objective function* $Z = 10x_1 + 12x_2$ |
|---|---|
| (2, 0) | $Z = 10\,(2) + 12\,(0) = 20$ |
| (8, 0) | $Z = 10\,(8) + 12\,(0) = 80$ |
| (6.67, 1.33) | $Z = 10\,(6.67) + 12\,(1.33) = 82.67$ |
| (0.57, 2.86) | $Z = 10\,(0.57) + 12\,(2.86) = 40$ |

$$\text{Minimize } (Z) = 20 \text{ at } x_1 = 2, x_2 = 0$$

**Type III : Linear programming problems having Infeasible solutions**

If in a LPP, there is no feasible solution area, i.e. there are no. points that satisfy all constraints of the problem, then it is said to have no solution or infeasible solution. An infeasible LPP with two decision variables can be identified through its graph.

**Example 2.11** *Solve the LPP*

$$\text{Maximize } (z) = 5x_1 + 8x_2$$

*Subject to*

$$x_1 + x_2 \geq 8$$
$$4x_1 + x_2 \leq 2$$
$$x_1, x_2 \geq 0$$

***Solution*** Plotting the lines $x_1 + x_2 = 8$ and $4x_1 + x_2 = 2$, there is no common region which satisfies the above constraint inequations in the first quadrant, as shown below :

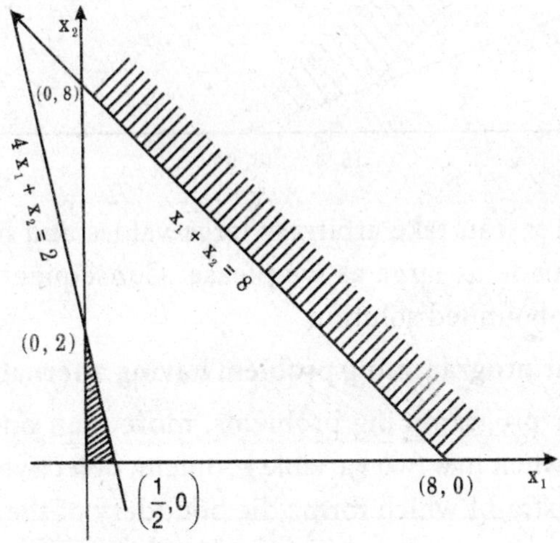


Clearly, The given LPP has no solution.

**Type IV : Linear programming problems having unbounded solution**

If in a maximization problem, the feasible region is unbounded, then one or more decision variables will increase indefinitely without violating feasibility conditions and objective function can be made arbitrarily large which is practically not possible.

**Example 2.12** *Solve the LPP*

$$\text{Maximize } (z) = 5x_1 + 6x_2$$

*Subject to*

$$x_1 + x_2 \geq 5$$
$$x_1 + 3x_2 \geq 6$$
$$x_1, x_2 \geq 0$$

***Solution*** Identifying the common region of the given constraint inequations, as shown in given figure.

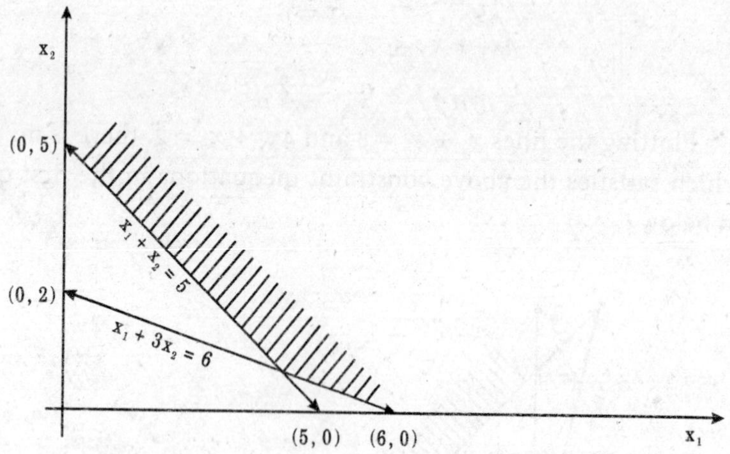


Clearly $x_1$ and $x_2$ can take arbitrary large values and hence the objective function can be made as large as we please. Consequently we say that the problem has an unbounded solution.

**Type V : Linear programming problem having alternate optimal solution**

In some linear programming problems, more than one optimal solution exists. It happens when in a two variable problem, objective function equation is parallel to a constraint which forms the boundary of the feasible region.

**Example 2.13** *Solve the LPP*

$$\text{Maximize } (z) = 6x_1 + 4x_2$$

*Subject to*

$$x_1 + x_2 \leq 5$$
$$3x_1 + 2x_2 \leq 12$$
$$x_1, x_2 \geq 0$$

***Solution*** Plotting the lines $x_1 + x_2 = 5$ and $3x_1 + 2x_2 = 12$ and identifying the common region.

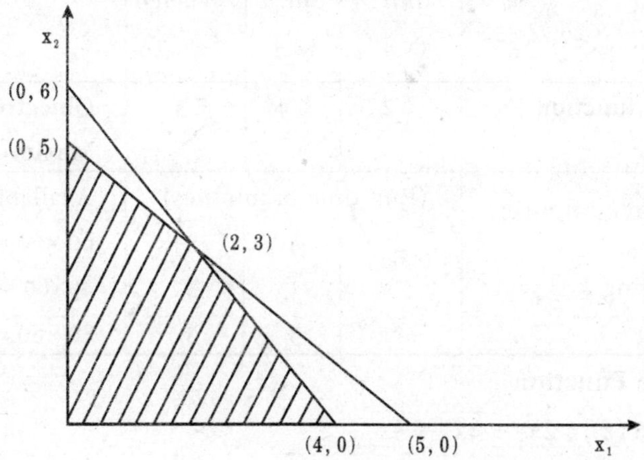


| *Point* $(x_1, x_2)$ | *Value of the objective function* $Z = 6x_1 + 4x_2$ |
|---|---|
| (4, 0) | $Z = 6(4) + 4(0) = 24$ |
| (2, 3) | $Z = 6(2) + 4(3) = 24$ |
| (0, 5) | $Z = 6(0) + 4(5) = 20$ |

$$\text{Maximize } (z) = 24 \text{ at } x_1 = 4, x_2 = 0 \text{ and } x_1 = 2, x_2 = 3$$

## PREVIOUS YEARS QUESTIONS

**Example 2.14** *A Company has three operational departments (Weaving, processing and packing) with capacity to produce three different types of clothes namely suiting, shirting and woolen yielding the profit of ₹ 2 ₹ 4 and ₹ 3 per metre respectively. One metre suiting requires 3 minutes in weaving, 2 minutes in processing and 1 minute in packing. Similarly one metre of shirting requires 4 minutes in weaving, 1 minute in processing and 3 minutes in packing while one metre woolen requires 3 minutes in each department. In a week, total run time of each department is 60, 40 and 80 hours for weaving, processing and packing departments respectively.*

*Formulate the linear programming problem to find the product mix to maximize the profit.* **[GGSIPU, I-Term, 2008]**

***Solution*** Let

$x_1$ – Number of suit

$x_2$ – Number of shirt

$x_3$ – Number of woolen

| | *Suit* ($x_1$) | *Shirt* ($x_2$) | *Woolen* ($x_3$) | |
|---|---|---|---|---|
| **Objective function Profit** | ₹ 2 | ₹ 4 | ₹ 3 | **Objective Maximization** |
| Constraints | (Run time in minutes) | | | Availability |
| 1. Weaving | 3 | 4 | 3 | 60 ×60 = 3600 minutes |
| 2. Processing | 2 | 1 | 3 | 40×60 = 2400 minutes |
| 3. Packing | 1 | 3 | 3 | 80 ×60 = 4800 minutes |

**Objective Function**

Maximize $(z) = 2x_1 + 4x_2 + 3x_3$

Constraints :

$$3x_1 + 4x_2 + 3x_3 \leq 3600 \text{ (Weaving department)}$$
$$2x_1 + 1x_2 + 3x_3 \leq 2400 \text{ (Processing department)}$$
$$1x_1 + 3x_2 + 3x_3 \leq 4800 \text{ (Packing department)}$$

Also since number of different types of clothes are non-negative.

$$x_1 \geq 0, x_2 \geq 0, x_3 \geq 0$$

∴ LPP model is

$$\text{Maximize } (z) = 2x_1 + 4x_2 + 3x_3$$

Subject to

$$3x_1 + 4x_2 + 3x_3 \leq 3600$$
$$2x_1 + x_2 + 3x_3 \leq 2400$$
$$x_1 + 3x_2 + 3x_3 \leq 4800$$
$$x_1, x_2, x_3 \geq 0$$

**Example 2.15** *A company has two grades of inspectors I and II to undertake quality control inspection. At least 1500 pieces must be inspected in an 8 hour day. Grade I inspector can check 20 pieces in an hour with an accuracy of 96%. Grade II inspector checks 14 pieces an hour with an accuracy of 92%. Wages of grade I inspector are ₹ 5 per hour while those of grade II inspector are ₹ 4 per hour. Any error made by an inspector costs ₹ 3 to the company. If there are in all 10 grade I inspectors and 15 grade II inspectors in the company, find the optimal assignment of inspectors that minimizes the daily inspection cost.*

**[GGSIPU, End-Term, 2012]**

***Solution*** Let $x_1$ – Grade I inspector

$x_2$ - Grade II inspectors

| | *Grade (I)* $x_1$ | *Grade (II)* $x_2$ | |
|---|---|---|---|
| **Objective Function : Cost** | | | **Objective : Minimization** |
| Wages Rate per hour | ₹ 5 | ₹ 4 | |
| Cost of inaccurate checking | ₹ 3 | ₹ 3 | |
| **Constraints** | | | **Availability** |
| No. of inspectors | 10 | 15 | 10 Grade I, 15 Grade II |
| Pieces checked per hour | 20 | 14 | 1500 pieces |
| Inaccuracy in checking | 1 – 0.96 = 0.04 | 1 – 0.92 = 0.08 | |

Duration of inspection = 8 hr per day

Total number of pieces which must be inspected = 1500

Hourly cost of grade I inspectors can be calculated as

$$₹\ (5 + 20 \times 0.04 \times 3) \Rightarrow ₹\ 7.40$$

Hourly cost of grade II inspectors can be calculated as

$$₹\ (4 + 14 \times 0.08 \times 3) \Rightarrow ₹\ 7.36$$

**Objective Function**

$$\text{Minimize } (z) = 8\,(7.40\,x_1 + 7.36\,x_2)$$

**Constriants**

$$8 \times 20x_1 + 8 \times 14\,x_2 \geq 1500$$

$$x_1 \leq 10$$

$$x_2 \leq 15$$

Since number of inspectors are non-negative,

$\therefore$ $$x_1 \geq 0, x_2 \geq 0$$

$\therefore$ **LPP model is**

$$\text{Minimize } (z) = 59.2\,x_1 + 58.88\,x_2$$

Subject to

$$116\,x_1 + 112\,x_2 \geq 1500$$

$$x_1 \leq 10$$

$$x_2 \leq 15$$

$$x_1, x_2 \geq 0$$

**Example 2.16** *Avisha wants to invest at most ₹25,000 in postal life insurance and public provident funds. According to rules, she has to invest at least ₹4,000 in postal life insurance and at least ₹8,000 in public provident funds. If the rate of interest on postal life insurance is 8% per annum and the rate of interest on public provident funds is 10% per annum, how much money should she invest to earn maximum yearly income. Formulate a linear programming model.*

***Solution*** Let

$x_1$ – Investment in PLI

$x_2$ – Investment in PPF

| | *PLI*<br>$x_1$ | *PPF*<br>$x_2$ | |
|---|---|---|---|
| **Objective Function : Profit** | | | **Objective : Minimization** |
| Rate of Interest | 8% | 10% | |
| **Constraints** | | | **Availability** |
| Investment | 4000 min | 8000 min | ₹ 25,000 |

**Objective Function**

$$\text{Maximize } (Z) = \frac{8}{100}x_1 + \frac{10}{100}x_2$$

**Constraint**

$$x_1 + x_2 \leq 25{,}000$$

$$x_1 \geq 4{,}000 \text{ (PLI)}$$

$$x_2 \geq 8{,}000 \text{ (PPF)}$$

Also investments in PLI and PPF are non negative

$\therefore$ $$x_1 \geq 0, x_2 \geq 0$$

$\therefore$ **LPP model is**

$$\text{Maximize } (z) = 0.08\, x_1 + 0.10\, x_2$$

Subject to

$$x_1 + x_2 \leq 25{,}000$$

$$x_1 \geq 4{,}000$$

$$x_2 \geq 8{,}000$$

$$x_1, x_2 \geq 0$$

**Example 2.17** *An Indian Airline aeroplane can carry a maximum of 400 passenger. A profit of ₹600 is made on each Business class ticket and a profit of ₹ 400 is made on each economy class ticket. The airline reserves at least 30 seats for business class. However, at least 5 times as many passengers perfer to travel by economy class to the Business class. Formulate the linear programming problem to determine how many each type of tickets must be sold in order to maximize the profit for the Indian Airlines.*

**Solution** Let $x_1$ – Tickets for Business class

$x_2$ – Tickets for Economy class

| | *Business class* | *Economy class* | |
|---|---|---|---|
| | $x_1$ | $x_2$ | |
| **Objective Function : Profit** | ₹ 600 | ₹ 400 | **Objective : Minimization** |
| **Constraints** | | | **Availability** |
| Seat reserved business class | 30 | — | |
| Seats | 1 | 1 | |
| Seats Ratio | 1 | Min 5 times | 400 passengers |

**Objective Function**

$$\text{Maximize } (z) = 600\, x_1 + 400\, x_2$$

**Constraint**

$$x_1 + x_2 \leq 400 \text{ (Passengers)}$$
$$x_1 \geq 30 \text{ (Reservation)}$$
$$x_2 \geq 5x_1 \text{ (Condition)}$$

Also since number of tickets to be sold are non negative,

$\therefore$ $$x_1 \geq 0, x_2 \geq 0$$

$\therefore$ **LPP model is**

$$\text{Maximize } (z) = 600\, x_1 + 400\, x_2$$

Subject to

$$x_1 + x_2 \leq 400$$
$$x_1 \geq 30$$
$$x_2 \geq 5x_1$$
$$x_1, x_2 \geq 0$$

**Example 2.18** *A firm manufactures two items. It purchases castings which are then machined, bored and polished. Costing for items A and B cost ₹ 2 and ₹ 3 respectively and are sold at ₹5 and ₹ 6 each respectively. Running* costs of the three machines are ₹ 20, ₹ 14 and ₹ 17.5 per hour respectively. Capacity of the machines are

| | *Part A* | *Part B* |
|---|---|---|
| *Machinery Capacity* | *25/hr* | *40/hr* |
| *Boring Capacity* | *28/hr* | *35/hr* |
| *Polishing Capacity* | *35/hr* | *25/hr* |

*Formulate the linear programming model to determine the product mix that maximizes the profit.*

***Solution*** Let $x_1$ – Number of units of item A

$x_2$ – Number of units of item B

| | *Item (A)* $x_1$ | *Item (B)* $x_2$ | |
|---|---|---|---|
| **Objective Function : Profit** | | | **Objective : Maximization** |
| Cost Price (CP) per unit | ₹ 2 | ₹ 3 | |
| Cost of Machinery (per unit) | $\frac{20}{25}=0.8$ | $\frac{20}{40}=0.5$ | * Cost of Machinery of 25 units of item $A$ = ₹ 20 |
| Cost of Boring (per unit) | $\frac{14}{28}=0.5$ | $\frac{14}{35}=0.4$ | ∴ Cost of Machinery of 1 unit of |
| Cost of Polishing (per unit) | $\frac{17.5}{35}=0.5$ | $\frac{17.5}{25}=0.7$ | item $A = \frac{20}{25}$ |
| Total cost | 3.8 | 4.6 | Similary for $B$ |
| Sale Price (per unit) | 5 | 6 | |
| Profit (S.P. – C.P.) | 1.2 | 1.4 | |
| **Constraints** | | | Availability = 1 hour |
| No. of items machined per hour | 25 | 40 | Either item $A$ or $B$ |
| No. of items Bored per hour | 28 | 35 | Either item $A$ or $B$ |
| No. of items Polished per hour | 35 | 25 | Either item $A$ or $B$ |

**Objective function**

$$\text{Profit} = \text{S.P.} - (\text{C.P.} + \text{Machining Cost} + \text{Boring Cost} + \text{Polishing Cost})$$

$\therefore$ Profit per unit of item $A = 5 - (2 + 0.8 + 0.5 + 0.5) =$ ₹ 1.2

Profit per unit of item $B = 6 - (3 + 0.5 + 0.4 + 0.7) =$ ₹ 1.4

Objective Function is :

$$\text{Max}(z) = 1.2\,x_1 + 1.4\,x_2$$

**Constraints**

$$\frac{x_1}{25} + \frac{x_2}{40} \leq 1$$

(As both items $A$ and $B$ are to be machined each hour)

$$\frac{x_1}{25} + \frac{x_2}{35} \leq 1$$

(As both items $A$ and $B$ are to be bored each hour)

$$\frac{x_1}{35} + \frac{x_2}{25} \leq 1$$

(As both items $A$ and $B$ are to be polished each hour)

$$x_1\,x_2 \geq 0$$

## MULTIPLE CHOICE QUESTIONS

2.1. A feasible solution to lpp is also

(a) an optimal solution

(b) a solution to the lpp without non-negativety restrictions

(c) a solution to the lpp with non-negative restrictions

(d) non of the above

2.2. An optimal solution to the lpp is also

(a) a feasible solution

(b) a solution to the lpp without non-negativity restrictions

(c) a solution to the lpp with non-negativity restrictions

(d) all of the above

2.3. A feasible solution to a lpp

(a) need not satisfy all the constraints at the same time

(b) must optimize the value of the objetive function

(c) must be a corner point of the feasible region

(d) must satisfy all the constraints simutlaneously

2.4. If an iso-profit line yielding the optimal solution is parallel to a constraint line, then

(a) the solution is unbounded
(b) the solution is infeasible
(c) an alternate optimal solution exists
(d) none of the above

2.5. If two constraints do not intersect in the positive quadrant (first quadrant) of the graph, then

(a) the solution is unbounded
(b) the problem is having an infeasible solution
(c) one of the constraints is redundant
(d) none of the above

2.6. The solution of the lpp becomes unbounded when

(a) in a minimization lpp, feasible region is unbounded
(b) in a maximization lpp, feasible region is unbounded
(c) the lpp model is incorrectly formulated
(d) the objective function is unbounded.

2.7. An alternative optimal solution exists its a linear programming problem, when

(a) two constraints are parallel
(b) one of the constraints is redundant
(c) the objective function equation is parallel to one of the constraints
(d) none of the above.

2.8. While solving a lpp, infeasibility may be removed by

(a) removing a constraint
(b) removing a variable
(c) adding another constraint
(d) adding another variable

2.9. While solving a lpp graphically the area bounded by the constraints

(a) is a convex polygon
(b) is a non-convex polygon
(c) may be circular region
(d) none of the above

2.10. Infeasibility means that the number of solutions to the linear programming problem satisfying all the constraints is

(a) at least 1 (b) none
(c) at least 2 (d) none of the above

## ANSWERS

1. (c) 2. (d) 3. (d) 4. (c) 5. (b) 6. (b) 7. (c)
8. (a) 9. (a) 10. (b)

## SUMMARY

1.A general linear programming problem (lpp) with $m$ constrants and $n$ variables is :

Maximize (or Minimize) $Z = C_1 x_1 + C_2 x_2 + ..... + C_n x_n$ ... (A)

Subject to :

$a_{11}x_1 + a_{12}x_2 + ..... + a_{1n} x_n \leq (or \geq or =) b_1$ ... (1)

$a_{11} x_1 + a_{22} x_2 + ....... + a_{2n} x_n \leq (or \geq or =) b_2$ ... (2)

$\vdots$

$a_{m1} x_1 + a_{m2} x_2 + ...... + a_{mn} x_n \leq (or \geq or =) b_m$ ... (m)

$x_1, x_2 ........ x_n \geq 0$

Here

(i) $x_1, x_2 ......., x_n$ are the $n$ decision variables whose values are to be determined.

(ii) Equation (A) is the objective function 'Z' whose value is to be maximized or minimized.

(iii) The inequations (1), (2) ...... ($m$) are $m$ constraints which determine the feasible region of the lpp.

(iv) $b_i$ ($i = 1, 2, ...... m$) represents the requirement or availability of the $i^{th}$ constraint.

(v) $C_j$ ($j = 1, 2, ....., n$) represents the contribution of variable $x_j$ to the objective function $Z$.

(vi) The coefficients $a_{ij}$ ($i = 1, 2, .... m; j = 1, 2, ..... n$) are substitution coefficients.

2. **Solution of the LPP.** The set of values of the decision variables which satisfy the constraints of the given lpp is called solution of the lpp.

3. **Feasible Solution.** A solution of the given lpp, which also satisfies its non-negativety restrictions is called feasible solution.

4. **Optimum Solution.** A feasible solution which optimizes (maximizes or minimizes) the objective function of the lpp is called its optimum solution.

CHAPTER 3

# Linear Programming (Simplex Method)

## 3.1 INTRODUCTION

Graphical methods are impracticable for solving linear programming problems having more than two variables. Simplex method is the most versatile method applied to LPPs having two or more decision variables. A problem having $n$ variable can be thought of in terms of $n$-dimensional space, where feasible region is a convex set (region) having corner points or vertices. Simplex method provides an algorithm which consists of moving from one vertex of a basic feasible solution to another vertex (another basic feasible solution) in such a manner that the value of objective function at the succeeding vertex is greater than value at the preceding vertex in a maximization problem and vice-versa in a minimization problem.

## 3.2 IMPORTANT DEFINITIONS AND RESULTS

### 3.2.1 Slack, Surplus and Artificial Variables

A non-negative variable which is added to L.H.S. of the constraint inequation of the type

$$a_{11} x_1 + a_{12} x_2 + ..... + a_{1n} x_n \leq b$$

to convert it into equality is called a slack variable.

The constraint inequation can hence be rewritten as

$$a_{11} x_1 + a_{12} x_2 + ..... + a_{1n} x_n + s_1 = b_1$$

Here $s_1$ represents the slack variable.

Surplus variable is a non-negative variable subtracted from the constraint inequation of the type :

$$a_{21}x_1 + a_{22}x_2 + \ldots\ldots + a_{2n}x_n \geq b_2$$

The equation after subtracting surplus variable $s_2$ is :

$$a_{21}x_1 + a_{22}x_2 + \ldots\ldots + a_{2n}x_n - s_2 = b_2$$

Artifical variable is temporarily added to a constraint equation so that it may act as a basic variable in the initial simplex table. It is to be noted that artificial variables are required only in constraint equation having '=' sign and not in those having '≤' or '≥' signs. Artificial variables must be driven to zero to obtain a basic solution to the original constraints.

### 3.2.2 Basic Solution

If there are $(m + n)$ variables and $m$ constraints, only $m$ variables can form a basic solution taking remaining $n$ variables as zero. The $m$ variables forming the solution are called basic variables and remaining variables are called non-basic variables. Total number of basic solutions is ${}^{m+n}C_m$.

### 3.2.3 Basic Feasible Solution

If the solution yeilds all non-negative basic variables, it is called basic feasible solution, otherwise it is infeasible.

### 3.2.4 Non-Degenerate Basic Feasible Solution

A basic solution is said to be non-degenerate if it yields exactly $m$ positive (non-zero) values. The solution is degenerate if one or more of the $m$ basic varibales vanish. 'Degenerate' literally means that the problem will not generate new solution in further iteration.

**Example 3.1** *Find all basic feasible solutions of the equations.*

$$2x_1 + 6x_2 + 2x_3 + x_4 = 3$$

$$6x_1 + 4x_2 + 4x_3 + 6x_4 = 2$$

***Solution*** Total number of basic sclutions

$$= {}^4C_2 = 6$$

**Ist Solution.** Let $x_1$ and $x_2$ be basic variables, i.e., $x_3 = 0$ and $x_4 = 0$.

∴ System reduces to

$$2x_1 + 6x_2 = 3$$

$$6x_1 + 4x_2 = 2$$

Solving these two, we get $x_1 = 0, x_2 = \frac{1}{2}$

$\therefore$ Solution is basic feasible but degenerate.

**2nd Solution.** Let $x_1$ and $x_3$ be basic variables i.e., $x_2 = x_4 = 0$

System reduces to

$$2x_1 + 2x_3 = 3$$
$$6x_1 + 4x_3 = 2$$

Solving, we get

$$x_1 = -2, x_3 = \frac{7}{2}$$

$\therefore$ Solution is not feasible and degenerate .

**3rd Solution.** Let $x_1$ and $x_4$ be basic variables

System reduces to

$$2x_1 + x_4 = 3$$
$$6x_1 + 6x_4 = 2$$

Solving, we get

$$x_1 = -\frac{8}{3}, x_4 = -\frac{7}{3}$$

$\therefore$ Solution is not feasible and degenerate.

**4th Solution.** Let $x_2$ and $x_3$ be basic varibales

System reduces to

$$6x_2 + 2x_3 = 3$$
$$4x_2 + 4x_3 = 2$$

Solving, we get

$$x_2 = \frac{1}{2}, x_3 = 0$$

$\therefore$ Solution is basic feasible and degenerate.

**5th Solution.** Let $x_2$ and $x_4$ be basic variables

System reduces to

$$6x_2 + x_4 = 3$$
$$4x_2 + 6x_4 = 2$$

Solving, we get

$$x_2 = \frac{1}{2}, x_4 = 0$$

$\therefore$ Solution is basic feasible and degenerate.

**6th Solution.** Let $x_3$, $x_4$ be basic variables

$$\Rightarrow \quad \begin{aligned} 2x_3 + x_4 &= 3 \\ 4x_3 + 6x_4 &= 2 \end{aligned}$$

Solving, we get

$$x_3 = 2, x_4 = -1$$

$\therefore$ Solution is not feasible and degenerate.

## 3.3 CANONICAL AND STANDARD FORMS OF LPP

By using elementary tansformations, general form of lpp can always be put in following form :

$$\text{Maximize } (Z) = C_1 x_1 + C_2 x_2 + \ldots\ldots + C_n x_n$$

Subject to

$$a_{i1} x_1 + a_{i2} x_2 + \ldots\ldots + a_{in} x_n \leq b_i \; (i = 1, 2, \ldots m)$$

$$x_1, x_2, \ldots\ldots, x_n \geq 0$$

This is known as canonical form of lpp canonical form of lpp must have following characteristics

(i) Objective function should be of maximization type.

(ii) All the constraints should be of less than ($\leq$) type.

(iii) All the decision variables should be non-negative standard form of lpp is of the form :

$$\text{Maximize } (Z) = C_1 x_1 + C_2 x_2 + \ldots.. + C_n x_n$$

Subject to

$$a_{i1} x_1 + a_{i2} x_2 + \ldots\ldots + a_{in} x_n = b_i \; (i = 1, 2, \ldots., m)$$

$$x_{11} \; x_{21} \; x_n \geq 0$$

Here all the constraints should be of equality type.

## 3.4 SIMPLEX METHOD (ALGORITHM)

Consider the lpp

$$\text{Maximize (or Minimize) } Z = C_1 x_1 + C_2 x_2 + \ldots. + C_n X_n$$

Subject to

$$\sum_{j=1}^{n} a_{ij}x_j \leq \text{(or = or} \geq\text{)}\ b_i,\ (i = 1, 2, \ldots\ldots, m)$$

$$x_j \geq 0,$$
$$(j = 1, 2, \ldots\ldots, n)$$

**Step 1 :** If the problem is of minimization, change it to maximization one by putting Min. Z = Max. $Z'$, where $Z' = -Z$.

**Step 2 :** Make all $b_i$'s $(i = 1, 2, \ldots., m)$ positive if not originally.

**Step 3 :** Keeping all $b_i$'s positive; if all the constraints have '$\leq$' sign, proceed to Step–4, otherwise use big–$M$, two-phase or dual simplex method.

**Step 4 :** Convert all the constraint inequation into equation by adding slack variables.

**Step 5 :** Construct the initial table by assuming slack variables as basic variables i.e., values of decision variables are assumed to be zero intially. Note that basic variables form unit matrix in the simplex table. Decision variables may be taken as basic variables if they give unit matrix contribution in simplex table.

**Step 6 :** Calculate $\Delta_j = z_j - C_j = C_B\, x_j - C_j$ for all basic and non-basic varibales, $\Delta_j$ values for all basic variabes must be zero.

**Step 7 :** If all $\Delta_j \geq 0$, current solution is the optimal solution, otherwise proceed to Step 8.

**Step 8 :** Find the entering variable, whose coefficient in $\Delta_j$ row is most negative.

**Step 9 :** Find the departing variable i.e., basic variable (say $x_k$) in the row whose ratio $V_B/x_k$ $(x_k > 0)$ is minimum.

**Step 10 :** Find the pivotal entry at the intersection of entering variable column and departing variable row.

**Step 11 :** Make pivotal element '1' and remaining element in the entering variable column '0'. Return to Step 6.

**Notations used in Simplex Method**

$Z_j = C_B\, x_j$

$C_B$ = Cost vector (coefficient of basic variable)

$x_j$ = Coefficient of constraints

$C_j$ = Coefficient of objective function

$B.V.$ = Basic variable

$V_B$ = Solution (RHS data in conetrainet equations)

**Example 3.2** *Solve the lpp*

$$\text{Maximize } (Z) = 3x_1 + 2x_2$$

*Subject to*

$$x_1 + x_2 \leq 4$$
$$x_1 - x_2 \leq 2$$
$$x_1, x_2 \geq 0$$

***Solution*** Adding slack variables $s_1$ and $s_2$

$$\text{Maximize } (Z) = 3x_1 + 2x_2 + 0s_1 + 0s_2$$

Subject to

$$x_1 + x_2 + s_1 = 4$$
$$x_1 - x_2 + s_2 = 2$$
$$x_1, x_2, s_1, s_2 \geq 0$$

| | $C_j$ | | 3 | 2 | 0 | 0 | |
|---|---|---|---|---|---|---|---|
| B.V. | $C_B$ | $V_B$ | $x_1$ | $x_2$ | $s_1$ | $s_2$ | Min Ratio $V_B/x_k$, $x_k > 0$ |
| $s_1$ | 0 | 4 | 1 | 1 | 1 | 0 | 4 |
| $s_2$ | 0 | 2 | [1] | −1 | 0 | 1 | 2 → |
| | Z = 0 | | −3↑ | −2 | 0 | 0 | |
| | | | $x_1$ enters, $s_2$ leaves | | | | |
| $s_1$ | 0 | 2 | 0 | [2] | 1 | −1 | → |
| $x_1$ | 3 | 2 | 1 | −1 | 0 | 1 | |
| | Z = 6 | | 0 | −5↑ | 0 | 3 | |
| | | | $x_2$ enters, $s_1$ leaves | | | | |
| $x_2$ | 2 | 1 | 0 | 1 | $\frac{1}{2}$ | $-\frac{1}{2}$ | |
| $x_1$ | 3 | 3 | 1 | 0 | $\frac{1}{2}$ | $\frac{1}{2}$ | |
| | Z = 1 | | 0 | 0 | $\frac{5}{2}$ | $\frac{1}{2}$ | All $\Delta_j \geq 0$ |

Optimal Solution is

$$\text{Maximum } (Z) = 11$$

at

$$x_1 = 3$$

$$x_2 = 1$$

**Example 3.3** *Solve the lpp*

$$\text{Minimize } (Z) = x_1 - 2x_2$$

Subject to

$$x_1 + x_2 \leq 24$$

$$x_1 + 2x_2 \leq 32$$

$$-2x_1 + x_2 \geq -24$$

$$x_1, x_2 \geq 0$$

***Solution*** Changing the lpp into standard form :

$$\text{Maximize } (Z') = -x_1 + 2x_2$$

Subject to

$$x_1 + x_2 \leq 24$$

$$x_1 + 2x_2 \leq 32$$

$$2x_1 - x_2 \leq 24$$

$$x_1, x_2 \geq 0$$

Adding slack variables $s_1, s_2, s_3$

$$\text{Maximize } (Z') = -x_1 + 2x_2$$

Subject to

$$x_1 + x_2 + s_1 = 24$$

$$x_1 + 2x_2 + s_2 = 32$$

$$2x_1 - x_2 + s_3 = 24$$

$$x_1, x_2, s_1, s_2, s_3 \geq 0$$

| | $C_j$ | | -1 | 2 | 0 | 0 | 0 | |
|---|---|---|---|---|---|---|---|---|
| B.V. | $C_B$ | $V_B$ | $x_1$ | $x_2$ | $s_1$ | $s_2$ | $s_3$ | $\frac{V_B}{x_k}, x_k > 0$ *Min. ratio* |
| $s_1$ | 0 | 24 | 1 | 1 | 1 | 0 | 0 | $\frac{24}{1} = 24$ |

*Contd.*

*Contd.*

| | $C_j$ | | −1 | 2 | 0 | 0 | 0 | |
|---|---|---|---|---|---|---|---|---|
| B.V. | $C_B$ | $V_B$ | $x_1$ | $x_2$ | $s_1$ | $s_2$ | $s_3$ | $\frac{V_B}{x_k}, x_k > 0$ *Min. ratio* |
| $s_2$ | 0 | 32 | 1 | [2] | 0 | 1 | 0 | $\frac{32}{2} = 16 \rightarrow$ |
| $s_3$ | 0 | 24 | 2 | −1 | 0 | 0 | 1 | |
| | | | 1 | −2↑ | 0 | 0 | 0 | |
| | | $z = 0$ | $x_2$ enters, $s_2$ leaves | | | | | |
| $s_1$ | 0 | 8 | $\frac{1}{2}$ | 0 | 1 | $-\frac{1}{2}$ | 0 | |
| $x_2$ | 2 | 16 | $\frac{1}{2}$ | 1 | 0 | $\frac{1}{2}$ | 0 | |
| $s_3$ | 0 | 40 | $\frac{5}{2}$ | 0 | 0 | $\frac{1}{2}$ | 1 | |
| | $Z' = 32$ | | 2 | 0 | 0 | 1 | 0 | All $\Delta_j \geq 0$ |

Optimal solution is

Maximum

$$Z = -32$$

at

$$x_1 = 0$$

$$x_2 = 2$$

### 3.4.1 Multiple Optima (Alternative Optimal Solution)

For optimal solution $\Delta_j \geq 0$ and $\Delta_j = 0$ for basic variables. But if $\Delta_j = 0$ for some non-basic variable column in the optimal simplex table then alternative optimal solution exists.

**Example 3.4** *Solve the given lpp. Also find on alternative optimal solution, if it exists.*

$$\text{Maximize } (z) = 6x_1 + 10x_2 + 2x_3$$

*Subject to*

$$2x_1 + 4x_2 + 3x_3 \leq 40$$

$$x_1 + x_2 \leq 10$$

$$2x_2 + x_3 \leq 12$$

$$x_1, x_2, x_3 \geq 0$$

***Solution*** Adding slack varibales $s_1$, $s_2$ and $s_3$

$$\text{Maximize } (z) = 6x_1 + 10x_2 + 2x_3 + 0s_1 + 0s_2 + 0s_3$$

Subject to

$$2x_1 + 4x_2 + 3x_3 + s_1 = 40$$

$$x_1 + x_2 + s_2 = 10$$

$$2x_2 + x_3 + s_3 = 12$$

$$x_1, x_2, x_3, s_1, s_2, s_3 \geq 0$$

| | | $C_j$ | 6 | 10 | 2 | 0 | 0 | 0 | |
|---|---|---|---|---|---|---|---|---|---|
| *B.V.* | $C_B$ | $V_B$ | $x_1$ | $x_2$ | $x_3$ | $s_1$ | $s_2$ | $s_3$ | Min. $V_B/x_k$, $x_k > 0$ |
| $s_1$ | 0 | 40 | 2 | 4 | 3 | 1 | 0 | 0 | 10 |
| $s_2$ | 0 | 10 | 1 | 1 | 0 | 0 | 1 | 0 | 10 |
| $s_3$ | 0 | 12 | 0 | [2] | 1 | 0 | 0 | 1 | 6 → |
| | | $z = 0$ | − 6 | − 10↑ | − 2 | 0 | 0 | 0 | |
| | | | $x_2$ enters, $s_3$ leaves | | | | | | |
| $s_1$ | 0 | 16 | 2 | 0 | 1 | 1 | 0 | − 2 | 8 |
| $s_2$ | 0 | 4 | [1] | 0 | − 1/2 | 0 | 1 | − 1/2 | 4 → |
| $x_2$ | 10 | 6 | 0 | 1 | 1/2 | 0 | 0 | 1/2 | |
| | | $z = 60$ | − 6↑ | 0 | 3 | 0 | 0 | 5 | |
| | | | $x_1$ enters, $s_2$ leaves | | | | | | |
| $s_1$ | 0 | 8 | 0 | 0 | [2] | 1 | − 2 | − 1 | 4 → |
| $x_1$ | 6 | 4 | 1 | 0 | − 1/2 | 0 | 1 | − 1/2 | |
| $x_2$ | 10 | 6 | 0 | 1 | 1/2 | 0 | 0 | 1/2 | 12 |
| | | $z = 84$ | 0 | 0 | 0↑ | 0 | 6 | 2 | $\Delta_j \geq 0$ |
| | | | Optimal solution $x_1 = 4$, $x_2 = 6$, $x_3 = 0$ | | | | | | |
| | | | $\Delta_3 = 0$ ∴ forcing in $x_3$, $s_1$ leaves | | | | | | |
| $x_3$ | 2 | 4 | 0 | 0 | 1 | 1/2 | − 1 | − 1/2 | |
| $x_1$ | 6 | 6 | 1 | 0 | 0 | 1/4 | 1/2 | − 3/4 | |
| $x_2$ | 10 | 4 | 0 | 1 | 0 | − 1/4 | 1/2 | 3/4 | |
| | | $z = 84$ | 0 | 0 | 0 | 0 | 6 | 2 | $\Delta_j \geq 0$ |

∴ Alternate optimal solution is

$$x_1 = 6$$
$$x_2 = 4$$
$$x_3 = 4$$
$$\text{Maximum } z = 84$$

### 3.4.2 Unbounded Solution

If in a mximization problem, corresponding to most negative $\Delta_j$, all the entries in the corresponding column are negative or zero, then the problem has an unbounded solution.

**Example 3.5** *Solve the lpp*

$$\text{Maximize } (z) = 4x_1 + x_2 + 3x_3 + 5x_4$$

*Subject to*

$$4x_1 - 6x_2 - 5x_3 - 4x_4 \geq -20$$
$$-3x_1 - 2x_2 + 4x_3 + x_4 \leq 10$$
$$-8x_1 - 3x_2 + 3x_3 + 2x_4 \leq 20$$
$$x_1, x_2, x_3, x_4 \geq 0$$

***Solution*** Changing its standard form and adding slack varibales

$$\text{Maximize } (z) = 4x_1 + x_2 + 3x_3 + 5x_4 + os_1 + os_2 + os_3$$

Subject to

$$-4x_1 + 6x_2 + 5x_3 + 4x_4 + s_1 = 20$$
$$-3x_1 - 2x_2 + 4x_3 + x_4 + s_2 = 10$$
$$-8x_1 - 3x_2 + 3x_3 + 2x_4 + s_3 = 20$$
$$x_1, x_2, x_3, x_4, s_1, s_2, s_3 \geq = 0$$

| | | $C_j \rightarrow$ | 4 | 1 | 3 | 5 | 0 | 0 | 0 | Min. Ratio |
|---|---|---|---|---|---|---|---|---|---|---|
| B.V. | $C_B$ | $V_B$ | $x_1$ | $x_2$ | $x_3$ | $x_4$ | $s_1$ | $s_2$ | $s_3$ | $V_B/x_k,\ x_k > 0$ |
| $s_1$ | 0 | 20 | – 4 | 6 | 5 | [4] | 1 | 0 | 0 | 5 → |
| $s_2$ | 0 | 10 | – 3 | – 2 | 4 | 1 | 0 | 1 | 0 | 10 |
| $s_3$ | 0 | 20 | – 8 | – 3 | 3 | 2 | 0 | 0 | 1 | 10 |
| | | z = 0 | – 4 | – 1 | – 3 | – 5 ↑ | 0 | 0 | 0 | |
| | | | $x_4$ enters, $s_1$ leaves | | | | | | | |
| $x_4$ | 5 | 5 | – 1 | 3/2 | 5/4 | 1 | 1/4 | 0 | 0 | |
| $s_2$ | 0 | 5 | – 2 | – 7/2 | 11/4 | 0 | – 1/4 | 1 | 0 | |
| $s_3$ | 0 | 10 | – 6 | – 6 | 1/2 | 0 | – 1/2 | 0 | 1 | |
| | | | – 9 | 13/2 | 13/4 | 0 | 5/4 | 0 | 0 | $\Delta_j$ |

All the elements in $\Delta_1$ column are negative

∴ No vector can leave the basis.

∴ Given problem is having an unbounded solution.

## 3.5 SIMPLEX ALGORITHM (INQUALITIES OF ≤, ≥ OR = TYPE)

So far we have discussed simplex algorithm where all inequalities were of '≤' type keeping all $b_i$'s positive.

In this section we will discuss the lpp's having inequalities of mixed type i.e. '≤' or '≥' or '='.

### 3.5.1 Big M Method or Charne's Penalty Method

Big $M$ method is used in lpp's in which constraints may also have '≥' or '=' signs after ensuring that all $b_{i's}$ are +ve.

**Algorithm (Big M Method)**

**Step 1 :** Express lpp in standard form.

**Step 2 :** After making all $b_i$ +ve, assign slack variable to constraints having '≤' type inequality. Also assign surplus and artificial variables to constraints having '≥' or '=' signs. Artificial variables are added to ensure unit matrix contributions from basic variables.

**Step 3 :** Assign a significantly high –ve cost to artificial variables.

**Step 4 :** Solve the problem with usual simplex method until the optimal solution is reached.

Whenever if a artifical variable happens to leave the basis, we drop that variable and omit all the entries corresponding to its coloum from Simple Table.

**Example 3.6** *Solve the following lpp using Big–M method*

$$Maximize\ (z) = 2x_1 + x_2 - 3x_3$$

*Subject to*

$$x_1 + x_2 + x_3 \geq 6$$
$$2x_1 + x_2 = 14$$
$$x_1, x_2, x_3 \geq 0$$

***Solution*** Problem is of maximization type and all $b_i$'s are +ve, adding surplus and artificial variables to constraints and assigning high –ve cost to artificial variable :

$$\text{Maximize } (z) = 2x_1 + x_2 - 3x_3 + os_1 - Ma_1$$

Subject to

$$x_1 + x_2 + x_3 - s_1 = 6$$
$$2x_1 + x_2 + a_1 = 14$$
$$x_1, x_2, s_1, a_1 \geq 0$$

**Note.** Artificial variable is not added to first constraint as varaible $x_3$ can provide unit matrix contribution in initial simplex table.

| | $C_j \rightarrow$ | | 2 | 1 | $-3$ | 0 | $-M$ | |
|---|---|---|---|---|---|---|---|---|
| B.V. | $C_B$ | $V_B$ | $x_1$ | $x_2$ | $x_3$ | $s_1$ | $a_1$ | *Min. Ratio* $\frac{V_B}{x_k}, x_k > 0$ |
| $x_3$ | $-3$ | 6 | $\boxed{1}$ | 1 | 1 | $-1$ | 0 | $\frac{6}{1} = 6 \rightarrow$ |
| $a_1$ | $-M$ | 14 | 2 | 1 | 0 | 0 | 1 | $\frac{14}{2} = 7$ |
| | $Z = -18-14M$ | | $-5-2M\uparrow$ | $-4-M$ | 0 | 3 | 0 | $\Delta_j$ |
| | | | $x_1$ enters, $x_3$ leaves | | | | | |
| $x_1$ | 2 | 6 | 1 | 1 | 1 | $-1$ | 0 | |
| $a_1$ | $-M$ | 2 | 0 | $-1$ | $-2$ | $\boxed{2}$ | 1 | $\rightarrow$ |
| | $Z = 12-2M$ | | 0 | $1+M$ | $5+2M$ | $-2-2M\uparrow$ | 0 | $\Delta_j$ |
| | | | $s_1$ enters, $a_1$ leaves | | | | | |
| $x_1$ | 2 | 7 | 1 | $\boxed{\frac{1}{2}}$ | 0 | 0 | × | $\rightarrow$ |
| $s_1$ | 0 | 1 | 0 | $-\frac{1}{2}$ | $-1$ | 1 | × | |
| | $Z = 14$ | | 0 | $0\uparrow$ | 3 | 0 | × | All $\Delta_j \geq 0$ |
| | | | Solution optimal, Multiple optimal solution exist $x_2$ enters, $x_1$ leaves | | | | | |
| $x_2$ | 1 | 14 | 2 | 1 | 0 | 0 | × | |
| $s_1$ | 0 | 8 | 1 | 0 | $-1$ | 1 | × | |
| | $Z = 14$ | | 0 | 0 | 3 | 0 | × | All $\Delta_j \geq 0$ |

Optimal solution is

$$\text{Maximum (Z)} = 14$$

at $x_1 = 7, x_2 = 0; x_3 = 0$

or $x_1 = 0, x_2 = 14, x_3 = 0$

**Example 3.7** *Solve the following lpp using Big–M method :*

$$\text{Maximum } (Z) = 4x_1 + x_2$$

*Subject to*

$$x_1 + 2x_2 \leq 4$$
$$3x_1 + x_2 = 3$$
$$4x_1 + 3x_2 \geq 6$$
$$x_1, x_2 \geq 0$$

***Solution*** Problem is of maximization type and each $b_i$ is positive, adding slack, surplus and artificial variables and assigning high –ve cost to artificial variables :

$$\text{Max. (Z)} = 4x_1 + x_2 + os_1 + os_2 - Ma_1 - Ma_2$$

Subject to

$$x_1 + 2x_2 + s_1 = 4$$
$$3x_1 + x_2 + a_1 = 3$$
$$4x_1 + 3x_2 - s_2 + a_2 = 6$$
$$x_1, x_2, s_1, s_2, a_1, a_2 \geq 0$$

| | | $C_j \rightarrow$ | 4 | 1 | 0 | 0 | – M | – M | |
|---|---|---|---|---|---|---|---|---|---|
| B.V | $C_B$ | $V_B$ | $x_1$ | $x_2$ | $s_1$ | $s_2$ | $a_1$ | $a_2$ | Min. Ratio $\frac{V_B}{x_k}, x_k > 0$ |
| $s_1$ | 0 | 4 | 1 | 2 | 1 | 0 | 0 | 0 | $\frac{4}{1} = 4$ |
| $a_1$ | – M | 3 | [3] | 1 | 0 | 0 | 1 | 0 | $\frac{3}{3} = 1 \rightarrow$ |
| $a_2$ | – M | 6 | 4 | 3 | 0 | – 1 | 0 | 1 | $\frac{6}{4} = \frac{3}{2}$ |
| | Z = – 9M | | –7M–4↑ | – 4M–1 | 0 | M | 0 | 0 | $\Delta_j$ |
| | | | $x_1$ enters $a_1$ leaves | | | | | | |

| | | | | | | | | | |
|---|---|---|---|---|---|---|---|---|---|
| $s_1$ | 0 | 3 | 0 | $\frac{5}{3}$ | 1 | 0 | × | 0 | $\frac{9}{5}$ |
| $x_1$ | 4 | 1 | 1 | $\frac{1}{3}$ | 0 | 0 | × | 0 | 3 |
| $a_2$ | – M | 2 | 0 | $\boxed{\frac{5}{3}}$ | 0 | – 1 | × | 1 | $\frac{6}{5}\rightarrow$ |
| | $Z = 4-2M$ | | 0 | $-\frac{5}{3}M+\frac{1}{3}\uparrow$ | 0 | $M$ | × | 0 | $\Delta_j$ |
| | | | | $x_2$ enters, $a_2$ leaves | | | | | |
| $s_1$ | 0 | 1 | 0 | 0 | 1 | 1 | × | × | |
| $x_1$ | 4 | $\frac{3}{5}$ | 1 | 0 | 0 | $\frac{1}{5}$ | × | × | |
| $x_2$ | 1 | $\frac{6}{5}$ | 0 | 1 | 0 | $-\frac{3}{5}$ | × | × | |
| | $Z = \frac{18}{5}$ | | 0 | 0 | 0 | $\frac{1}{5}$ | × | × | All $\Delta_j \geq 0$ |

Optimal solution is

$$\text{Maximum } (Z) = \frac{18}{5}$$

at $$x_1 = \frac{3}{5}, x_2 = \frac{6}{5}, x_3 = 0$$

**Note.** Artifical variables once leave the basis, do not re-enter the basis.

### 3.5.2 Two-phase Method

Two-phase method is also used to solve lpp's having '≥' or '=' signs in the constraints.

**Algorithm (Two-phase Method)**

**Phase I**

Simplex method is applied to a specially constructed auxiliary lpp.

**Step 1 :** Assign a cost '– 1' to artificial variables and a cost '0' to all other variables (in place of original cost) in the objective function say $z^*$.

**Step 2 :** Solve the auxiliary lpp by simplex method. Two cases arise :

(i) **Maximum $z^* < 0$.**

In this case at least one artificial variable appears at +ve level.

Any feasible solution does not exist to given lpp in this case.

(ii) **Maximum $z^* = 0$.**

In this case an artifical variable may occur at '0' level or there may be no artificial variable in the basis. Proceed to Phase–II in such case.

**Phase II**

Assign actual costs to variables in the objective function and '0' cost to every slack, surplus or artifical variable (which is at zero level) in the basis. The artificial variables which are non-basic are removed from the solution solve the lpp by usual simple method.

**Example 3.8** *Solve the given lpp using two-phase method.*

$$\text{Maximize } (Z) = x_1 + x_2$$

*Subject to*

$$3x_1 + 2x_2 \leq 12$$

$$x_1 \geq 5$$

$$x_1, x_2 \geq 0$$

***Solution*** Given lpp is of maximization type and also each $b_i$ is +ve. Adding slack, surplus and artificial variables to the constraints inequations :

$$\text{Maximize } (Z) = x_1 + x_2$$

Subject to

$$3x_1 + 2x_2 + s_1 = 12$$

$$x_1 - s_2 + a_1 = 5$$

$$x_1, x_2, s_1, s_2, a_1 \geq 0$$

**Phase I.** Assiging a cost '– 1' to artificial variable $a_1$ and cost '0' to all other varibales,

The new objective function of auxiliary problem becomes

$$\text{Maximize } (Z^*) = 0x_1 + 0x_1 + 0s_1 + 0s_2 - 1a_1$$

Subject to given constraints.

| | $C_j \rightarrow$ | | 0 | 0 | 0 | 0 | − 1 | |
|---|---|---|---|---|---|---|---|---|
| | | | | | | | | *Min. Ratio* |
| *B.V.* | $C_B$ | $V_B$ | $x_1$ | $x_2$ | $s_1$ | $s_2$ | $a_1$ | $\frac{V_B}{x_k}, x_k > 0$ |
| $s_1$ | 0 | 12 | [3] | 2 | 1 | 0 | 0 | $\frac{12}{3} = 4 \rightarrow$ |
| $a_1$ | − 1 | 5 | 1 | 0 | 0 | − 1 | 1 | $\frac{5}{1} = 5$ |
| | $Z^* = -5$ | | $-1\uparrow$ | 0 | 0 | 1 | 0 | $\Delta_j$ |
| | | | $x_1$ enters, $s_1$ leaves | | | | | |
| $x_1$ | 0 | 4 | 1 | $\frac{2}{3}$ | $\frac{1}{3}$ | 0 | 0 | |
| $a_1$ | − 1 | 1 | 0 | $-\frac{2}{3}$ | $-\frac{1}{3}$ | − 1 | 1 | |
| | $Z^* = -1$ | | 0 | $\frac{2}{3}$ | $\frac{1}{3}$ | 1 | 0 | $\Delta_j \geq 0$ |

All $\Delta_j \geq 0$ and maximum $Z^* < 0$ with an artificial variable $a_1$ in the basis at +ve level, therefore there is no feasible solution to given lpp.

**Example 3.9** *Solve the lpp given in Example 3.6 using Two-phase method.*

$$\textit{Maximize } (Z) = 2x_1 + x_2 - 3x_3$$

*Subject to*

$$x_1 + x_2 + x_3 \geq 6$$
$$2x_1 + x_2 = 14$$
$$x_1, x_2, x_3 \geq 0$$

**Solution** Adding surplus and artificial variables :

$$\text{Maximize } (Z) = 2x_1 + x_2 - 3x_3$$

Subject to

$$x_1 + x_2 + x_3 - s_1 = 6$$
$$2x_1 + x_2 + a_1 = 14$$
$$x_1, x_2, x_3, s_1, a_1 \geq 0$$

**Phase I.** Assining a cost '− 1' to artificial variable $a_1$ and cost '*0*' its all other varibales, the new objective function of auxiliary problem becomes.

$$\text{Maximize } Z^* = 0x_1 + 0x_2 + 0x_3 + 0s_1 - 1a_1$$

Subject to given constraints

| | | $C_j \rightarrow$ | 0 | 0 | 0 | 0 | −1 | Min. Ratio |
|---|---|---|---|---|---|---|---|---|
| B.V. | $C_B$ | $V_B$ | $x_1$ | $x_2$ | $x_3$ | $s_1$ | $a_1$ | Min. $\frac{V_B}{x_k}, x_k > 0$ |
| $x_3$ | 0 | 6 | [1] | 1 | 1 | −1 | 0 | $\frac{6}{1} = 6 \rightarrow$ |
| $a_1$ | −1 | 14 | 2 | 1 | 0 | 0 | 1 | $\frac{14}{2} = 7$ |
| | $Z^* = -14$ | | $-2\uparrow$ | −1 | 0 | 0 | 0 | $\Delta_j$ |
| | | | $x_1$ enters, $x_3$ leaves | | | | | |
| $x_1$ | 0 | 6 | 1 | 1 | 1 | −1 | 0 | |
| $a_1$ | −1 | 2 | 0 | −1 | −2 | [2] | 1 | $\rightarrow$ |
| | $Z^* = -2$ | | 0 | 1 | 2 | $-2\uparrow$ | 0 | $\Delta_j$ |
| | | | $s_1$ enters, $a_1$ leaves | | | | | |
| $x_1$ | 0 | 7 | 1 | $\frac{1}{2}$ | 0 | 0 | × | |
| $s_1$ | 0 | 1 | 0 | $-\frac{1}{2}$ | −1 | 1 | × | |
| | $Z^* = 0$ | | 0 | 0 | 0 | 0 | | $\Delta_j \geq 0$ |

All $\Delta_j \geq 0$ and Max. $Z^* = 0$ with no artificial variable in the basis, proceeding to Phase II.

| | | $C_j \rightarrow$ | 0 | 0 | 0 | 0 | Min. Ratio |
|---|---|---|---|---|---|---|---|
| B.V. | $C_B$ | $V_B$ | $x_1$ | $x_2$ | $x_3$ | $s_1$ | $\frac{V_B}{x_k}, x_k > 0$ |
| $x_1$ | 2 | 7 | 1 | [$\frac{1}{2}$] | 0 | 0 | $\rightarrow$ |
| $s_1$ | 0 | 1 | 0 | $-\frac{1}{2}$ | −1 | 1 | |
| | $Z = 14$ | | 0 | $0\uparrow$ | 3 | 9 | All $\Delta_j \geq 0$ |
| | | | Multiple optimal solution exists | | | | |

| | $C_j \rightarrow$ | | 0 | 0 | 0 | 0 | |
|---|---|---|---|---|---|---|---|
| B.V. | $C_B$ | $V_B$ | $x_1$ | $x_2$ | $x_3$ | $s_1$ | Min. Ratio $\frac{V_B}{x_k}, x_k > 0$ |
| $x_2$ | 1 | 14 | 2 | 1 | 0 | 0 | |
| $s_1$ | 0 | 8 | 1 | 0 | −1 | 1 | |
| | $Z = 14$ | | 0 | 0 | 3 | 0 | $\Delta_j \geq 0$ |

Optimal solution is

$$\text{Maximum } (Z) = 14$$

at $x_1 = 7, x_2 = 0, x_3 = 0$

or $x_1 = 0, x_2 = 14, x_3 = 0$

## 3.6 DEGENERACY IN LINEAR PROGRAMMING PROBLEMS

Degeneracy in lpps arise if one of the basic variable takes a zero value in a basic feasible solution. Degeneracy if induced may cost extra iterations which unnecessarily increases manual efforts. Degeneracy may be caused due to redundant constraints or due to wrong choice of tied entering or leaving variables.

### 3.6.1 Steps to Resolve Tie in Entering Variables

(i) Tie between two decision variables, choose arbitrarily

(ii) Tie between two slack or surplus variables, choose arbitraily.

(iii) Tie between decision and slack/surplus variable, choose decision variable.

### 3.6.2 Steps to Resolve Tie in Leaving Variables

(i) Tie between two decision variables, choose arbitrarely.

(ii) Tie between two slack/surplus variables, choose arbitrarily.

(iii) Tie between decision and slack/surplus variable, choose slack/surplus variable.

(iv) In case of tie between slack/surplus/decision variable and artificial variable, choose artificial variable to leave.

#### 3.6.2.1 *Perturbation Technique*

We may also use perturbation technique to resolve tie between two slack/surplus leaving variables.

According to perturbation technique, consider two tied rows of slack/surplus variables and take corresponding column from unit matrix. Divide each such element in tied rows by same $\frac{V_B}{x_k}$ value. The one with first minimum ratio leaves the basis.

**Example 3.10** *Solve the lpp*

$$Max.\ (Z) = 3x_1 + 2x_2$$

*Subject to*

$$2x_1 + x_2 \leq 1$$
$$3x_1 + 4x_2 \geq 4$$
$$x_1, x_2 \geq 0$$

**Solution** Adding slack and surplus variables and solving using Big–$M$ method :

$$\text{Max. } (Z) = 3x_1 + 2x_2 + 0s_1 + 0s_2 - Ma_1$$

Subject to

$$2x_1 + x_2 + s_1 = 1$$
$$3x_1 + 4x_2 - s_2 + a_1 = 4$$
$$x_1, x_2, s_1, s_2, a_1 \geq 0$$

$$C_j \rightarrow \quad 3 \quad 2 \quad 0 \quad 0 \quad -M$$

| *B.V* | $C_B$ | $V_B$ | $x_1$ | $x_2$ | $s_1$ | $s_2$ | $a_1$ | | *Min. Ratio* $\frac{V_B}{x_k}, x_k > 0$ |
|---|---|---|---|---|---|---|---|---|---|
| $s_1$ | 0 | 1 | 2 | 1 | 1 | 0 | 0 | | 1 } Tie |
| $a_1$ | $-M$ | 4 | 3 | [4] | 0 | $-1$ | 1 | $\rightarrow$ | 1 } Tie |
| | $Z = -4M$ | | $-3M-3$ | $-4M-2\uparrow$ | 0 | $M$ | 0 | | $\Delta_j$ |
| | | | | $x_2$ enters, $a_1$ leaves | | | | | |
| $s_1$ | 0 | 0 | 0 | [$\frac{5}{4}$] | 0 | 1 | $\frac{1}{4}$ | $\times$ | $0 \rightarrow$ |
| $x_2$ | 2 | 1 | 1 | $\frac{3}{4}$ | 1 | 0 | $-\frac{1}{4}$ | $\times$ | $\frac{4}{3}$ |
| | $Z = 2$ | | $-\frac{3}{2}\uparrow$ | 0 | 0 | $-\frac{1}{2}$ | $\times$ | | |
| | | | | $x_1$ enters, $s_1$ leaves | | | | | |

| | $C_j \rightarrow$ | | 3 | 2 | 0 | 0 | $-M$ | |
|---|---|---|---|---|---|---|---|---|
| B.V | $C_B$ | $V_B$ | $x_1$ | $x_2$ | $s_1$ | $s_2$ | $a_1$ | Min. Ratio $\frac{V_B}{x_k}, x_k > 0$ |
| $x_1$ | 3 | 0 | 1 | 0 | $\frac{4}{5}$ | $\boxed{\frac{1}{5}}$ | × | → |
| $x_2$ | 2 | 1 | 0 | 1 | $-\frac{3}{5}$ | $-\frac{2}{5}$ | × | |
| | $Z = 2$ | | 0 | 0 | $\frac{6}{5}$ | $-\frac{1}{5}\uparrow$ | × | $\Delta_j$ |
| | | | $s_2$ enters, $x_1$ leaves | | | | | |
| $s_2$ | 0 | 0 | 5 | 0 | 4 | 1 | × | |
| $x_2$ | 2 | 1 | 2 | 1 | 1 | 0 | × | |
| | $Z = 2$ | | 1 | 0 | 2 | 0 | × | All $\Delta_j \geq 0$ |

**Note.** Solution does not improve in 3 iterations as basic variable $s_1$ is at '*0*' level in 2nd iteration. Hence the solution is degenerate.

Optimal solution is

$$\text{Maximum } (Z) = 2$$

at
$$x_1 = 0$$
$$x_2 = 1$$

Also not that we have chosen artificial variable $a_1$ to leave the basis in the 1st iteration. Suppose we apply perturbation technique to choose leaving variable.

| | $s_1$ | $a_1$ | |
|---|---|---|---|
| $s_1$ | $\frac{1}{1} = 1$ | $\frac{0}{1} = 0$ | |
| $a_1$ | $\frac{0}{4} = 0$ | $\frac{1}{4}$ | → |

According to perturbation technique, the one with first minimum ratio leaves the basis. Hence $a_1$ leaves the basis.

## PREVIOUS YEARS QUESTIONS

**Example 3.11** *Solve by Simplex Method the following LP Problem.*

(GGSIPU, 2010)

$$\text{Minimize } (z) = x_1 - 3x_2 + x_3$$

S.T.

$$3x_1 - x_2 + 2x_3 \le 7$$
$$-2x_1 + 4x_2 \le 12$$
$$-4x_1 + 3x_2 + 8x_3 \le 10$$
$$x_1, x_2, x_3 \ge 0$$

**Solution** *Std. form*

$$\text{Max. } (-z') = -x_1 + 3x_2 - x_3 + 0s_1 + 0s_2 + 0s_3$$

$$3x_1 - x_2 + 2x_3 + s_1 = 7$$
$$-2x_1 + 4x_2 + s_2 = 12$$
$$-4x_1 + 3x_2 + 8x_3 + s_3 = 10$$
$$x_1, x_2, x_3, s_1, s_2, s_3 \ge 0$$

| | | $C_j \rightarrow$ | $-1$ | $3$ | $-1$ | $0$ | $0$ | $0$ | |
|---|---|---|---|---|---|---|---|---|---|
| B.V. | $C_B$ | $X_B$ | $x_1$ | $x_2$ | $x_3$ | $S_1$ | $S_2$ | $S_3$ | Min. ratio $x_B/x_k,\ x_k > 0$ |
| $S_1$ | 0 | 7 | 3 | $-1$ | 2 | 1 | 0 | 0 | –ve |
| $S_2$ | 0 | 12 | $-2$ | $\boxed{4}$ | 0 | 0 | 1 | 0 | $\frac{12}{4} = 3 \rightarrow$ |
| $S_3$ | 0 | 10 | $-4$ | 3 | 8 | 0 | 0 | 1 | $\frac{10}{3} = 3.3$ |
| | $z' = 0$ | | 1 | $-3\uparrow$ | 1 | 0 | 0 | 0 | $\Delta_j = Z_j - C_j$ |
| $S_1$ | 0 | 10 | $\boxed{\frac{5}{2}}$ | 0 | 2 | 1 | $\frac{1}{4}$ | 0 | $\frac{10}{5} \times 2 = 4 \rightarrow$ |
| $x_2$ | 3 | 3 | $-\frac{1}{2}$ | 1 | 0 | 0 | $\frac{1}{4}$ | 0 | –ve |
| $S_3$ | 0 | 1 | $-\frac{5}{2}$ | 0 | 8 | 0 | $-\frac{3}{4}$ | 1 | –ve |
| | $z' = 9$ | | $-\frac{1}{2}\uparrow$ | 0 | 1 | 0 | $\frac{3}{4}$ | 0 | $\Delta_j = Z_j - C_j$ |

| | $C_j \rightarrow$ | | $-1$ | $3$ | $-1$ | $0$ | $0$ | $0$ | |
|---|---|---|---|---|---|---|---|---|---|
| B.V. | $C_B$ | $X_B$ | $x_1$ | $x_2$ | $x_3$ | $S_1$ | $S_2$ | $S_3$ | Min. ratio $x_B/x_k, x_k > 0$ |
| $x_1$ | $-1$ | 4 | 1 | 0 | $\frac{4}{5}$ | $\frac{2}{5}$ | $\frac{1}{10}$ | 0 | |
| $x_2$ | 3 | 5 | 0 | 1 | $\frac{2}{5}$ | $\frac{1}{5}$ | $\frac{3}{10}$ | 0 | |
| $s_3$ | 0 | 11 | 0 | 0 | 10 | 1 | $-\frac{1}{2}$ | 1 | |
| | $z' = 11$ | | 0 | 0 | $\frac{7}{5}$ | $\frac{1}{5}$ | $\frac{4}{5}$ | 0 | $\Delta_j \geq 0$ |

Hence all $\Delta_j \geq 0$

$\therefore$ This is the optimal solution

$\therefore$ Min. $z' = -z' = -11$

at $x_1 = 4$

$x_2 = 5$

$x_3 = 0$

**Example 3.12** $\quad Min. (z) = 8x_1 + 4x_2$

*S.t.*

$$x_1 + x_2 \geq 3$$
$$4x_1 + 2x_2 \geq 9$$
$$x_1, x_2 \geq 0$$

**(GGSIPU, 2012)**

***Solution*** *Std. form*

$$Max. (-z') = -8x_1 - 4x_2 - 0S_1 - 0S_2 - Ma_1 - Ma_2$$

*S.t.*

$$x_1 + x_2 - S_1 + a_1 = 3$$
$$4x_1 + 2x_2 - S_2 + a_2 = 9$$
$$x_1, x_2, S_1, S_2, a_1, a_2 \geq 0$$

| | $C_j \rightarrow$ | | $-1$ | $3$ | $-1$ | $0$ | $0$ | $0$ | |
|---|---|---|---|---|---|---|---|---|---|
| B.V. | $C_B$ | $x_B$ | $x_1$ | $x_2$ | $S_1$ | $S_2$ | $a_1$ | $a_2$ | Min ratio $x_B/x_k, x_k > 0$ |
| $a_1$ | $-M$ | 3 | 1 | 1 | $-1$ | 0 | 1 | 0 | $\frac{3}{1} = 3$ |
| $a_2$ | $-M$ | 9 | [4] | 2 | 0 | $-1$ | 0 | 1 | $\frac{9}{4} = 2.25 \rightarrow$ |
| | $z' = -12M$ | | $-5M+8\uparrow$ | $-3M+4$ | $M$ | $M$ | 0 | 0 | $\Delta_j = Z_j - C_j$ |

| | | $C_j \rightarrow$ | -1 | 3 | -1 | 0 | 0 | 0 | |
|---|---|---|---|---|---|---|---|---|---|
| B.V. | $C_B$ | $x_B$ | $x_1$ | $x_2$ | $S_1$ | $S_2$ | $a_1$ | $a_2$ | Min ratio $x_B/x_k$, $x_k > 0$ |
| $a_1$ | $-M$ | $\frac{3}{4}$ | 0 | $\boxed{\frac{1}{2}}$ | $-1$ | $\frac{1}{4}$ | 1 | × | $\frac{3}{4} \times 2 = 1.5 \rightarrow$ |
| $x_1$ | $-8$ | $\frac{9}{4}$ | 1 | $\frac{1}{2}$ | 0 | $-\frac{1}{4}$ | 0 | × | $\frac{9}{4} \times 2 = 4.5$ |
| | $z' = -18 - \frac{3M}{4}$ | | 0 | $-\frac{1}{2}M \uparrow$ | $M$ | $-\frac{M}{4}+2$ | | × | |
| $x_2$ | $-4$ | $\frac{3}{2}$ | 0 | 1 | $-2$ | $\frac{1}{2}$ | × | × | |
| $x_1$ | $-8$ | $\frac{3}{2}$ | 1 | 0 | 1 | $-\frac{1}{2}$ | × | × | |
| | $Z' = -18$ | | 0 | 0 | 0 | 2 | × | × | $\Delta_j \geq 0$ |

Hence all $\Delta_j \geq 0$

∴ This is the optimal solution

∴ $$\text{Miz. } z = -z' = 18$$

at $$x_1 = \frac{3}{2}$$

$$x_2 = \frac{3}{2}$$ **Ans.**

**Example 3.13** $$Miz. (z) = 4x_1 + 3x_2$$

S.t.

$$2x_1 + x_2 \geq 10$$
$$-3x_1 + 2x_2 \leq 6$$
$$x_1 + x_2 \geq 6$$
$$x_1, x_2 \geq 0$$

**(GGSIPU, Feb. 2012)**

***Solution*** *Std. form*

$$Max. (-z') = -4x_1 - 3x_2 - 0S_1 + 0S_2 - 0S_3 - Ma_1 - Ma_2$$

S.t.

$$2x_1 + x_2 - S_1 + a_1 = 10$$

$$-3x_1 + 2x_2 + S_2 = 6$$
$$x_1 + x_2 - S_3 + a_2 = 6$$
$$a_1, a_2, x_1, x_2, S_1, S_2 \geq 0$$

| | $C_j \rightarrow$ | | $-4$ | $-3$ | $0$ | $0$ | $0$ | $-M$ | $-M$ | |
|---|---|---|---|---|---|---|---|---|---|---|
| B.V. | $C_B$ | $x_B$ | $x_1$ | $x_2$ | $S_1$ | $S_2$ | $S_3$ | $a_1$ | $a_2$ | Min ratio $x_B/x_k, x_k > 0$ |
| $a_1$ | $-M$ | 10 | [2] | 1 | $-1$ | 0 | 0 | 1 | 0 | $\frac{10}{2} = 5 \rightarrow$ |
| $S_2$ | 0 | 6 | $-3$ | 2 | 0 | 1 | 0 | 0 | 0 | $-$ve |
| $a_2$ | $-M$ | 6 | 1 | 1 | 0 | 0 | $-1$ | 0 | 1 | $\frac{6}{1} = 6$ |
| | $z = (-16M)$ | | $-3M+4\uparrow$ | $-2M+3$ | $M$ | 0 | $M$ | 0 | 0 | $\Delta_j = z_j - C_j$ |
| $x_1$ | $-4$ | 5 | 1 | $\frac{1}{2}$ | $-\frac{1}{2}$ | 0 | 0 | $\times$ | 0 | 10 |
| $S_2$ | 0 | 21 | 0 | $\frac{7}{2}$ | $-\frac{3}{2}$ | 1 | 0 | $\times$ | 0 | 6 |
| $a_2$ | $-M$ | 1 | 0 | $\left[\frac{1}{2}\right]$ | $\frac{1}{2}$ | 0 | $-1$ | $\times$ | 1 | $2 \rightarrow$ |
| | $z' = (-M-20)$ | | 0 | $\frac{-M}{2}+1\uparrow$ | $\frac{-M}{2}+2$ | 0 | $M$ | $\times$ | 0 | |
| $x_1$ | $-4$ | 4 | 1 | 0 | $-1$ | 0 | $-1$ | $\times$ | $\times$ | |
| $S_2$ | 0 | 14 | 0 | 0 | $-5$ | 1 | 7 | $\times$ | $\times$ | |
| $x_2$ | $-3$ | 2 | 0 | 1 | 1 | 0 | $-2$ | $\times$ | $\times$ | |
| | $z' = -22$ | | 0 | 0 | 1 | 0 | 10 | $\times$ | $\times$ | $\Delta_j \geq 0$ |

Hence all $\Delta_j \geq 0$

$\therefore$ This is the optimal solution

$\therefore$ $$\text{Min. } z = -z' = 22$$

at $$x_1 = 4$$
$$x_2 = 2$$

**Ans.**

**Example 3.14** $$Max\ (z) = 3x_1 + 2x_2$$

*S.t.*

$$2x_1 + x_2 \leq 2$$

$$3x_1 + 4x_2 \geq 12$$

$$x_1, x_2 \geq 0$$ **(GGSIPU, Feb. (Re), 2012)**

*Std. form*

$$Max.\ (z) = 3x_1 + 2x_2 + 0S_1 - 0S_2 - Ma_1$$

*S.t.*

$$2x_1 + x_2 + S_1 = 02$$

$$3x_1 + 4x_2 - S_2 + a_1 = 12$$

$$x_1, x_2, S_1, S_2, a_1 \geq 0$$

| | $C_j \rightarrow$ | | 3 | 2 | 0 | 0 | $-M$ | |
|---|---|---|---|---|---|---|---|---|
| B.V. | $C_B$ | $x_B$ | $x_1$ | $x_2$ | $s_1$ | $s_2$ | $a_1$ | Min. ratio $x_B/x_k, x_k > 0$ |
| $S_1$ | 0 | 2 | 2 | [1] | 1 | 0 | 0 | $\frac{2}{1} = 2 \rightarrow$ |
| $a_1$ | $-M$ | 12 | 3 | 4 | 0 | $-1$ | 1 | $\frac{12}{4} = 3$ |
| | $z = -12M$ | | $-3M-3$ | $-4M-2\uparrow$ | 0 | $M$ | 0 | $\Delta_j = z_i - C_j$ |
| $x_2$ | 2 | 2 | 2 | 1 | 1 | 0 | 0 | |
| $a_1$ | $-M$ | 4 | $-5$ | 0 | $-4$ | $-1$ | 1 | |
| | $z = -4M+4$ | | $+5M+1$ | 0 | $4M+2$ | $M$ | 0 | |

Artificial variable appears at +ve level

Hence, no feasible solution.

**Example 3.15** $$Max.\ (z) = 40x_1 + 88x_2$$

*Subject to*

$$x_1 + 4x_2 \leq 30$$

$$5x_1 + 2x_2 \leq 69$$

$$x_1, x_2 \geq 0$$ **(GGSIPU, 2013)**

**Solution** *Std. form*

$$Max.\ (z) = 40x_1 + 88x_2 + 0S_1 + 0S_2$$

*S.t.*

$$x_1 + 4x_2 + S_1 = 30$$

$$5x_1 + 2x_2 + S_2 = 60$$

$$x_1, x_2, S_2, S_2 \geq 0$$

| | $C_j \rightarrow$ | | 40 | 88 | 0 | 0 | |
|---|---|---|---|---|---|---|---|
| *B.V.* | $C_B$ | $x_B$ | $x_1$ | $x_2$ | $s_1$ | $s_2$ | *Min. ratio* $x_B/x_k, x_k > 0$ |
| $S_1$ | 0 | 30 | 1 | $\boxed{4}$ | 1 | 0 | $\frac{30}{4} = 7.5 \rightarrow$ |
| $S_2$ | 0 | 60 | 5 | 2 | 0 | 1 | 30 |
| | $z = 0$ | | $-40$ | $-88\uparrow$ | 0 | 0 | $\Delta_j = Z_j - C_j$ |
| $x_2$ | 88 | 75 | $\frac{1}{4}$ | 1 | $\frac{1}{4}$ | 0 | $\frac{30}{4} \times \frac{4}{1} = 30$ |
| $S_2$ | 0 | 45 | $\boxed{\frac{9}{2}}$ | 0 | $-\frac{1}{2}$ | 1 | $\frac{45}{5} \times 2 = 10 \rightarrow$ |
| | $z = 660$ | | $-18\uparrow$ | 0 | 22 | 0 | |
| $x_2$ | 88 | 5 | 0 | 1 | $\frac{5}{18}$ | $-\frac{1}{18}$ | |
| $x_1$ | 40 | 10 | 1 | 0 | $-\frac{1}{9}$ | $\frac{2}{9}$ | |
| | | $z = 840$ | 0 | 0 | 20 | 4 | |

All the $\Delta_j \geq 0$. Hence this is the optimal solution

$$\text{Max. } (z) = 840$$
$$x_1 = 10$$
$$x_2 = 5$$

**Ans.**

**Example 3.16** $\quad$ *Max.* $(z) = 5x_1 - 2x_2 + 3x_3$

*S.t.*

$$2x_1 + 2x_2 - x_3 \geq 2$$
$$3x_1 - 4x_2 \leq 3$$
$$x_1 + 3x_3 \leq 5$$
$$x_1, x_2, x_3 \geq 0$$

**(GGSIPU, Feb., 2013)**

***Solution*** *Std. form*

$$\text{Max. } (z) = 5x_1 - 2x_2 + 3x_3 - 0S_1 + 0S_2 + 0S_3 - Ma_1$$

*S.t.*

$$2x_1 + 2x_2 - x_3 - S_1 + a_1 = 2$$

$$3x_1 - 4x_2 + S_2 = 3$$
$$x_1 + 3x_3 + S_3 = 5$$
$$x_1, x_2, x_3 \geq 0$$
$$S_1, S_2, S_3, a_1 \geq 0$$

**Leaving →** **Note**

**1 2 3 (Preferance)** *A—Artificial*

***A S D*** *S — Slack/Surplus*

**3 2 1 (Preferance)** *D—Disicion Variable*

**←Entering**

| | $C_j \rightarrow$ | | 5 | −2 | 3 | 0 | 0 | 0 | −M | |
|---|---|---|---|---|---|---|---|---|---|---|
| B.V. | $C_B$ | $x_B$ | $x_1$ | $x_2$ | $x_3$ | $S_1$ | $S_2$ | $S_3$ | $a_1$ | Min ratio $x_B/x_k, x_k > 0$ |
| $a_1$ | −M | 2 | [2] | 2 | −1 | −1 | 0 | 0 | 1 | $\frac{2}{2}=1\rightarrow$ |
| $S_2$ | 0 | 3 | 3 | −4 | 0 | 0 | 1 | 0 | 0 | $\frac{3}{3}=1$ |
| $S_3$ | 0 | 5 | 1 | 0 | 3 | 0 | 0 | 1 | 0 | $\frac{5}{1}=5$ |
| | $z=-2M$ | | $-2M-5\uparrow$ | $-2M+2$ | $M-3$ | $M$ | 0 | 0 | 0 | $\Delta_j = z_j - C_j$ |
| $x_1$ | 5 | 1 | 1 | 1 | $-\frac{1}{2}$ | $-\frac{1}{2}$ | 0 | 0 | × | −ve |
| $S_2$ | 0 | 0 | 0 | −7 | [$\frac{3}{2}$] | $\frac{3}{2}$ | 1 | 0 | × | $0\rightarrow$ |
| $S_3$ | 0 | 4 | 0 | −1 | $\frac{7}{2}$ | $\frac{1}{2}$ | 0 | 1 | × | $\frac{8}{7}$ |
| | $z=5$ | | 0 | 7 | $-\frac{11}{2}\uparrow$ | $-\frac{5}{2}$ | 0 | 0 | × | |
| $x_1$ | 5 | 1 | 1 | $-\frac{4}{3}$ | 0 | 0 | $\frac{1}{3}$ | 0 | × | −ve |
| $x_3$ | 3 | 0 | 0 | $-\frac{14}{3}$ | 1 | 1 | $\frac{2}{3}$ | 0 | × | −ve |

| | $C_j \rightarrow$ | | 5 | −2 | 3 | 0 | 0 | 0 | −M | |
|---|---|---|---|---|---|---|---|---|---|---|
| B.V. | $C_B$ | $x_B$ | $x_1$ | $x_2$ | $x_3$ | $S_1$ | $S_2$ | $S_3$ | $a_1$ | Min ratio $x_B/x_k, x_k > 0$ |
| $S_3$ | 0 | 4 | 0 | $\boxed{\frac{46}{3}}$ | 0 | −3 | $-\frac{7}{3}$ | 1 | × | $\frac{12}{46}\ \frac{6}{23} \rightarrow$ |
| | | $z = 5$ | 0 | $-\frac{56}{3}\uparrow$ | 0 | 3 | $\frac{11}{3}$ | 0 | × | |
| $x_1$ | 5 | $\frac{31}{23}$ | 1 | 0 | 0 | $-\frac{6}{23}$ | $\frac{3}{23}$ | $\frac{2}{3}$ | × | −ve |
| $x_3$ | 3 | $\frac{28}{23}$ | 0 | 0 | 1 | $\boxed{\frac{2}{23}}$ | $-\frac{1}{23}$ | $\frac{7}{23}$ | × | $14\rightarrow$ |
| $x_2$ | −2 | $\frac{6}{23}$ | 0 | 1 | 0 | $-\frac{9}{46}$ | $-\frac{7}{46}$ | $\frac{3}{46}$ | × | −ve |
| | | $z = -\frac{227}{23}$ | 0 | 0 | 0 | $-\frac{15}{23}\uparrow$ | $\frac{437}{529}$ | $\frac{284}{69}$ | × | |
| $x_1$ | 5 | 5 | 1 | 0 | 3 | 0 | 0 | $\frac{109}{69}$ | × | |
| $S_1$ | 0 | 14 | 0 | 0 | $\frac{23}{2}$ | 1 | $-\frac{1}{2}$ | $\frac{7}{2}$ | × | |
| $x_2$ | −2 | 3 | 0 | 1 | $\frac{9}{4}$ | 0 | $-\frac{1}{4}$ | $\frac{3}{4}$ | × | |
| | | $z = 19$ | 0 | 0 | $\frac{15}{2}$ | 0 | $\frac{1}{2}$ | $\frac{883}{138}$ | × | $\Delta_j \geq 0$ |

Hence all $\Delta_j \geq 0$, therefore this is the optimal solution,

at,

$$x_1 = 5$$
$$x_2 = 3$$
$$x_3 = 0$$
$$\text{Max. } (z) = 19$$

**Example 3.17** $\quad$ *Max.* $(z) = 3x_1 + 2x_2$

*S.t.*

$$2x_1 + x_2 \leq 12$$
$$3x_1 + 4x_2 \geq 12$$
$$x_1, x_2 \geq 0$$

**(GGSIPU, Feb. 2014)**

**Solution** *Std. form*

$$Max.\ (z) = 3x_1 + 2x_2 + 0S_1 - 0S_2 - Ma_1$$

S.t.

$$2x_1 + x_2 + S_1 = 12$$
$$3x_1 + 4x_2 - S_2 + A_1 = 12$$
$$x_1, x_2, S_1, S_2, a_1 \geq 0$$

| | $C_j \rightarrow$ | | 3 | 2 | 0 | 0 | $-M$ | |
|---|---|---|---|---|---|---|---|---|
| B.V. | $C_B$ | $x_B$ | $x_1$ | $x_2$ | $S_1$ | $S_2$ | $a_1$ | Min. ratio $x_B/x_k$, $x_k > 0$ |
| $S_1$ | 0 | 12 | 2 | 1 | 1 | 0 | 0 | $\frac{12}{1} = 12$ |
| $a_1$ | $-M$ | 12 | 3 | [4] | 0 | $-1$ | 1 | $\frac{12}{4} = 3 \rightarrow$ |
| | $z = -12M$ | | $-3M-3$ | $-4M-2\uparrow$ | 0 | $M$ | 0 | $\Delta_j = Z_j - C_j = C_B x_j - C_j$ |
| $s_2$ | 0 | 9 | $\frac{5}{4}$ | 0 | 1 | $\frac{1}{4}$ | × | $\frac{9\times 4}{5} = 7.2$ |
| $x_2$ | 2 | 3 | [$\frac{3}{4}$] | 1 | 0 | $-\frac{1}{4}$ | × | $\frac{3\times 4}{3} = 4 \rightarrow$ |
| | $z = 6$ | | $-\frac{3}{2}\uparrow$ | 0 | 0 | $-\frac{1}{2}$ | × | |
| $S_1$ | 0 | 4 | 0 | $-\frac{5}{3}$ | 1 | [$\frac{2}{3}$] | × | $\rightarrow$ |
| $x_1$ | 3 | 4 | 1 | $\frac{4}{3}$ | 0 | $-\frac{1}{3}$ | × | $-$ve |
| | $z = 12$ | | 0 | 2 | 0 | $-1\uparrow$ | × | |
| $S_2$ | 0 | 6 | 0 | $-\frac{5}{2}$ | $\frac{3}{2}$ | 1 | × | $-$ve |
| $x_1$ | 3 | 6 | 1 | [$\frac{1}{2}$] | $\frac{1}{2}$ | 0 | × | $\rightarrow$ |
| | $z = 18$ | | 0 | $-\frac{1}{2}\uparrow$ | $\frac{3}{2}$ | 0 | × | |
| $S_2$ | 0 | 36 | 5 | 0 | 4 | 1 | × | |
| $x_2$ | 2 | 12 | 2 | 1 | 1 | 0 | × | |
| | $z = 24$ | | 1 | 0 | 2 | 0 | × | $\Delta_j \geq 0$ |

Hence, all the $\Delta_j \geq 0$

$\therefore$ This is the optimal solution,

at

$$x_1 = 0$$
$$x_2 = 12$$
$$\text{Max. } (z) = 24$$

## MULTIPLE CHOICE QUESTIONS

3.1. To covert the constraints of an LPP into equalities, Slack Variable are

(a) Subtracted from LHS (b) Added in LHS

(c) Added in RHS (c) None

3.2. Which variables are subtracted from LHS to convert constraints into equalities

(a) Surplus variables (b) Slack variables

(c) Artificial variables (d) Basic Variables

3.3. If one or more basic variables are zero, then a basic feasible solution is said to be

(a) Non degenerated solution (b) Optimal solution

(c) Degenerated solution (d) unbounded solution

3.4. What is the other name for Big–M Method.

(a) Panelty method (b) Simplex method

(d) Two phase method (d) All of these

3.5. When does the Simplex Method indicate that the LPP has unbounded solution

(a) All variables are negative in key column

(b) All variables are positive in key column

(c) Both (a and b)

(d) None

3.6. In Simplex Method the current feasible solution is said to be optimal if

(a) $Z_j - C_j < 0$ (b) $Z_j + C_j \geq 0$

(c) $Z_j + C_j \leq 0$ (d) $Z_j - C_j \geq 0$

3.7. The key row is selected by finding the ratio

(a) min. $\left(\frac{V_B}{x_k}, x_k > 0\right)$ (b) min. $\left(\frac{V_B}{x_k}, x_k < 0\right)$

(c) max. $\left(\frac{V_B}{x_k}, x_k > 0\right)$ (d) max. $\left(\frac{V_B}{x_k}, x_k < 0\right)$

3.8. For getting IBFS, any non-negative variable introduced in the constraint equation is called

(a) Surplus variable (b) Artificial variable

(c) Slack variable (d) Basic variable

3.9. What does it indicate in optimal Simplex table if $(Z_j - C_j) = 0$ for some non-basic varibales

(a) Degenerated solution exist

(b) Unbounded solution exist

(c) Alternate optimal solution exist

(d) No solution exist

3.10. The methods used for solving LPP with involvment of Artificial varibales are

(a) Charne's Penalty Method (b) Two-phase method

(c) Simplex method (d) both (a and b)

(e) both (a and c)

## ANSWERS

1. (b) 2. (a) 3. (c) 4. (a) 5. (a) 6. (d) 7. (a)
8. (b) 9. (c) 10. (d)

## SUMMARY

**Basic Solution.** If there are $(m + n)$ variables and $m$ constraints, only $m$ variables can form a basic solution taking remaining $n$ variables as zero. The $m$ variables forming the solution are called basic variables and remaining variables, are called non-basic variables. Total number of basic solutions in $^{m+n}C_m$.

**Basic Feasible Solution.** If the solution yeilds all non-negative basic variables, it is called basic feasible solution, otherwise it is infeasible.

**Non-Degenerate Basic Feasible Solution.** A basic solutions is said to be non-degenerate if it yields exactly $m$ positive (non-zero) values. The solution is degenerate if one or more of the $m$ basic varibales vanish. 'Degenerate' literally means that the problem will not generate new solution in further iterations.

**Simplex Method (Algoritm)**

Consider the lpp

$$\text{Maximmize (or Minimize) } (Z) = C_1 x_1 + C_2 x_2 + \dots + C_n X_n$$

Subject to

$$\sum_{j=1}^{n} a_{ij} x_j \le (\text{or} = \text{or} \ge)\ b_i,\ (i = 1, 2, \dots, m)$$

$$x_j \ge 0,\ (j = 1, 2, \dots, n)$$

**Step 1 :** If the problem is of minimization, change it to maximization one by putting Min. Z = Max. $Z'$, where $Z' = -Z$

**Step 2 :** Make all $b_i$'s ($i = 1, 2, ...., m$) positive, if not originally.

**Step 3 :** Keeping all $b_i$'s positive; if all the constraints have '$\leq$' sign proceed to Step–4, otherwise use big–$M$, two-phase or dual simplex method.

**Step 4 :** Convert all the constraint inequations into equations by adding slack variables.

**Step 5 :** Construct the initial table by assuming slack variables as basic variables i.e., values of decision variables are assumed to be zero initially. Note that basic variables form unit matrix in the simplex table. Decision variables may be taken as basic variables if they give unit matrix contribution in simplex table.

**Step 6 :** Calculate $\Delta_j = z_j - C_j = C_B x_j - C_j$ for all basic and non-basic varibales $\Delta_j$ values for all basic variables must be zero.

**Step 7 :** If all $\Delta_j \geq 0$, current solution is the optimal solution, otherwise proceed to Step 8.

**Step 8 :** Find the entering variable whose coefficient in $\Delta_j$ row is most negative.

**Step 9 :** Find the departing variable i.e., basic variable (say $x_k$) in the row whose ratio $V_B/x_k$ ($x_k > 0$) is minimum.

**Step 10 :** Find the pivotal entry at the intersection of entering variable column and departing variable row.

**Step 11 :** Make pivotal element '1' and remaining element in the entering variable column '0'. Return to Step 6.

**Big M Method or Charne's Penalty Method.** Big M method is used in lpps in which constraints may also have '$\geq$' or '=' signs after ensuring that all $b_i$'s are +ve.

**Algorithm (Big M Method)**

**Step 1 :** Express lpp in standard form.

**Step 2 :** After making all $b_i$ +ve, assign slack variable to constraints having $\leq$' type inequality. Also assign surplus and artificial variables to constraints haveing '$\geq$' or '=' signs. Artificial variables are added to ensure unit matric contributions from basic variables.

**Step 3 :** Assign a significantly high –ve cost to artificial variables.

**Step 4 :** Solve the problem with usual simplex method until the optimal solution is reached.

**Two-phase Method**

Two-phase method is also used to solve lpps having '$\geq$' or '=' signs in the constraints.

**Algorithm (Two-phase Method)**

**Phase I**

Simplex method is applied to a specially constructed auxiliary lpp.

**Step 1 :** Assign a cost '– 1' to artificial variables and a cost '0' to all other variables (in place of original cost) in the objective function say $z^*$.

**Step 2 :** Solve the auxiliary lpp by simplex method cases arise :

(i) **Maximum $z^* < 0$.** In this case at least one artificial variable appears at +ve level. Any feasible solution does not exist to given lpp in this case.

(ii) **Maximum $z^* = 0$.** In this case an artificial variable may occur at '0' level or there may be no artificial variable in the basis. Proceed to Phase–II in such case.

**Phase II**

Assign actual costs to variables in the objective function and '0' cost to every slack, surplus or artificial variable (which is at zero level) in the basis. The artificial variables which are non-basic are removed from the solution. Solve the lpp by usual simplex method.

CHAPTER 4

# Duality in Linear Programming

## 4.1 INTRODUCTION

Every linear programming problem (called the primal) is associated with another lpp (called its dual). Duality provides interesting insight into economic and sensitivity analysis.

## 4.2 PROCEDURE TO CONVERT PRIMAL INTO DUAL LPP

**Step 1.** If the constraints in primal have equality sign, change it into two inequality constraints.

$\therefore \qquad a = b$

$\Leftrightarrow \qquad a \leq b$

and $\qquad a \geq b$

**Step 2.** Unrestricted varibales if any are replaced by the difference of two non –ve variables.

**Step 3.** If the objective function in the primal is of maximization type, ensure all inequality signs are of '≤' type and if objective function is of minimization, make all inequality signs to '≥' if not originally.

**Step 4.** Finally dual of the given lpp is obtained by

(i) Changing the 'Maximize' function to 'Minimize' and vice-versa.

(ii) Transposing the rows and columns of constraint coefficients.

(iii) Changing the inequality sign of '≤' to '≥' and vice versa.

(iv) Transposing the coefficient ($C_1$, $C_2$, ......, $C_n$) of the objective function and right hand side constants ($b_1$, $b_2$, ...... $b_m$)

**Note.** (i) Number of dual varibales ($\omega_1, \omega_2, ..... \omega_m$) are equal to number of inequality constraints in the primal (after changing equality constraints if any into inequations).

(ii) Number of constraints in the dual are same as number of variables in the primal.

(iii) An *i*th '=' constraint in the primal implies *i*th unrestrict variable ($\omega_i = \omega_i' - \omega_i''$) in the dual.

(iv) *j*th unristricted variable in the primal implies *j*th equality constraint in the dual.

**Example 4.1** *Write dual of the given lpp.*

$$\text{Maximize } Z = x_1 + 3x_2 + 2x_3$$

*Subject to :*

$$x_1 + x_2 + x_3 \leq 15$$
$$2x_1 - x_3 \geq -1$$
$$3x_1 - x_2 + 2x_3 \leq 12$$
$$x_1, x_2, x_3 \geq 0$$

***Solution*** Since the objective function in the primal is of maximization type, changing all constraints to '$\leq$' type

$$\text{Maximize } Z = x_1 + 3x_2 + 2x_3$$

Subject to :

$$x_1 + x_2 + x_3 \leq 15$$
$$-2x_1 + x_3 \leq 1$$
$$3x_1 - x_2 + 2x_3 \leq 12$$
$$x_1, x_2, x_3 \geq 0$$

Dual of given lpp is

$$\text{Maximize } Z_w = 15w_1 + w_2 + 12w_3$$

Subject to

$$w_1 - 2w_2 + 3w_3 \geq 1$$
$$w_1 - w_3 \geq 3$$
$$w_1 + w_2 + 2w_3 \geq 2$$
$$w_1, w_2, w_3 \geq 0$$

**Example 4.2** *Find the dual of the given lpp*

$$\text{Minimize } Z = x_1 + 3x_2$$

*Subject to*

$$x_1 + x_2 \geq 5$$
$$5x_1 - 2x_2 \leq -3$$
$$x_1 \geq 0$$
$$x_2 \text{ unrestricted}$$

***Solution*** Writing unrestricted variable $x_2$ as $(x_2' - x_2'')$ and multiplying 2nd constraint by '– 1' so that '≤' sign is changed to '≥' as in primal objective function is of minimization type :

$$\text{Minimize } Z = x_1 + 3x_2' - 3x_2''$$

Subject to :

$$x_1 + x_2' - x_2'' \geq 5$$
$$-5x_1 + 2x_2' - 2x_2' \geq 3$$
$$x_1, x_2', x_2'' \geq 0$$

Dual of given lpp is

$$\text{Maximize } Z_w = 5w_1 + 3w_2$$

Subject to :

$$w_1 - 5w_2 \leq 1$$
$$w_1 + 2w_2 \leq 3$$
$$-w_1 - 2w_2 \leq -3$$

or

$$w_1 + 2w_2 \geq 3$$
$$w_1, w_2 \geq 0$$

Hence the dual is

$$\text{Maximize } Z_w = 5w_1 + 3w_2$$

Subject to :

$$w_1 - 5w_2 \leq 1$$
$$w_1 + 2w_2 = 3$$
$$w_1, w_2 \geq 0$$

**Note :** $2^{nd}$ unresticted variable in primal implies $2^{nd}$ equality consraint in dual.

**Example 4.3** *write the dual of the given lpp*

$$\text{Minimize } Z = x_1 + x_2$$

*Subject to*

$$2x_1 - x_2 \geq 1$$
$$x_1 + 2x_2 = 10$$
$$x_1, x_2 \geq 0$$

***Solution*** Since the 2nd constraint is of '=' type, rewriting it into two inequality constraints.

$$\text{Minimize } Z = 2x_1 + x_2$$

Subject to :

$$2x_1 - x_2 \geq 1$$
$$x_1 + 2x_2 \geq 10$$
$$x_1 + 2x_2 \leq 10$$
$$x_1, x_2 \geq 0$$

Multiplying constraint 3 by '–1' so that it is changed to '≥' type as objective function is of minimization type.

$$\text{Minimize } Z = 2x_1 + x_2$$

Subject to :

$$2x_1 - x_2 \geq 1$$
$$x_1 + 2x_2 \geq 10$$
$$-x_1 - 2x_2 \geq -10$$
$$x_1 x_2 \geq 0$$

As the 2nd constraint in of '=' type presuming 2nd variable in dual as unrestricted, dual of given lpp is

$$\text{Maximize } Z_w = w_1 + 10\,(w_2' - w_2'')$$

Subject to

$$2w_1 + (w_2' - w_2'') \leq 2$$
$$-w_1 + 2\,(w_2' - w_2'') \leq 1$$
$$w_1, w_2', w_2'' \geq 0$$

Hence dual of given lpp is

$$\text{Maximize } Z_w = w_1 + 10\,w_2$$

Subject to :

$$2w_1 + w_2 \leq 2$$
$$-w_1 + 2w_2 \leq 1$$
$$w_1 \geq 0,\ w_2 \text{ unrestricted}$$

## 4.3 RESULTS ON DUALITY (DUAL-PRIMAL RELATIONSHIP)

1. The dual of the dual lpp is again primal.

2. If primal has a finite optimal solution, then dual also possesses the same and similarly if a finite optimal solution of dual implies the same for primal problem. Also maximum value of objective function for one is same as minimum value of other, i.e. Max. $Z$ = Min. $Z_w$ and Min. $Z$ = Max. $Z_w$.

3. If either primal or the dual lpp has an unbounded solution then the other has an infeasible solution.

4. For a +ve basic variable (non-zero) in the primal, the corresponding dual variable will be zero and for a non-basic variable in the primal (which is zero), the corresponding dual variable will be basic and +ve. This is known as complimentary slackness property. Thus solution of dual may be read from primal simplex table and vice-versa. '$\Delta_j$' values of the optimal simplex table provide values of basic variables of the dual lpp.

**Example 4.4** *Write down the dual of following lpp and solve it. Hence find solution of primal lpp.*

$$\text{Minimize. } Z = 7x_1 + 12x_2 + 10x_3$$

*Subect to :*

$$3x_1 - 2x_2 - 4x_3 \geq -1$$
$$-x_1 + 4x_2 + 3x_3 \geq 3$$
$$2x_1 + 8x_3 \geq -2$$
$$x_1, x_2, x_3 \geq 0$$

***Solution*** Dual of given ;pp is

$$\text{Max. } Z_w = -w_1 + 3w_2 - 2w_3$$

Subject to

$$3w_1 - w_2 + 2w_3 \leq 7$$
$$-2w_1 + 4w_2 \leq 12$$
$$-4w_1 + 3w_2 + 8w_3 \leq 10$$
$$w_1, w_2, w_3 \geq 0$$

Adding slack variabales $s_1, s_2, s_3$ to constraints for solving the dual lpp.

$$\text{Max. } Z_w = -w_1 + 3w_2 - 2w_3 + os_1 + os_2 + os_3$$

Subject to

$$3w_1 - w_2 + 2w_3 + os_1 = 7$$

$$-2w_1 + 4w_2 + os_2 = 12$$
$$-4w_1 + 3w_2 + 8w_3 + os_3 = 10$$
$$w_1, w_2, w_3, s_1, s_2, s_3 \geq 0$$

| | | $C_j \rightarrow$ | $-1$ | $3$ | $-2$ | $0$ | $0$ | $0$ | |
|---|---|---|---|---|---|---|---|---|---|
| *B.V* | $C_B$ | $W_B$ | $w_1$ | $w_2$ | $w_3$ | $s_1$ | $s_2$ | $s_3$ | *Min. ratio* $W_B/w_k$, $w_k > 0$ |
| $s_1$ | 0 | 7 | 3 | $-1$ | 2 | 1 | 0 | 0 | |
| $s_2$ | 0 | 12 | $-2$ | $\boxed{4}$ | 0 | 0 | 1 | 0 | $\frac{12}{4} \rightarrow$ |
| $s_3$ | 0 | 10 | $-4$ | 3 | 8 | 0 | 0 | 1 | $\frac{10}{3}$ |
| | | $Z_w = 0$ | 1 | $-3\uparrow$ | 2 | 0 | 0 | 0 | $\Delta_j$ |
| | | | $w_2$ enters and $s_2$ leaves | | | | | | |
| $s_1$ | 0 | 10 | $\boxed{\frac{5}{2}}$ | 0 | 2 | 1 | $\frac{1}{4}$ | 0 | $\rightarrow$ |
| $w_2$ | 3 | 3 | $-\frac{1}{2}$ | 1 | 0 | 0 | $\frac{1}{4}$ | 0 | |
| $s_3$ | 0 | 1 | $-\frac{5}{2}$ | 0 | 8 | 0 | $-\frac{3}{4}$ | 1 | |
| | | $Z_w = 9$ | $-\frac{1}{2}\uparrow$ | 0 | 2 | 0 | $\frac{3}{4}$ | 0 | $\Delta_j$ |
| | | | $w_1$ enters and $s_1$ leaves | | | | | | |
| $w_1$ | $-1$ | 4 | 1 | 0 | $\frac{4}{5}$ | $\frac{2}{5}$ | $\frac{1}{10}$ | 0 | |
| $w_2$ | 3 | 5 | 0 | 1 | $\frac{2}{5}$ | $\frac{1}{5}$ | $\frac{3}{10}$ | 0 | |
| $s_3$ | 0 | 11 | 0 | 0 | 10 | 1 | $-\frac{1}{2}$ | 1 | |
| | | $Z_w = 11$ | 0 | 0 | $\frac{12}{5}$ | $\frac{1}{5}\downarrow$ $x_1$ | $\frac{4}{5}\downarrow$ $x_2$ | $0\downarrow$ $x_3$ | $\Delta_j > 0$ |

**Solution of dual**

$$w_1 = 4$$
$$w_2 = 5$$
$$w_3 = 0$$
$$\text{Max. } Z_w = 11$$

**Solution of primal**

$$x_1 = \frac{1}{5}$$

$$x_2 = \frac{4}{5}$$

$$x_3 = 0$$

$$\text{Min. } Z = 11$$

**Note.** $\Delta_j$ values of slack variables provide solution of decision varibales in primal lpp.

**Example 4.5** *Write down the dual of the following lpp and solve it. Hence find solution of the primal lpp.*

$$\text{Max. } Z = 3x_1 - 2x_2$$

*Subject to :*

$$x_1 + x_2 \leq 5$$

$$x_1 \leq 4$$

$$x_2 \leq 6$$

$$x_2 \geq 1$$

$$x_1, x_2 \geq 0$$

***Solution*** Given lpp can be rewritten as

$$\text{Max. } Z = 3x_1 - 2w_2$$

Subject to :

$$x_1 + x_2 \leq 5$$

$$x_1 \leq 4$$

$$x_2 \leq 6$$

$$-x_2 \leq -1$$

$$x_1, x_2 \geq 0$$

Dual of given lpp is

$$\text{Min. } Z_w = 5w_1 + 4w_2 + 6w_3 - w_4$$

Subject to

$$w_1 + w_2 + ow_3 + ow_4 \geq 3$$

$$w_1 + ow_2 + w_3 - w_4 \geq -2$$

$$w_1, w_2, w_3, w_4 \geq 0$$

Rewriting dual lpp as

$$\text{Max. } Z_{w'} = -5w_1 - 4w_2 - 6w_3 + w_4$$

Subject to

$$w_1 + w_2 + ow_2 + ow_4 \geq 3$$

$$-w_1 + ow_2 - w_3 + w_4 \leq 2$$

$$w_1, w_2, w_3, w_4 \geq 0$$

Adding and subtracting slack and surplus variables repectively in constraint 2 and constraint 1.

$$\text{Max. } Z_{w'} = -5w_1 - 4w_1 - 6w_3 + w_4 + os_1 + os_2$$

Subject to

$$w_1 + w_2 + ow_3 + ow_4 - s_1 = 3$$

$$-w_1 + ow_2 - w_3 + w_4 + s_2 = 2$$

$$w_1, w_2, w_3, w_4, s_1, s_2 \geq 0$$

**Note.** Here $w_2$ and $w_4$ can be taken as basic variables as they form unit matrix in initial simplex table, so no need to add artificial varibale in constraint 1.

| | $C_j \rightarrow$ | | $-5$ | $-4$ | $-6$ | $1$ | $0$ | $0$ | |
|---|---|---|---|---|---|---|---|---|---|
| B.V. | $C_B$ | $W_B$ | $w_1$ | $w_2$ | $w_3$ | $w_4$ | $s_1$ | $s_2$ | Min ratio $W_B/w_k,\ w_k > 0$ |
| $w_2$ | $-4$ | 3 | 1 | 1 | 0 | 0 | $-1$ | 0 | |
| $w_4$ | 1 | 2 | $-1$ | 0 | $-1$ | 1 | 0 | 1 | |
| | $Z_w' = -10$ | | 0 | 0 | 5 | 0 | 4↓ $x_1$ | 1↓ $x_2$ | $\Delta_j \geq 0$ |

$$\text{Max. } Z'_w = -10$$

$$\Rightarrow \quad \text{Min. } Z_w = 10$$

$$w_1 = 0$$

$$w_2 = 3$$

$$w_3 = 0$$

$$w_4 = 2$$

Solution of primal Lpp is

$$\text{Max. } Z = 10$$

at,

$$x_1 = 4$$

$$x_2 = 1$$

## 4.4 ECONOMIC INTERPRETATION OF DUALITY

Consider the following lpp

$$\text{Max. } (Z) = C_1x_1 + C_2x_2 + \ldots + C_n x_n$$

Subject to

$$a_{11} x_1 + a_{12} x_2 + \ldots + a_{1n} x_n \leq b_1$$

$$\vdots$$

$$a_{m1} x_1 + a_{m2} x_2 + \ldots + a_{mn} x_n \leq b_m$$

$$x_1, x_2, \ldots, x_n \geq 0$$

Here $C_j$ is revenue per unit of activity $j$

$b_i$ is maximum availability of resource $i$

$a_{ij}$ is consumption of resource $i$ per unit of activity $j$

Let the $m$ equations in the primal represent availability of $m$ resources $b_1, \ldots, b_m$.

Suppose an entrepreneur wants to purchase all the requisite resources $b_1$, ...., $b_m$ at minimum costing.

If $w_1, w_2, \ldots w_m$ represent dual variables showing fair prices of 1 unit of associated resource, also known as marginal prices or shadow prices, then dual lpp may be formed as :

$$\text{Min. } (Z_w) = b_1 w_1 + b_2 w_2 + \ldots + b_m w_m$$

Subject to

$$a_{11} w_1 + a_{21} w_2 + \ldots + a_{m1} w_m \geq C_1$$

$$\vdots$$

$$a_{1n} w_1 + a_{2n} w_2 + \ldots + a_{mn} w_m \geq C_n$$

$$w_1, w_2, \ldots, w_m \geq 0$$

When both primal and dual solutions are optimal, the objective function satisfy strict equality, i.e.,

$$\text{Max. } Z = \text{Min. } Z_w$$

## 4.5 ADVANTAGES OF DUALITY

The advantages of duality are summarised as follows :

1. Computational procedure may be considerably reduced by converting

into dual if primal contains a large number of constraints and small number of varibales.

2. Solution of dual checks the accuracy of the primal solution for computational errors.

3. Duality may be used to solve lpp in which the initial solution is infeasible.

4. Economic interpretation of duality helps the management in making future decisions.

## 4.6 DUAL SIMPLEX METHOD

The dual simplex method is similar to regular simplex method. The only difference lies in the criterion used for selecting entering variable and leaving variable. The main advantage of dual simplex method is that it can solve lpp involving '$\leq$' and '$\geq$' signs in a convenient way without the use of artificial varibales.

**Working Rule for Dual Simplex Method**

**Step 1.** If the problem is of minimization, change it to maximization by putting $Z' = -(Z)$.

**Step 2.** Change all inequality signs to '$\leq$' type if not originally, multiplying by '$-1$'.

**Step 3.** By introducing slack variables, change the lpp into standard form and construct an initial dual simplex table.

**Step 4.** Test the signs of $\Delta_j$ $(Z_j - C_j)$ in the simplex table.

(i) If all $\Delta_j \geq 0$ and $X_B$ are non-negative, optimum solution has been attained.

(ii) If all $\Delta_j \geq 0$ and at least one of the $X_B$ is negative, go to step 5.

(iii) If at least one $\Delta_j$ is negative, method is not applicable to given lpp.

**Step 5.** Select the most negative $X_B$, the corresponding vector leaves the basis.

**Step 6.** Check all $x_j$'s for leaving vector,

(i) if all $x_j$'s $\geq 0$, infeasible solution.

(ii) If at least one $x_j$ is –ve, calculate the ratio $\left\{\dfrac{\Delta_j}{x_j}; x_j < 0\right\}, j = 1, 2, ...., n.$

The vector corresponding to max. $\left\{\frac{\Delta_j}{x_j}; x_j < 0\right\}$ enters the basis.

**Step 7.** Find the pivotal entry at the intersection of entering variable column and departing variable row.

**Step 8.** Make the pivotal element 1 and remaining elements in the column zero. Return to Step 4.

**Example 4.6** *Solve the following Lpp using dual simplex method.*

$$Min.\ Z = 2x_1 + x_2$$

*Subject to :*

$$3x_1 + x_2 \geq 3$$
$$4x_1 + 3x_2 \geq 6$$
$$x_1 + 2x_2 \geq 3$$
$$x_1, x_2 \geq 0$$

***Solution*** Rewriting given lpp as

$$\text{Max. } (-Z) = -2x_1 - x_2$$

Subject to :

$$-3x_1 - x_2 \leq -3$$
$$-4x_1 - 3x_2 \leq -6$$
$$-x_1 - 2x_2 \leq -3$$
$$x_1, x_2 \geq 0$$

Adding slack varibales to constraints

$$\text{Max. } (-Z) = -2x_1 - x_2 + os_1 + os_2 + os_3$$

Subject to :

$$-3x_1 - x_2 + s_1 = -3$$
$$-4x_1 - 3x_2 + s_2 = -6$$
$$-x_1 - 2x_2 + s_3 = -3$$
$$x_1, x_2, s_1, s_2, s_3 \geq 0$$

| | | $C_j \rightarrow$ | $-2$ | $-1$ | $0$ | $0$ | $0$ | |
|---|---|---|---|---|---|---|---|---|
| B.V. | $C_B$ | $W_B$ | $x_1$ | $x_2$ | $s_1$ | $s_2$ | $s_3$ | Max. $\Delta_j/x_j,\ x_j < 0$ |
| $s_1$ | 0 | $-3$ | $-3$ | $-1$ | 1 | 0 | 0 | $\left(\text{Max.}\left\{\frac{2}{-4}, \frac{1}{3}\right\} = -\frac{1}{3}\right)$ |
| $s_2$ | 0 | $-6$ | $-4$ | $\boxed{-3}$ | 0 | 1 | 0 | $\longrightarrow$ |
| $s_3$ | 0 | $-3$ | $-1$ | $-2$ | 0 | 0 | 1 | |
| | | $-Z = 0$ | 2 | $1\uparrow$ | 0 | 0 | 0 | $\Delta_j$ |
| | | | $s_2$ leaves and $x_2$ enters | | | | | |
| $s_1$ | 0 | $-1$ | $\boxed{-\frac{5}{3}}$ | 0 | 1 | $-\frac{1}{3}$ | 0 | $\longrightarrow$ |
| $x_2$ | $-1$ | 2 | $\frac{4}{3}$ | 1 | 0 | $-\frac{1}{3}$ | 0 | $\text{Max.}\left\{\frac{\frac{2}{3}}{-\frac{5}{3}}, \frac{\frac{1}{3}}{-\frac{1}{3}}\right\} = -\frac{2}{5}$ |
| $s_3$ | 0 | 1 | $\frac{5}{3}$ | 0 | 0 | $-\frac{2}{3}$ | 1 | |
| | | $-Z = -2$ | $\frac{2}{3}\uparrow$ | 0 | 0 | $\frac{1}{3}$ | 0 | $\Delta_j$ |
| | | | $s_1$ leaves and $x_1$ enters | | | | | |
| $x_1$ | $-2$ | $\frac{3}{5}$ | 1 | 0 | $-\frac{3}{5}$ | $\frac{1}{5}$ | 0 | |
| $x_2$ | $-1$ | $\frac{6}{5}$ | 0 | 1 | $\frac{4}{5}$ | $-\frac{3}{5}$ | 0 | |
| $s_3$ | 0 | 0 | 0 | 0 | 1 | $-1$ | 1 | |
| | | $-Z = -\frac{12}{5}$ | 0 | 0 | $\frac{2}{5}$ | $\frac{1}{5}$ | 0 | $\Delta_j \geq 0$ and $X_B \geq 0$ |

Solutioin in Minimum $Z = -\left(-\frac{12}{5}\right) = \frac{12}{5}$

at $x_1 = \frac{3}{5}$ and $x_2 = \frac{6}{5}$

## PREVIOUS YEARS QUESTIONS

**Example 4.7** *Find out the solution of dual of the fllowing lpp.*

$$Min.\ (Z) = 15y_1 + 7y_2 + 100\,y_3$$

Subject to

$$y_1 + 7y_2 + 3y_3 \geq 4$$
$$y_1 + 5y_2 + 5y_3 \geq 5$$
$$y_1 + 3y_2 + 10y_3 \geq 9$$
$$y_1 + 2y_2 + 15y_3 \geq 11$$
$$y_1, y_2, y_3 \geq 0$$

by using principle of duality. **(GGSIPU, End Term, 2013)**

***Solution*** **Primal** $\text{Min}\ (Z) = 15y_1 + 7y_2 + 100y_3$

Subject to :

$$y_1 + 7y_2 + 3y_3 \geq 4$$
$$y_1 + 5y_2 + 5y_3 \geq 5$$
$$y_1 + 3y_2 + 10y_3 \geq 9$$
$$Y_1 + 2y_2 + 15y_3 \geq 11$$

Dual of given lpp is

$$\text{Max.}\ (Z) = 4x_1 + 5x_2 + 9x_3 + 11x_4$$

Subject to

$$x_1 + x_2 + x_3 + x_4 \leq 15$$
$$7x_1 + 5x_2 + 3x_3 + 2x_4 \leq 7$$
$$3x_1 + 5x_2 + 10x_3 + 15x_4 \leq 100$$
$$x_1, x_2, x_3, x_4 \geq 0$$

Adding slack varibales $S_1$, $S_2$, $S_3$ to constraints for solving the dual lpp, i.e. standard forms

$$\text{Max.}\ (Z) = 4x_1 + 5x_2 + 9x_3 + 11x_4 + os_1 + os_2 + os_3$$

Subject to :

$$x_1 + x_2 + x_3 + x_4 + s_1 = 15$$
$$7x_1 + 5x_2 + 3x_3 + 2x_4 + s_2 = 7$$
$$3x_1 + 5x_2 + 10x_3 + 15x_4 + s_3 = 100$$
$$x_1, x_2, x_3, s_1, s_2, s_3 \geq 0$$

| | $C_j \rightarrow$ | | 4 | 5 | 9 | 11 | 0 | 0 | 0 | |
|---|---|---|---|---|---|---|---|---|---|---|
| B.V. | $C_B$ | $x_B$ | $x_1$ | $x_2$ | $x_3$ | $x_4$ | $S_1$ | $S_2$ | $S_3$ | Min ratio $x_B/x_k,\ x_k > 0$ |
| $S_1$ | 0 | 15 | 1 | 1 | 1 | 1 | 1 | 0 | 0 | $\frac{15}{1}=15$ |
| $S_2$ | 0 | 7 | 7 | 5 | 3 | [2] | 0 | 1 | 0 | $\frac{7}{2}=3.5 \rightarrow$ |
| $S_3$ | 0 | 100 | 3 | 5 | 10 | 15 | 0 | 0 | 1 | $\frac{100}{15}=6.6$ |
| | | $z = 0$ | $-4$ | $-5$ | $-9$ | $-11\uparrow$ | 0 | 0 | 0 | $\Delta_j = z_j - C_j$ |
| $S_1$ | 0 | $11.5=\frac{23}{2}$ | $-\frac{5}{2}$ | $-\frac{3}{2}$ | $-\frac{1}{2}$ | 0 | 1 | $-\frac{1}{2}$ | 0 | |
| $x_4$ | 11 | $3.5=\frac{7}{2}$ | $\frac{7}{2}$ | $\frac{5}{2}$ | $\frac{3}{2}$ | 1 | 0 | $\frac{1}{2}$ | 0 | |
| $S_3$ | 0 | $42.5=\frac{95}{2}$ | $-\frac{99}{2}$ | $-\frac{65}{2}$ | $-\frac{25}{2}$ | 0 | 0 | $-\frac{15}{2}$ | 1 | |
| | | $z = 38.5$ | $\frac{69}{2}$ | $\frac{45}{2}$ | $\frac{15}{2}$ | 0 | 0 | $\frac{11}{2}$ | 0 | $\Delta_j = z_j - C_j$ |

Hence all the $Z_j - C_j \geq 0$, this is the optimal solution

**Dual Solution**

$$\text{Max. } (Z) = 38.5$$
$$x_1 = 0$$
$$x_2 = 0$$
$$x_3 = 0$$
$$x_4 = 3.5$$

**Ans.**

**Primal Solution**

$$\text{Min. } (Z) = 38.5$$
$$y_1 = 0$$
$$y_2 = 5.5$$
$$y_3 = 0$$

**Ans.**

***Example* 4.8** *Find out the dual of the following Lpp.*

$$\text{Min. } Z = 3x_1 + 10x_2 + 2x_3$$

*Subject to :*

$$2x_1 + 3x_2 + 2x_3 \leq 7$$
$$3x_1 - 2x_2 + 4x_3 = 3$$
$$x_1, x_2, x_3 \geq 0$$

**(GGSIPU, Feb. 2013)**

***Solution***

**Primal**

$$\text{Min. } Z = 3x_1 + 10x_2 + 2x_3$$

Subject to :

$$-2x_1 - 3x_2 - 2x_3 \geq -7$$
$$-3x_1 + 2x_2 - 4x_3 \geq -3$$
$$3x_1 - 2x_2 + 4x_3 \geq 3$$
$$x_1, x_2, x_3 \geq 0$$

**Dual**

$$\text{Max. } (Z) = -7w_1 - 3w_2' + 3w_2''$$

Subject to :

$$-2w_1 - 3w_2' + 3w_2'' \leq 3$$
$$-3w_1 + 2w_2' - 2w_2'' \leq 10$$
$$-2w_1 - 4w_2' + 4w_2'' \leq 2$$

Hence

$$\text{Max. } (Z) = -7w_1 - 3w_2$$

Subject to :

$$-2w_1 - 3w_2 \leq 3$$
$$-3w_1 + 2w_2 \leq 10$$
$$-2w_1 - 4w_2 \leq 2$$
$$w_1 \geq 0, \quad w_2 \rightarrow \text{unrestricted}$$
$$w_2 = w_2' - w_2''$$

**Example 4.9** *Obtain dual of the following Lpp*

$$\text{Min. } (Z) = x_1 + x_2 + x_3$$

*Subject to :*

$$x_1 - 3x_2 + 4x_3 = 5$$
$$x_1 - 2x_2 \leq 3$$

$$2x_1 - x_3 \geq 4$$
$$x_1, x_2 \geq 0,$$
$x_3$ *is unrestricted* **(GGSIPU, Feb. 2014)**

***Solution*** **Primal**

$$\text{Min. } (Z) = x_1 + x_2 + x_3$$

Subject to :

$$x_1 - 3x_2 + 4x_3 = 5$$
$$x_1 - 2x_2 \leq 3$$
$$2x_2 - x_3 \geq 4$$
$$x_1, x_2 \geq 0, \quad x_3 \quad \text{unrestricted}$$

$$\therefore \qquad \text{Min. } Z = x_1 + x_2 + x_3' - x_3''$$

Subject to :

$$x_1 - 3x_2 + 4x_3' - 4x_3'' \geq 5$$
$$-x_1 + 3x_2 - 4x_3' + 4x_3'' \geq -5$$
$$-x_1 + 2x_2 \geq -3$$
$$2x_2 - x_3' + x_3'' \geq 4$$

**Dual**

$$\text{Max. } (Z) = 5w_1' - 5w_1'' - 3w_2 + 4w_3$$

Subject to :

$$w_1' - w_1'' - w_2 \leq 1$$
$$-3w_1' + 3w_1'' + 2w_2 + 2w_3 \leq 1$$
$$4w_1' + 4w_1'' - 1w_3 \leq 1$$
$$-4w_1' + 4w_1'' + w_3 \leq -1$$

Now

$$\text{Max. } (Z) = 5(w_1' - w_1'') - 3w_2 + 4w_3$$

Subject to :

$$(w_1' - w_1'') - w_2 \leq 1$$
$$-3(w_1' - w_1'') + 2w_2 + 2w_3 \leq 1$$
$$4(w_1' - w_1'') - w_3 \leq 1$$
$$-4(w_1' - w_1'') + w_3 \leq -1$$
$$\text{Max. } (Z) = 5w_1 - 3w_2 + 4w_3$$

Subject to

$$w_1 - w_2 \leq 1$$
$$-3w_1 + 2w_2 + 2w_3 \leq 1$$
$$4w_1 - w_3 \leq 1$$
$$4w_1 - w_3 \geq 1$$

$\therefore$
$$\text{Max. }(Z) = 5w_1 - 3w_2 + 4w_3$$

Subject to

$$w_1 - w_2 \leq 1$$
$$-3w_1 + 2w_2 + 2w_3 \leq 1$$
$$4w_1 - w_3 = 1$$
$$w_2 \cdot w_3 \geq 0$$
$$w_1 - \text{unrestricted}$$

**Example 4.10** *Obtain the dual of the following linear programming problem :*

$$\text{Min. }(Z) = x_1 + 2x_2$$
$$2x_1 + x_2 \leq 160$$
$$x_1 - x_2 = 30$$
$$x_1 \geq 10$$
$$x_1, x_2 \geq 10$$

**(GGSIPU, Feb. 2012, Reappear)**

***Solution*** **Primal**

$$\text{Min. }(Z) = x_1 + 2x_2$$

Subject to :

$$2x_1 + x_2 \leq 160$$
$$x_1 - x_2 = 30$$
$$x_1 \geq 10$$
$$x_1, x_2 \geq 0$$

$$\text{Min. }(Z) = x_1 + 2x_2$$

Subject to ;

$$-2x_1 - x_2 \geq -160$$
$$-x_1 + x_2 \geq -30 \qquad a \leq b$$
$$x_1 - x_2 \geq 30 \qquad a \geq b$$
$$x_1 \geq 10$$

$$x_1, x_2 \geq 0$$

**Dual**

$$\text{Max. } (Z) = -160\, w_1 - (30w_2' - 30w_2'') + 10w_3$$

Subject to :

$$-2w_1 - (w_2' - w_2'') + w_3 \leq 1$$
$$-w_1 + (w_2' - w_2'') \leq 2$$
$$w_1, w_2', w_2'', w_3 \geq 0$$

**Dual**

$$\text{Max. } (Z^*) = -160w_1 - 30w_2 + 10w_3$$

Subject to :

$$-2w_1 - w_2 + w_3 \leq 1$$
$$-w_1 + w_2 \leq 2$$
$$w_1, w_3 \geq 0$$
$$w_2 \text{ unrestricted}$$

**Example 4.11** *Obtain the dual of the following linear programming problem*

$$\text{Min. } (Z) = x_1 - 3x_2 - 2x_3$$

*Subject to :*

$$3x_1 - x_2 + 2x_3 \leq 7$$
$$2x_1 - 4x_2 \geq 12$$
$$-4x_1 + 3x_2 + 8x_3 \leq 10$$
$$x_1, x_2 \geq 9$$

*$x_3$ is unrestricted in sign* **(GGSIPU, Feb. 2012)**

***Solution*** **Primal**

$$\text{Min. } (Z) = x_1 - 3x_2 - 2x_3$$

Subject to :

$$3x_1 - x_2 + 2x_3 \leq 7$$

{Minimize problem hence all the sign are $\geq$ type}

$$2x_1 - 4x_2 \geq 12$$
$$-4x_1 + 3x_2 + 8x_3 \leq 10$$
$$x_1, x_2 \geq 0$$

$x_3$ is unrestricted in sign

**For Dual**

$$\text{Min. } (Z) = x_1 - 3x_2 - 2x_3' + 2x_3''$$

Subject to

$$-3x_1 + x_2 - 2x_3' + 2x_3'' \geq -7 \qquad (x_3 = x_3' - x_3'')$$
$$2x_1 - 4x_2 > 12$$
$$4x_1 - 3x_2 - 8x_3' + 8x_3'' \geq -10$$
$$x_1, x_2, xx_3', xx_3'' \geq 0$$

**Dual**

$$\text{Max. } (Z) = -7w_1 + 12w_2 - 10w_3$$

Subject to

$$-3w_1 + 2w_2 + 4w_3 \leq 1$$
$$1w_1 - 4w_2 - 3w_3 \leq -3$$
$$\begin{bmatrix} -2w_1 - 8w_3 \leq -2 \\ 2w_1 + 8w_3 \leq 2 \end{bmatrix} \qquad \begin{matrix} -a \leq -b \Rightarrow a \geq b \\ a \leq b \end{matrix}$$
$$2w_1 + 8w_3 = 2 \qquad \left.\begin{matrix} a \geq b \\ a \leq b \end{matrix}\right] a = b$$
$$w_1, w_2, w_3 \geq 0$$

## MULTIPLE CHOICE QUESTIONS

4.1. Duality may be used to solve lpp in which the initial solution is

(a) Bounded (b) Infeasible
(c) Unbounded (d) Feasible

4.2. In primal form if the *j*th variable is unsestricted in sign, then the corresponding dual constraint is

(a) ≤ sign (b) ≥ sign
(c) = sign (d) None

4.3. The dual of dual is

(a) Primal (b) Feasible
(c) Optimal (d) Dual

4.4. If either primal or the dual lpp has an unbounded solution then the other has

(a) no solution (b) a bounded solution
(c) alternate optimal solution (d) infeasible solution

4.5. What is the relation between number of constraints in the dual and number of varibales in the primal?

(a) Both are same
(b) Both are independent
(c) Constraints are more than variables

(d) None

4.6. If the dual has alternate optimal solution then the given problem has

(a) degenerated optimal solution (b) no solution

(c) alternate optimal solution (d) bounded solution

## ANSWERS

1. (b) 2. (c) 3. (a) 4. (d) 5. (a) 6. (a)

## SUMMARY

Every linear programming problem (called the primal) is associated with another Lpp (called its dual). Duality provides interesting insight into economic and sensitivity analysis.

**Procedure to convert primal into dual LPP**

**Step 1.** If the constraints in primal have equality sign, change it into two inequality constraints.

$$\therefore \quad a = b$$

$$\Leftrightarrow \quad a \leq b$$

$$\text{and} \quad a \geq b$$

**Step 2.** Unrestricted varibales if any are replaced by the difference of two non –ve variables.

**Step 3.** If the objective function in the primal is of maximization type, ensure all inequality signs are of '$\leq$' type and if objective function is of minimization, make all inequality signs to '$\geq$' if not original.

**Step 4.** Finally dual of the given lpp is obtained by

(i) changing the 'Maximize' function to 'Minimize' and vice-versa.

(ii) transposing the rows and column of constraint coefficients.

(iii) changing the inequality sign of '$\leq$' to '$\geq$' and vice versa.

(iv) transposing the coefficient ($C_1$, $C_2$, ...... $C_n$) of the objective function and right hand side constraints ($b_1$, $b_2$, ...... $b_m$)

**Advantages of duality**

The advantages of duality are summarised as follows :

1. Computational procedure may be considerably reduced by converting into dual if primal contains a large number of constraints and small number of varibales.
2. Solution of dual checks the accuracy of the primal solution for computational error.
3. Duality may be used to solve lpp in which the initial solution is infeasible.
4. Economic interpretation of duality helps the management in making future decisions.

CHAPTER 5

# Transportation Problems

## 5.1 INTRODUCTION

Transportation model is one of the most useful application of linear programming techniques. It is apparently used when we want to determine the minimum cost at which goods can be transported from given sources to specified destinations.

Let there be $m$ sources $S_i$, $i = 1, \ldots .. m$ and $n$ destinations $D_j$, $j = 1, \ldots ., n$. Also let $a_i$ denote availability in $i^{th}$ source, $b_j$ be the requirement in $j^{th}$ destination and the cost of transportation of one unit of commodity from source $S_i$ to destination $D_j$ be $C_{ij}$.

If $x_{ij}$ denotes quality transported from origin $S_i$ to destination $D_j$, then a general transportation problem may be formulated as

$$\text{Minimize } Z = \sum_{i=1}^{m} \sum_{j=1}^{n} C_{ij} x_{ij}$$

Subject to

$$\sum_{j=1}^{n} x_{ij} = a_i, \; i = 1, 2, \ldots, m$$

$$\sum_{i=1}^{m} x_{ij} = b_j, \; j = 1, 2, \ldots n$$

$$x_{ij} \geq 0$$

For a balanced transportation problem

$$\sum_{i=1}^{m} a_i = \sum_{j=1}^{n} b_j$$

## 5.2 INITIAL BASIC FEASIBLE SOLUTION

A general transportation problem can be represented in tabular form as shown below

**Destinations**

| | | $D_1$ | $D_2$ | | $D_j$ | | $D_n$ | Availability↓ |
|---|---|---|---|---|---|---|---|---|
| | $S_1$ | $C_{11}$ | $C_{12}$ | | $C_{ij}$ | | $C_{1n}$ | $a_1$ |
| Sources | $S_2$ | $C_{21}$ | $C_{22}$ | | $C_{2j}$ | | $C_{2n}$ | $a_2$ |
| | $S_i$ | $C_{i1}$ | $C_{i2}$ | | $C_{ij}$ | | $C_{in}$ | $a_i$ |
| | | | | | | | | |
| | $S_m$ | $C_{m1}$ | $C_{m2}$ | | $C_{mj}$ | | $C_{mn}$ | $a_m$ |
| Requirement → | | $b_1$ | $b_2$ | | $b_j$ | | $b_n$ | |

Problem aspires to find positive allocations $x_{ij}$, so that the total cost of transportation is minimized. Since the mathematical model of a transportation problem involves all equality constraints, the traditional methods of solving the problem using simplex technique may not be so convenient to use.

So some more practical methods are designed to find an initial basic feasible solution for a balanced transportation problem. If the problem is not balanced (i.e. sum of availabilities ≠ sum of requirements), it can be balanced by taking a fictitious source or destination as required and then an initial basic feasible relation may be found using any of the following methods :

1. North-West Corner Rule
2. Matrix-Minima or Least-Cost Method
3. Row-Minima Method
4. Column-Minima Method
5. Vogel's Approximation Method (VAM)

The basic feasible solution obtained by using any of the above methods is said to be non-degenerate if

1. total number of positive allocations is exactly equal to $(m + n - 1)$.

2. these allocations are in independent positions i.e., there is no loop formation by vertical or horizontal movement between the allocated cells.

**Independent positions**

| * | * | | |
|---|---|---|---|
| | | * | * |
| | * | | * |

**Non-independent positions**

| * | | * | |
|---|---|---|---|
| | * | | * |
| * | | * | * |

### 5.2.1 North-West Corner Rule

*Algorithm*

**Step 1.** Balance the transportation problem if not balanced initially.

**Step 2.** Allocate $x_{11} = \min(a_1, b_1)$ in the cell (1, 1) of the table, which is the north-west corner

**Step 3.** If $x_{11} = a_1$, the remaining cells of first row can not be allocated with any quantity, since the availability in $S_1$ has been exhausted. Replace $b_1$ by $b_1 - a_1$. Similarly if $x_{11} = b_1$, first column need not be allocated with any more amount since requirement of $D_1$ has been fulfilled. Replace $a_1$ by $a_1 - b_1$. If $a_1 = b_1$, then $x_{11} = a_1 = b_1$ and simultaneously remaining cells of first row and first column are crossed out. This gives an indication that a degenerate basic feasible solution will be produced i.e., number of allocated cells will be less than $m + n - 1$.

**Step 4.** Move to right hand cell (1, 2), if there is still any available quantity left, otherwise move down to cell (2, 1).

**Step 5.** Repeat step 4 until all quantity has been exhausted.

**Example 5.1** *Find the initial basic feasible solution to the following transportation problem by North-West corner rule.*

| | $D_1$ | $D_2$ | $D_3$ | $D_4$ | |
|---|---|---|---|---|---|
| $S_1$ | 2 | 3 | 4 | 1 | 10 |
| $S_2$ | 1 | 4 | 3 | 5 | 13 |
| $S_3$ | 3 | 2 | 1 | 4 | 12 |
| | 7 | 11 | 8 | 9 | |

*Solution* Given transportation problem is balanced one since

$$\sum_{i=1}^{3} a_i = \sum_{j=1}^{4} b_j = 35$$

Taking $x_{11} = \min(10, 7) = 7$, $\therefore$ $b_1 = 0$ and $a_1 = 10-7 = 3$

| | $D_1$ | $D_2$ | $D_3$ | $D_4$ | |
|---|---|---|---|---|---|
| $S_1$ | 2(7) | 3 | 4 | 1 | 3 |
| $S_2$ | 1(×) | 4 | 3 | 5 | 13 |
| $S_3$ | 3(×) | 2 | 1 | 4 | 12 |
| | 0 | 11 | 8 | 9 | |

Now moving to cell (1, 2), allocating an amount min. (3, 8) = 3 and crossing out remaining cells in the row

| | $D_1$ | $D_2$ | $D_3$ | $D_4$ | |
|---|---|---|---|---|---|
| $S_1$ | 2(7) | 3(3) | 4(×) | 1(×) | 0 |
| $S_2$ | 1(×) | 4 | 3 | 5 | 13 |
| $S_3$ | 3(×) | 2 | 1 | 4 | 12 |
| | 0 | 8 | 8 | 9 | |

Repeating the procedure with cell (2, 2)

| | $D_1$ | $D_2$ | $D_3$ | $D_4$ | |
|---|---|---|---|---|---|
| $S_1$ | 2(7) | 3(3) | 4(×) | 1(×) | 0 |
| $S_2$ | 1(×) | 4(8) | 3 | 5 | 5 |
| $S_3$ | 3(×) | 2(×) | 1 | 4 | 12 |
| | 0 | 0 | 8 | 9 | |

Now moving to right hand side cell (2, 3), allocating an amount min. {5, 8} = 5

| | $D_1$ | $D_2$ | $D_3$ | $D_4$ | |
|---|---|---|---|---|---|
| $S_1$ | 2(7) | 3(3) | 4(×) | 1(×) | 0 |
| $S_2$ | 1(×) | 4(8) | 3(5) | 5(×) | 0 |
| $S_3$ | 3(×) | 2(×) | 1 | 4 | 12 |
| | 0 | 0 | 3 | 9 | |

Now allocating 3 and 9 respectively to cells (3, 3) and (3, 4), final transportation table is given by

| | $D_1$ | $D_2$ | $D_3$ | $D_4$ | |
|---|---|---|---|---|---|
| $S_1$ | 2(7) | 3(3) | 4 | 1 | 10 |
| $S_2$ | 1 | 4(8) | 3(5) | 5 | 13 |
| $S_3$ | 3 | 2 | 1(3) | 4(9) | 12 |
| | 7 | 11 | 8 | 9 | |

$$\begin{aligned}\text{Total transportation cost} &= 2(7) + 3(3) + 4(8) + 3(5) + 1(3) + 4(9) \\ &= 14 + 9 + 32 + 15 + 3 + 36 \\ &= 109 \text{ units}\end{aligned}$$

***Note.*** Given example has been elaborated for demonstration purpose, practically all the steps can be performed in a single table.

**Example 5.2** *Following table depicts cost of transportation from four factory outlets $F_1$, $F_2$, $F_3$ and $F_4$ to three warehouses $W_1$, $W_2$ and $W_3$. Find an initial basic feasible solution using North-West corner rule.*

| | $W_1$ | $W_2$ | $W_3$ | *Availability* ↓ |
|---|---|---|---|---|
| $F_1$ | 2 | 7 | 4 | 5 |
| $F_2$ | 3 | 3 | 1 | 8 |
| $F_3$ | 5 | 4 | 7 | 7 |
| $F_4$ | 1 | 6 | 2 | 14 |
| *Requirement* → | 7 | 9 | 18 | |

***Solution*** Starting from cell (1, 1) and proceeding right (if any availability is left) and downwards to make suitable allocations ;

| | $W_1$ | $W_2$ | $W_3$ | *Availability* |
|---|---|---|---|---|
| $F_1$ | 2(5) | 7× | 4× | ~~5~~ |
| $F_2$ | 3(2) | 3(6) | 1× | ~~8~~ ~~6~~ |
| $F_3$ | 5 | 4(3) | 7(4) | ~~7~~ ~~4~~ |
| $F_4$ | 1× | 6× | 2(14) | ~~14~~ |
| *Requirement* → | ~~7~~ ~~2~~ | ~~9~~ ~~3~~ | ~~18~~ ~~4~~ | |

Total transportation cost $= 2(5) + 3(2) + 3(6) + 4(3) + 7(4) + 2(14)$

$= 102$ units

### 5.2.2 Matrix-Minima or Least-Cost Method

In this method, we look for the minimum cost element in the entire matrix and allocate as much as possible. The procedure is given below.

**Step 1.** Balance the transportation problem if not balanced initially.

**Step 2.** Locate the cell having minimum cost element $C_{ij}$ and allocate maximum possible quantity by taking $x_{ij} = \min(a_i, b_j)$. If such cell of minimum cost is not unique i.e. if there is a tie, select the cell where we can allocate more amount.

**Step 3.** If $x_{ij} = a_i$, cross out the remaining cells in the $i^{th}$ row, and if $x_{ij} = b_j$, remaining cells of $j^{th}$ column are crossed-out.

**Step 4.** Repeat steps 2 and 3 untill all allocations are made.

**Example 5.3** *Find an initial basic feasible solution for the transportation schedule given in example 5.1 using least cost method.*

***Solution*** There is tie among the least cost cells at (1, 4), (2, 1) and (3, 3). Thereby looking for a cell, which may be allocated with more quantity, which is (1, 4) initially.

| | $D_1$ | $D_2$ | $D_3$ | $D_4$ | |
|---|---|---|---|---|---|
| $S_1$ | 2× | 3(1) | 4× | 1(9) | ~~10~~ ~~1~~ |
| $S_2$ | 1(7) | 4(6) | 3× | 5× | ~~13~~ ~~6~~ |
| $S_3$ | 3× | 2(4) | 1(8) | 4× | ~~12~~ ~~4~~ |
| | ~~7~~ | ~~A~~ ~~7~~ ~~6~~ | ~~8~~ | ~~9~~ | |

Transportation cost $= 3(1) + 1(9) + 1(7) + 4(6) + 2(4) + 1(8)$

$= 59$ units

**Example 5.4** *Find an initial basic feasible solution for a transportation schedule given in example 5.2, using least cost method.*

***Solution*** Choosing the least cost cell (2, 3) initially, as we can allocate it with more quantity.

| | $W_1$ | $W_2$ | $W_3$ | |
|---|---|---|---|---|
| $F_1$ | 2× | 7(2) | 4(3) | ~~5~~ ~~2~~ |
| $F_2$ | 3× | 3× | 1(8) | ~~8~~ |
| $F_3$ | 5× | 4(7) | 7× | ~~7~~ |
| $F_4$ | 1(7) | 6× | 2(7) | ~~14~~ ~~7~~ |
| | ~~7~~ | ~~9~~ ~~2~~ | ~~18~~ ~~10~~ ~~3~~ | |

$$\text{Transportation cost} = 7(2) + 4(3) + 1(8) + 4(7) + 1(7) + 2(7)$$
$$= 83 \text{ units}$$

### 5.2.3 Row-Minima Method

We choose the cell in the first row with lowest cost and allocate as much amount as possible. This allocation is repeated for remaining cells of the first row till $a_1$ is exhausted. Procedure is continued by taking up next rows one at a time, till all $a_i$'s are exhausted.

**Example 5.5** *Find an initial basic feasible solution for a transportation schedule given in example 5.1, using Row-Minima method.*

***Solution*** Least cost cell in the first row is at the position (1, 4). Allocating $\min(a_1, b_4)$ i.e. min (10, 9) = 9 to (1, 4) and subsequently allocating min (1, 7) = 1 to (1, 1).

| | $D_1$ | $D_2$ | $D_3$ | $D_4$ | |
|---|---|---|---|---|---|
| $S_1$ | 2(1) | 3× | 4× | 1(9) | ~~10~~ ~~1~~ |
| $S_2$ | 1(6) | 4× | 3(7) | 5× | ~~13~~ ~~7~~ |
| $S_3$ | 3× | 2(11) | 1(1) | 4× | ~~12~~ ~~11~~ |
| | ~~7~~ ~~6~~ | ~~11~~ | ~~8~~ | ~~9~~ | |

$$\text{Transportation cost} = 2(1) + 1(9) + 1(16) + 3(7) + 2(11) + 1(1)$$
$$= 61 \text{ units}$$

**Example 5.6** *Find the initial basic feasible solution for the transportation schedule given in example 5.2, using Row-Minima method.*

***Solution*** Allocating quantity 5 to minimum cost cell at (1, 1)

| | $D_1$ | $D_2$ | $D_3$ | |
|---|---|---|---|---|
| $S_1$ | 2(5) | 7× | 4× | ~~5~~ |
| $S_2$ | 3× | 3× | 1(8) | ~~8~~ |
| $S_3$ | 5× | 4(7) | 7× | ~~7~~ |
| $S_4$ | 1(2) | 6(2) | 2(10) | ~~14~~ ~~12~~ ~~10~~ |
| | ~~7~~ ~~2~~ | ~~9~~ ~~2~~ | ~~18~~ ~~10~~ | |

$$\text{Transportation cost} = 2(5) + 1(8) + 4(7) + 1(2) + 6(2) + 2(10)$$
$$= 80 \text{ units}$$

### 5.2.4 Column-Minima Method

Procedure is similar to Row-Minima method. Here the minimum cost is selected in the first column and allocations are made till the requirement of first column is exhausted. Procedure is repeated till the allocations in the last column are made.

### 5.2.5 Vogel Approximation Method (VAM) or Unit Cost Penalty Method

**Step 1.** Balance the problem if not initially balanced.

**Step 2.** Write the difference of smallest and second smallest cost in each row and column on the right of corresponding row and below the corresponding column. Each difference is named as penalty.

**Step 3.** Row or column having largest penalty is allocated by maximum possible amount to the cell with lowest cost. In case of tie i.e., if two or more row/columns having equal penalities, choose arbitrarily.

**Step 4.** Recompute the row and column differences (penalties) and repeat the procedure explained in step 3. Cells in which allocations have been made or crossed cell are not taken into account.

**Step 5.** Continue the procedure till all allocations have been made.

**Note.** Final transportation cost may vary while finding initial basic feasible solution using VAM as allocations are made arbitrarily in a tie.

**Example 5.7** *Find the initial basic feasible solution for example 5.1 using VAM.*

***Solution*** Minimum cost and next minimum cost in first column are 1 and 2 whose difference is 1. Writing 1 below column one and similarly writing penalities for all rows and columns :

| | $D_1$ | $D_2$ | $D_3$ | $D_4$ | | |
|---|---|---|---|---|---|---|
| $S_1$ | 2× | 3(1) | 4× | 1(9) | ~~10~~ ~~1~~ | (1) (1) (1) |
| $S_2$ | 1(7) | 4(6) | 3× | 5× | ~~13~~ ~~6~~ | (2) (2) ③ |
| $S_3$ | 3× | 2(4) | 1(8) | 4× | ~~12~~ ~~4~~ | (1) (1)(1) |
| | ~~7~~ | ~~11~~ | ~~8~~ | ~~9~~ | | |
| | (1) | (1) | (2) | ③ | | |
| | (1) | (1) | ② | × | | |
| | (1) | (1) | × | × | | |

In the first round, maximum penalty '3' is for fourth column, so allocating maximum possible amount '9' to the minimum cost cell (1, 4).

In the second round, maximum penalty '2' is in 3rd column and 2nd row, arbitrarily choosing 3rd column and allocating maximum possible amount '8' to minimum cost cell (3, 3).

In the third round, maximum penalty '3' is ag'inst 2nd row, so allocating maximum possible amount '7' to cell (2, 1), which is minimum cost cell in 2nd row. Finally remaining availbilities 1, 6 and 4 in respective resources $S_1$, $S_2$ and $S_3$ are filled in 2nd column.

$$\text{Transportation cost} = 3(1) + 1(9) + 1(7) + 4(6) + 2(4) + 1(8)$$
$$= 59 \text{ units}$$

**Example 5.8** *Find the initial basic feasible solution using VAM for the problem 5.2 .*

**Solution** Minimum and next minimum cost cells in the first row are '2' and '4' so penalty for first row is 4 – 2= 2. Writing all the penalties on right of each row and below each column.

| | $D_1$ | $D_2$ | $D_3$ | | |
|---|---|---|---|---|---|
| $S_1$ | 2(3) | 7(2) | 4× | ~~5~~ ~~2~~ | (2) (2) (5) ⑤ |
| $S_2$ | 3× | 3× | 1(8) | ~~8~~ | ② × × × × |
| $S_3$ | 5× | 4(7) | 7× | ~~7~~ | (1) (1)(1)(1) |
| $S_4$ | 1(4) | 6× | 2(10) | ~~14~~ ~~4~~ | (1) (1)⑤ × |
| | ~~7~~ ~~3~~ | ~~9~~ | ~~18~~ ~~10~~ | 34 | |
| | (1) | (1) | (1) | | |
| | (1) | (2) | ② | | |
| | (1) | (2) | × | | |
| | (3) | (3) | × | | |

In round one maximum panalties are for Row 1 and Row 2. Choosing row 2 arbitrarily and alloting maximum possible amount 8 to cell (2, 3) which is the minimum cost cell.

In round two Row 1, Column 2 and Column 3 are having maximum penalties. Choosing column 3 arbitrarily and alloting maximum possible amount '10' to minimum cost cell (4, 3), requirement in $D_3$ has been met.

In round three maximum panalties are for $R_1$ and $R_4$. Choosing $R_4$ arbitrarily and alloting an amount '4' minimum cost cell (4, 1), availability with $S_4$ is over.

In round four, maximum penalty is against Row 1. So alloting the maximum possible amount '3' to (1, 1), requirement in $D_1$ is over.

Cells (1, 2) and (3, 2) are alloted with remaining amounts '2' and '7' respectively.

$$\text{Transportation cost} = 2(3) + 7(2) + 1(8) + 4(7) + 1(4) + 2(10)$$
$$= 80 \text{ units}$$

## 5.3 MODI (MODIFIED DISTRIBUTION) METHOD—OPTIMAL SOLUTION

An initial basic feasible solution is obtained by using any of the methods discussed in previous section. Most commonly used method for finding an optimal solution is MODI (Modified Distribution) method also known as U–V method.

### Working Rule for Finding Optimal Solution using MODI Method

**Step 1.** Find an inital basic feasible solution using any appropriate method preferably VAM.

**Step 2.** Find set of numbers $u_i$ ($i = 1, 2, ....., m$) and $v_j$ ($j = 1, 2,......, n$) for each occupied cell $(r, s)$ such that $C_{rs} = u_r + v_s$ by putting any of $v_i$ or $v_j$ zero initially.

**Step 3.** Enter $C_{ij}$ for each unoccupied cell $(i, j)$ in the upper left corner and enter $(u_i + v_j)$ for same cell in the upper right corner.

**Step 4.** Enter $d_{ij} = C_{ij} - (u_i + v_j)$ for each unoccupied cell $(i, j)$ at the lower right corner.

**Step 5.** Examine $d_{ij}$ for occupied cells.

(i) If all $d_{ij} > 0$, solution is optimal and unique.

(ii) If all $d_{ij} \geq 0$, solution is optimal and an alternative optimal solution exists.

(iii) If at least one $d_{ij} < 0$, solution is not optimal and go to step 6.

**Step 6.** Form a new basic feasible solution by giving an allocation to the cell for which $d_{ij}$ is most –ve by forming a loop with occupied cells and vcacating one of them. Amount to be allocated is the minimum allocation in occupied cells from where amount is to be subtracted.

**Step 7.** Repeat steps 2 to 5, until optimal solution is attained.

**Example 5.9** *Solve the following transportation problem.*

| | $D_1$ | $D_2$ | $D_3$ | |
|---|---|---|---|---|
| $S_1$ | 2 | 7 | 4 | 5 |
| $S_2$ | 3 | 3 | 1 | 8 |
| $S_3$ | 5 | 4 | 7 | 7 |
| $S_4$ | 1 | 6 | 2 | 14 |
| | 7 | 9 | 18 | |

***Solution*** ***Step 1.*** Initial basic feasible solution of the given problem using VAM as solved in example 5.8 is :

| | $D_1$ | $D_2$ | $D_3$ |
|---|---|---|---|
| $S_1$ | 2(3) | 7(2) | 4 |
| $S_2$ | 3 | 3 | 1(8) |
| $S_3$ | 5 | 4(7) | 7 |
| $S_4$ | 1(4) | 6 | 2(10) |

***Step 2.*** To find is $u_i$ and $v_j$, marking occupied cells with dots and entering associated cost in the upper left corner as follows :

| | | | |
|---|---|---|---|
| 2 • | 7 • | | $u_1 = 1$ |
| | | 1 • | $u_2 = -1$ |
| | 4 • | | $u_3 = -2$ |
| 1 • | | 2 • | $u_4 = 0$ |
| $v_1 = 1$ | $v_2 = 6$ | $v_3 = 2$ | |

To find $u_i$ and $v_j$'s, put $u_i$ corresponding to row having maximum allocations as zero. In case of tie, choose arbitrarily. Here we put $u_4 = 0$, alternatively one can also take $u_1$ as zero.

$$
\begin{aligned}
\text{Now} \quad & u_4 = 0 \\
\Rightarrow \quad & v_1 = 1 \text{ and } v_3 = 2 \\
& \text{as } C_{41} = u_4 + v_1 \text{ and } C_{43} = u_4 + v_3 \\
\Rightarrow \quad & u_1 = 1 \text{ and } u_2 = -1 \\
& \text{as } C_{11} = u_1 + v_1 \text{ and } C_{23} = u_2 + v_3 \\
\Rightarrow \quad & v_2 = 6 \text{ as } C_{12} = u_1 + v_2 \\
\Rightarrow \quad & u_3 = -2 \text{ as } C_{32} = u_3 + v_2
\end{aligned}
$$

Hence all $u_i$ and $v_j$ have been found out.

***Step 3 and Step 4.*** Putting $C_{ij}$ for unoccupied cells in top left corner and $u_i+v_j$ in top right corner, $d_{ij} = C_{ij} - (u_i + v_j)$ in lower right corner.

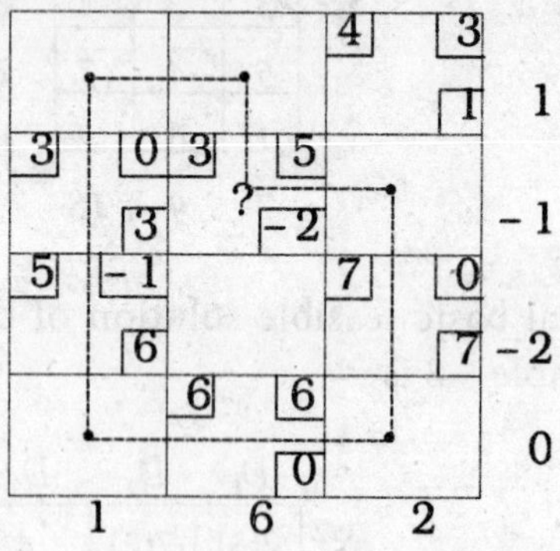


***Step 5.*** $d_{22} = -2 < 0$ $\therefore$ Solution is not optimal.

***Step 6.*** Since $d_{22}$ is –ve, making an allocation in (2, 2) by adding min{2, 4, 8} = 2 and then alternatively subtracting and adding same amount in occupied cells lieing on corner points of the loop as shown in given figure.

3+2 (1, 1) (1, 2) 2–2

0+2 8–2

(2, 2) (2, 3)

4–2 (4, 1) (4, 3) 10+2

***Step 7.*** The new basic feasible solution is

| | | |
|---|---|---|
| 2(5) | 7 | 4 |
| 3 | 3(2) | 1(6) |
| 5 | 4(7) | 7 |
| 1(2) | 6 | 2(12) |

New transportation cost $= 2(5) + 3(2) + 1(6) + 4(7) + 1(2) + 2(12)$

$= 76$ units

Checking for optimality : find $u_i$ and $v_j$.

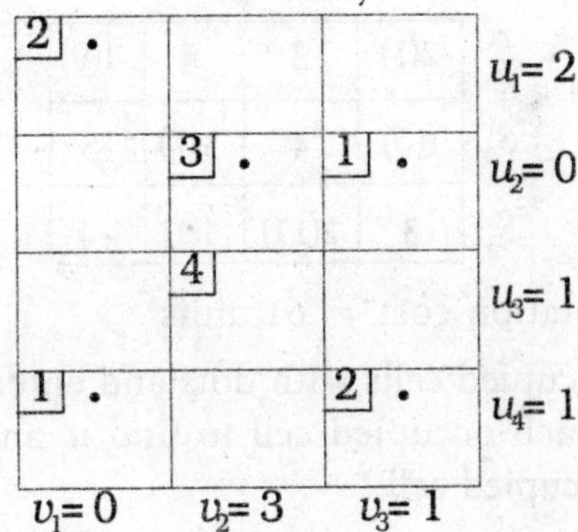

Checking $d_{ij}$ for each unoccupied cell

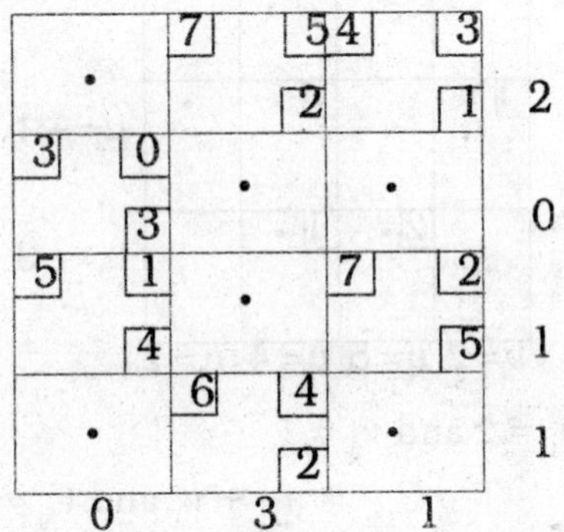

All $d_{ij} > 0$, Therefore solution is optimal and unique

Hence the current solution is the optimal solution

**Example 5.10** *Find an initial basic feasible solution of the following transportation problem using Row-Minima method. Hence find the optimal solution. Also find alternate optimal solution if it exist.*

| | $D_1$ | $D_2$ | $D_3$ | $D_4$ | |
|---|---|---|---|---|---|
| $S_1$ | 2 | 3 | 4 | 1 | 10 |
| $S_2$ | 1 | 4 | 3 | 5 | 13 |
| $S_3$ | 3 | 2 | 1 | 4 | 12 |
| | 7 | 11 | 8 | 9 | |

***Solution*** ***Step 1.*** Initial basic feasible solution using Row-Minima method as given in example 5.5 is :

| | $D_1$ | $D_2$ | $D_3$ | $D_4$ |
|---|---|---|---|---|
| $S_1$ | 2(1) | 3 | 4 | 1(9) |
| $S_2$ | 1(6) | 4 | 3(7) | 5 |
| $S_3$ | 3 | 2(11) | 1(1) | 4 |

Transprotation cost = 61 units

***Step 2.*** Marking occupied cells with dots and entering associated costs in upper left corner for each occupied cell to find $u_i$ and $v_j$ using the relation $C_{ij} = u_i + v_j$ for each occupied cell.

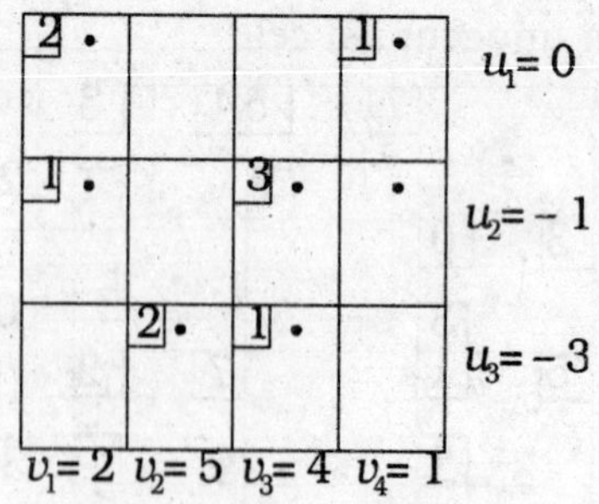


Putting $u_1 = 0 \Rightarrow v_1 = 2$ and $v_4 = 1$

as $\qquad C_{11} = u_1 + v_1$ and $C_{14} = u_1 + v_4$

Again $\qquad u_2 = -1$ as $C_{21} = u_2 + v_1$

$\Rightarrow \qquad v_3 = 4$ as $C_{23} = u_2 + v_3$

$\Rightarrow \qquad u_3 = -3$ as $C_{33} = u_3 + v_3$

$\Rightarrow \qquad v_2 = 5$ as $C_{32} = u_3 + v_2$

***Step 3 and Step 4.*** Putting $C_{ij}$ for unoccupied cells in top left corner, ($u_i$ + $v_j$) in top right corner.

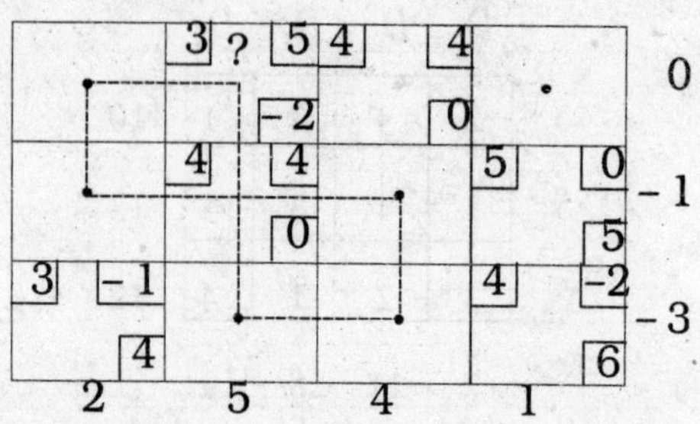

***Step 5.*** $d_{12} = 2,$

∴ we need an allocation in cell (1, 2)

***Step 6.*** Since $d_{12}$ is −ve making an allocation is (1, 2) by adding min(1, 11, 7) = 1 to it and alternatively subtracting and adding same amount in occupied cells in the loop.

1 − 1 (1, 1) (1, 2) 0 + 1
(2, 1)
6 + 1 (2, 3) 7 − 1
11 − 1 (3, 2) (3, 3) 1 + 1

***Step 7.*** The new basic feasible solution is

| | $D_1$ | $D_2$ | $D_3$ | $D_4$ |
|---|---|---|---|---|
| $S_1$ | 2 | 3(1) | 4 | 1(9) |
| $S_2$ | 1(7) | 4 | 3(6) | 5 |
| $S_3$ | 3 | 2(10) | 1(2) | 4 |

$$\text{New transprotation cost} = 3(1) + 1(9) + 1(7) + 3(6) + 2(10) + 1(2)$$
$$= 59 \text{ units}$$

Checking for optimality : finding $u_i$ and $v_j$ for occupied cells.

| | | | | |
|---|---|---|---|---|
| | 3 • | | 1 • | $u_1 = 1$ |
| 1 • | | 3 • | | $u_2 = 2$ |
| | 2 • | 1 • | | $u_3 = 0$ |
| $v_1 = -1$ | $v_2 = 2$ | $v_3 = 1$ | $v_4 = 0$ | |

Checking $d_{ij} = [C_{ij} - (u_i + v_j)]$ for each unoccupied cell.

| | | | | |
|---|---|---|---|---|
| 2 0 2 | • | 4 2 2 | • | 1 |
| • | 4 4 ? 0 | | 5 2 3 | 2 |
| 3 −1 4 | | | 4 0 4 | 0 |
| −1 | 2 | 1 | 0 | |

All $d_{ij} \geq 0$, $d_{22} = 0$, which implies that an alternate optimal solution exists.

To find alternate optimal solution, making an allocation in cell (2, 2) which is minimum of (6, 10) = 6.

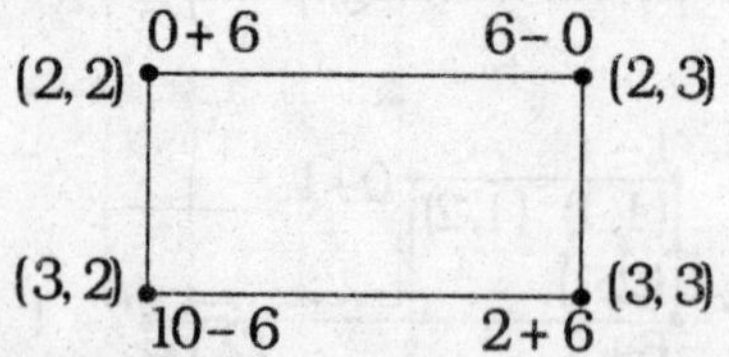


The new feasible alternate optimal solution is

| | $D_1$ | $D_2$ | $D_3$ | $D_4$ |
|---|---|---|---|---|
| $S_1$ | 2 | 3(1) | 4 | 1(9) |
| $S_2$ | 1(7) | 4(6) | 3 | 5 |
| $S_3$ | 3 | 2(4) | 1(8) | 4 |

$$\text{Transportation cost} = 3(1) + 1(9) + 1(7) + 4(6) + 2(4) + 1(8)$$
$$= 59 \text{ units}$$

## 5.4 DEGENERACY IN TRANSPORTATION PROBLEMS

Degeneracy in transportation problem occurs when,

1. Number of occupied cells is less than $m + n - 1$, where $m$ and $n$ respectively denote number of rows and columns in the transportation table.
2. Occupied cells form a loop by vertical or horizontal movements in between them.

Degeneracy can occur in two ways

(i) Initial basic feasible solution may be degenerate

(ii) The degeneracy develops at some intermediate stage while solving for an optimal solution.

### 5.4.1 Resolution of Degeneracy During Initial Stage

An extremely small quantity usually denoted by $\Delta$ or $\in$ is introduced in the least cost independent cell. If necessary two or more $\Delta s$ can be introduced in the least and second least cost cells in such a way that occupied cells do not form a loop.

**Example 5.11** *Determine your transportation plan for the given table. Also give minimum distribution cost.*

| | X | Y | Z | |
|---|---|---|---|---|
| A | 4 | 5 | 3 | 60 |
| B | 1 | 9 | 4 | 70 |
| C | 8 | 3 | 5 | 80 |
| | 50 | 80 | 80 | |

***Solution*** Finding an initial basic feasible solution using VAM :

| | X | Y | Z | | | | |
|---|---|---|---|---|---|---|---|
| A | 4× | 5× | 3(60) | ~~60~~ | 1 | 2 | ② |
| B | 1(50) | 9× | 4(20) | ~~70~~ ~~20~~ | 3 | ⑤ | × |
| C | 8× | 3(80) | 5× | ~~80~~ | 2 | 2 | 2 |
| | ~~50~~ | ~~80~~ | ~~80~~ ~~60~~ | | | | |
| | ③ | 2 | 1 | | | | |
| | × | 2 | 1 | | | | |
| | | 2 | 2 | | | | |

Total number of occupied cells are 4 which is less than 5 ($m+n-1$). Now if we allocate Δ to least cost unoccupied cell (1, 1), a loop is formed in between (1, 1), (1, 3), (2, 3) and (2, 1). So we can allocate Δ to next minimum cost cell (1, 2) or (3, 3) which do not form a loop within occupied cells. Choosing (3, 3) arbitrarily for allocation of Δ which is very small quantity and checking for optimality using MODI method :

***Step 1.*** Finding $u_i$ and $v_j$.

| | | | |
|---|---|---|---|
| | | 3 (60) | $u_1 = -2$ |
| 1 (50) | | 4 (20) | $u_2 = -1$ |
| | 3 (80) | 5 Δ | $u_3 = 0$ |
| $v_1 = 2$ | $v_2 = 3$ | $v_3 = 5$ | |

***Step 2.*** Checking for $d_{ij} = [C_{ij} - (u_i + v_j)]$

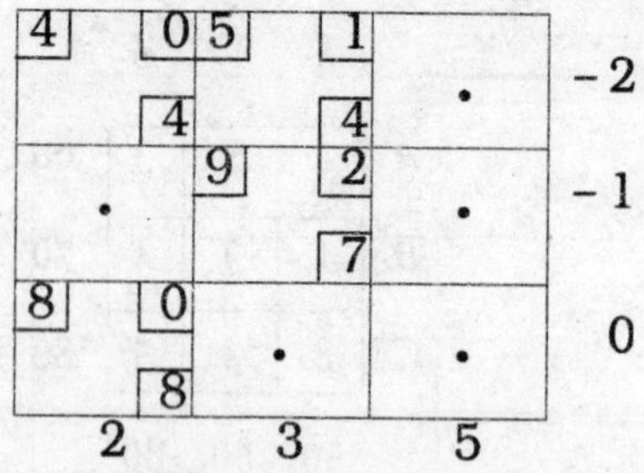

All $d_{ij} > 0$, therefore solution is optimal and unique.

Minimum transportation cost $= 3(60) + 1(50) + 4(20) + 3(80)$

$= 550$ units

**Note.** Δ plays only an auxilary role and has no contribution in final transportation cost, since it is very small quantity nearly zero. It is just introduced to find values of $u_i$ and $v_j$ which oterwise are not possible to find as number of allocated cells are less than $m+n-1$.

### 5.4.2 Resolution of Degeneracy During Solution Stage

While finding optimal solution, if simultaneously two or more occupied cells are vacated when a positive allocation is made to unoccupied cell having most negative $d_{ij}$ value, then degeneracy occurs.

To resolve degeneracy, allocate Δ to one or more of recently vacated cell, so that number of occupied cells is $m + n - 1$ in the new solution.

## 5.5 UNBALANCED TRANSPORTATION PROBLEM

If sum of availability is not equal to sum of requirements

i.e. $$\sum_{i=1}^{m} a_i \neq \sum_{j=1}^{n} b_j \text{, two cases arise :}$$

(i) $\Sigma a_i > \Sigma b_j$

To balance this type of transportation problem, add a dummy destination, '$D_j$'.

(ii) $\Sigma b_j > \Sigma a_i$

In this case add a dummy source, '$S_i$' to balance the problem.

**Example 5.12** *A potato chip manufacturer has two plants and three warehouses. Transportation costs for shipping from plants to warehouses, the plant availability and the warehouse requirements are as under :*

| *Plant* | *Warehouses* | | | *Supply* |
|---|---|---|---|---|
| | $W_1$ | $W_2$ | $W_3$ | |
| $P_1$ | 22 | 15 | 11 | 400 |
| $P_2$ | 20 | 12 | 10 | 400 |
| *Demand* | 250 | 250 | 500 | |

*Find the optimum shipping schedule.*

**Solution** Since total demand is 1000, whereas total supply is 800 units i.e.

$$\Sigma b_j > \Sigma a_i$$

Adding a dummy plant $P_3$ with zero transportation costs to all warehouses

| | $W_1$ | $W_2$ | $W_3$ | |
|---|---|---|---|---|
| $P_1$ | 22 | 15 | 11 | 400 |
| $P_2$ | 20 | 12 | 10 | 400 |
| $P_3$ | 0 | 0 | 0 | 200 |
| | 250 | 250 | 500 | 1000 |

Find initial basic feasible solution using VAM.

| | $W_1$ | $W_2$ | $W_3$ | | | |
|---|---|---|---|---|---|---|
| $P_1$ | 22× | 15× | 11(400) | ~~400~~ | 4 | ④ |
| $P_2$ | 20(50) | 12(250) | 10(100) | ~~400~~ | 2 | 2 |
| $P_3$ | 0(200) | 0× | 0× | ~~200~~ | 0 | × |
| | ~~250~~ ~~50~~ | ~~250~~ | ~~500~~ ~~100~~ | | | |
| | ⑳ | 12 | 10 | | | |
| | 2 | 3 | 1 | | | |

$$\text{Transportation cost} = 11(400) + 20(50) + 12(250) + 10(100) + 0(200) = 9400 \text{ units}$$

Total number of occupied cells = 5

∴ Solution is non-degenerate.

Checking for optimality using MODI method :

***Step 1.*** Finding $u_i$ and $v_j$ values :

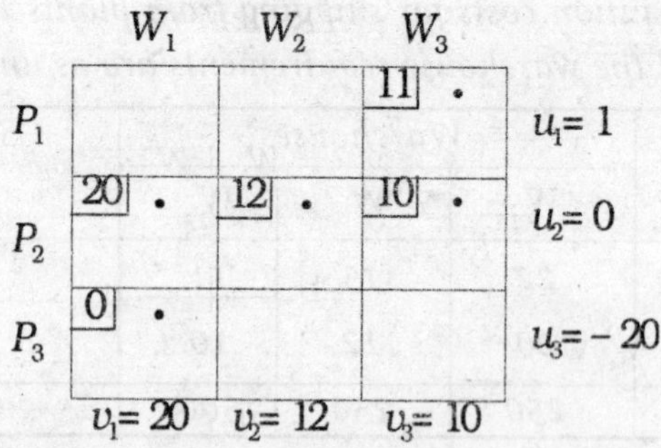


***Step 2.*** Checking $d_{ij}$ values for unoccupied cells :

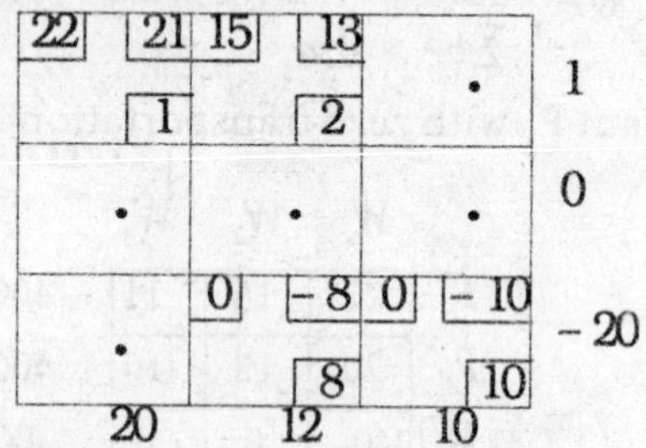


All $d_{ij} > 0$, Therefore solution is optimal and unique

## 5.6 MAXIMIZATION TRANSPORTATION PROBLEM

In certain transportation problem, objective function is of maximization instead of minimization.

To solve such type of transportation problem, each transportation cost is substracted from maximum transportation cost so that maximization transportation problem is converted into minimization transportation problem.

**Example 5.13** *Certain type of stationary items are produced in three factors and shipped to three warehouses. If the entries in the given table represent profit per unit produced at factory $F_i$ to warehouse $W_j$.*

| *Factory* | *Warehouses* | | | *Capacity* |
|---|---|---|---|---|
| | $W_1$ | $W_2$ | $W_3$ | |
| $F_1$ | 18 | 21 | 14 | 150 |
| $F_2$ | 20 | 25 | 16 | 200 |
| $F_3$ | 12 | 15 | 17 | 150 |
| *Demand* | 130 | 180 | 190 | |

*Solve the model for maximum profit.*

***Solution*** The revised transportation problem obtained by subtracting each profit from maximum profit '25' is :

| | $W_1$ | $W_2$ | $W_3$ | *Capacity* |
|---|---|---|---|---|
| $F_1$ | 7 | 4 | 11 | 150 |
| $F_2$ | 5 | 0 | 9 | 200 |
| $F_3$ | 13 | 10 | 8 | 150 |
| Demand | 130 | 180 | 190 | |

Finding an initial basic feasible solution using VAM.

| | $W_1$ | $W_2$ | $W_3$ | *Capacity* | |
|---|---|---|---|---|---|
| $F_1$ | [7] (110) | [4] | [11] (40) | ~~150~~ | 3 4 4 |
| $F_2$ | [5] (20) | [0] (180) | [9] × | ~~200~~ ~~20~~ | (5) 4 (4) |
| $F_3$ | [31] × | [10] × | [8] (150) | ~~150~~ | 2 (5) × |
| Demand | ~~130~~ ~~110~~ | ~~180~~ | ~~190~~ ~~40~~ | | |
| | 2 | 4 | 1 | | |
| | 2 | × | 1 | | |
| | 2 | × | 2 | | |

$$\text{Profit} = 18(110) + 14(40) + 20(20) + 25(180) + 17(150)$$

$$= 9990 \text{ units}$$

Checking for optimality using MODI method.

Step I Find $u_i$ and $v_j$ values

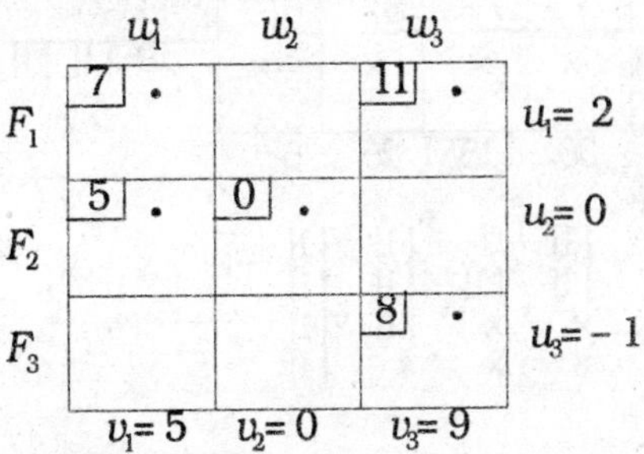

Step II Checking '$d_{ij}$' values for unoccupied cells

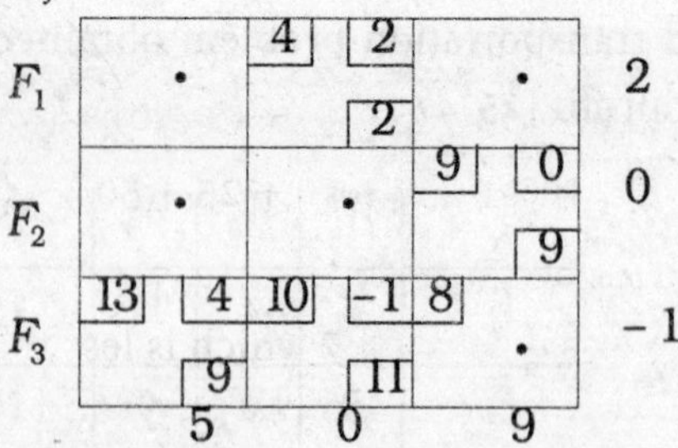

All $d_{ij} > 0$, solution is optimal

## PREVIOUS YEARS QUESTIONS

**Example 5.14** *Find the optimum solution to the following transportation problem in which the cells contain the transportation cost in rupees.*

| | $W_1$ | $W_2$ | $W_3$ | $W_4$ | $W_5$ | *Available* |
|---|---|---|---|---|---|---|
| $F_1$ | 7 | 6 | 4 | 5 | 9 | 40 |
| $F_2$ | 8 | 5 | 6 | 7 | 8 | 30 |
| $F_3$ | 6 | 8 | 9 | 6 | 5 | 20 |
| $F_4$ | 5 | 7 | 7 | 8 | 6 | 10 |
| *Required* | 30 | 30 | 15 | 20 | 5 | 100 (Total) |

**[GGSIPU, End Term - 2010]**

***Solution*** ***Step I.*** Initial basic feasible solution of the given problem using VAM.

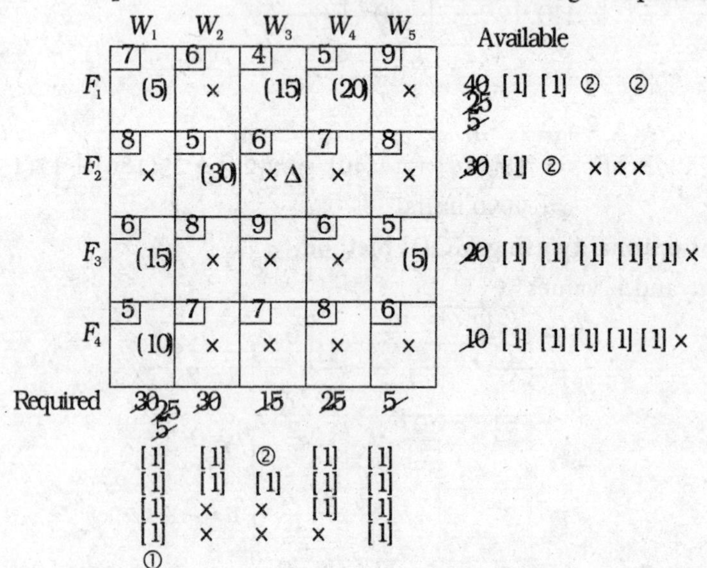

Number of occupied cell $= 7 \neq (m+n-1 \Rightarrow 8)$

Now transportation cost by VAM

$= (7\times5) + (4\times15) + (5\times20) + (5\times30) + (6\times15) + (5\times5) + (5\times10)$

$= 35 + 60 + 100 + 150 + 90 + 25 + 50$

$= 510$

Total number of occupied cells are 7 which is less than 8 (5 + 4 – 1). Now, if we allocate Δ to least cost unoccupied cell (4, 5) and (3.4) a loop is formed. So we can allocate Δ to next minimum cost cell (1, 2) and (2, 3) which do not form a loop with in occupied cells. Choosing (2, 3) arbitrarily for allocation of Δ which is very small quantity and checking for optimality using MODI's method.

***Step II.*** Finding $u_i$ and $v_j$ $(u_i + v_j = C_{ij})$

| | $W_1$ | $W_2$ | $W_3$ | $W_4$ | $W_5$ | $u_i$ |
|---|---|---|---|---|---|---|
| $F_1$ | 7 • | | 4 • | 5 • | | $u_1 = 0$ |
| $F_2$ | | 5 • | 6 • | | | $u_2 = 2$ |
| $F_3$ | 6 • | | | | 5 • | $u_3 = -1$ |
| $F_4$ | 5 • | | | | | $u_4 = -2$ |
| $v_j$ | $v_1 = 7$ | $v_2 = 3$ | $v_3 = 4$ | $v_4 = 5$ | $v_5 = 6$ | |

***Step III.*** Checking for $d_{ij} = C_{ij} - (u_i + v_j)$

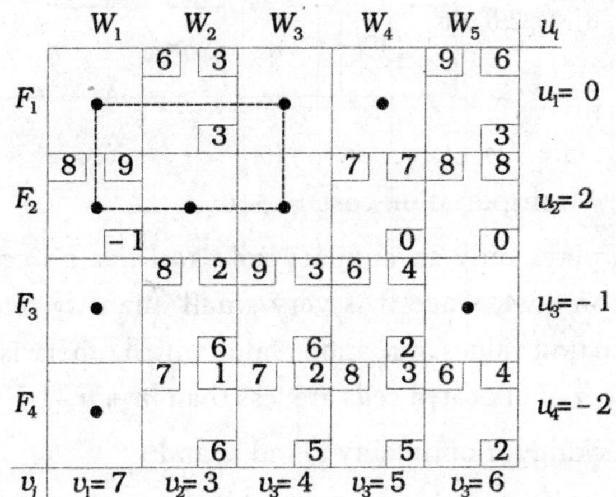

Now $d_{21} = -1 < 0$

∴ Solution is not optimal.

Since $d_{21}$ is –ve, making an allocation in (2, 1) by adding min {Δ, 5} = Δ and alternatively subtracting and adding same amount in occupied cells forming loop as shown below.

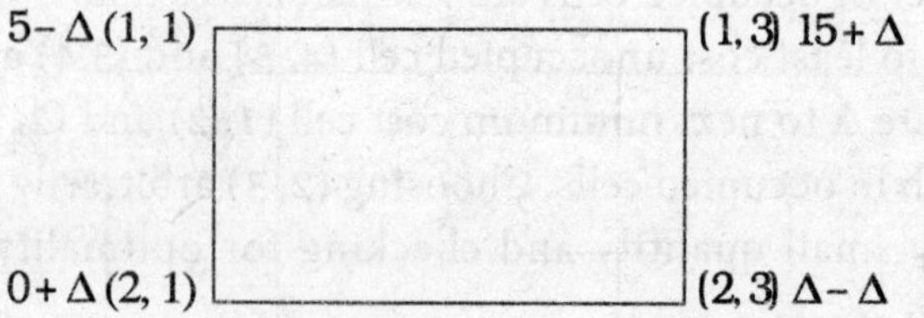

$5 - \Delta$ (1, 1) ——— (1, 3) $15 + \Delta$

$0 + \Delta$ (2, 1) ——— (2, 3) $\Delta - \Delta$

The new basic feasible solution is

| | $W_1$ | $W_2$ | $W_3$ | $W_4$ | $W_5$ |
|---|---|---|---|---|---|
| $F_1$ | 7 (5– Δ) | 6 × | 4 (15+ Δ) | 5 (20) | 9 × |
| $F_2$ | 8 (Δ) | 5 (30) | 6 | 7 × | 8 × |
| $F_3$ | 6 (15) | 8 × | 9 × | 6 × | 5 (5) |
| $F_4$ | 5 (10) | 7 × | 7 × | 8 × | 6 × |

The new transportation cost = 510

**Note.** Δ plays only an auxilary role and has no contribution in final transportation cost, since it is very small quantity nearly zero. It is just introduced to find values of $u_i$ and $v_j$ value which otherwise are not possible to find as number of allocated cells are less than $m + n - 1$.

Now checking for optimality : Find $u_i$ and $v_j$

| | $W_1$ | $W_2$ | $W_3$ | $W_4$ | $W_5$ | $u_i$ |
|---|---|---|---|---|---|---|
| $F_1$ | 7 • | | 4 • | 5 • | | $u_1 = 0$ |
| $F_2$ | 8 • | 5 • | | | | $u_2 = 1$ |
| $F_3$ | 6 • | | | | 5 • | $u_3 = -1$ |
| $F_4$ | 5 • | | | | | $u_4 = -2$ |
| $v_j$ | $v_1 = 7$ | $v_2 = 4$ | $v_3 = 4$ | $v_4 = 5$ | $v_5 = 6$ | |

Checking $d_{ij}$ for each unoccupied cell

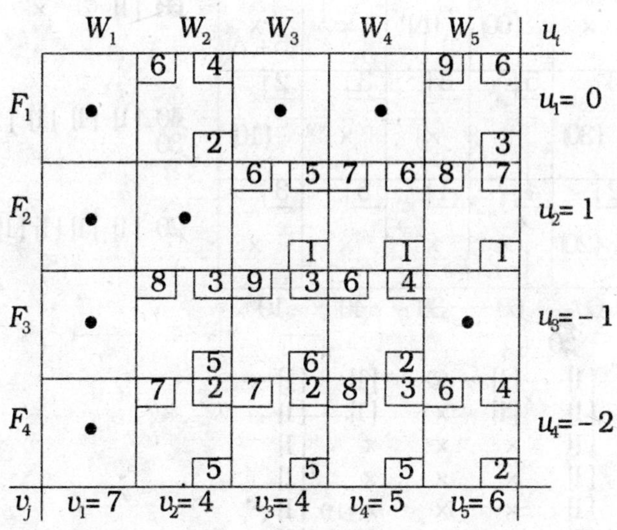

| | $W_1$ | $W_2$ | $W_3$ | $W_4$ | $W_5$ | $u_i$ |
|---|---|---|---|---|---|---|
| $F_1$ | • | 6 4 2 | • | • | 9 6 3 | $u_1 = 0$ |
| $F_2$ | • | • | 6 5 1 | 7 6 1 | 8 7 1 | $u_2 = 1$ |
| $F_3$ | • | 8 3 5 | 9 3 6 | 6 4 2 | • | $u_3 = -1$ |
| $F_4$ | • | 7 2 5 | 7 2 5 | 8 3 5 | 6 4 2 | $u_4 = -2$ |
| $v_j$ | $v_1 = 7$ | $v_2 = 4$ | $v_3 = 4$ | $v_4 = 5$ | $v_5 = 6$ | |

All $d_{ij} > 0$, therefore, solution is optimal and unique.

Hence, current solution is the optimal solution.

**Example 5.15** *Find the basic feasible solution of the following transportation problem. Also find the optimal transprotation plan.*

**[GGSIPU, End Term, 2011]**

| | 1 | 2 | 3 | 4 | 5 | *Available* |
|---|---|---|---|---|---|---|
| *A* | 4 | 3 | 1 | 2 | 6 | 80 |
| *B* | 5 | 2 | 3 | 4 | 5 | 60 |
| *C* | 3 | 5 | 6 | 3 | 2 | 40 |
| *D* | 2 | 4 | 4 | 5 | 3 | 20 |
| *Required* | 60 | 60 | 30 | 40 | 10 | |

***Solution Step I.*** Initial basic feasible solution of the given problem using VAM.

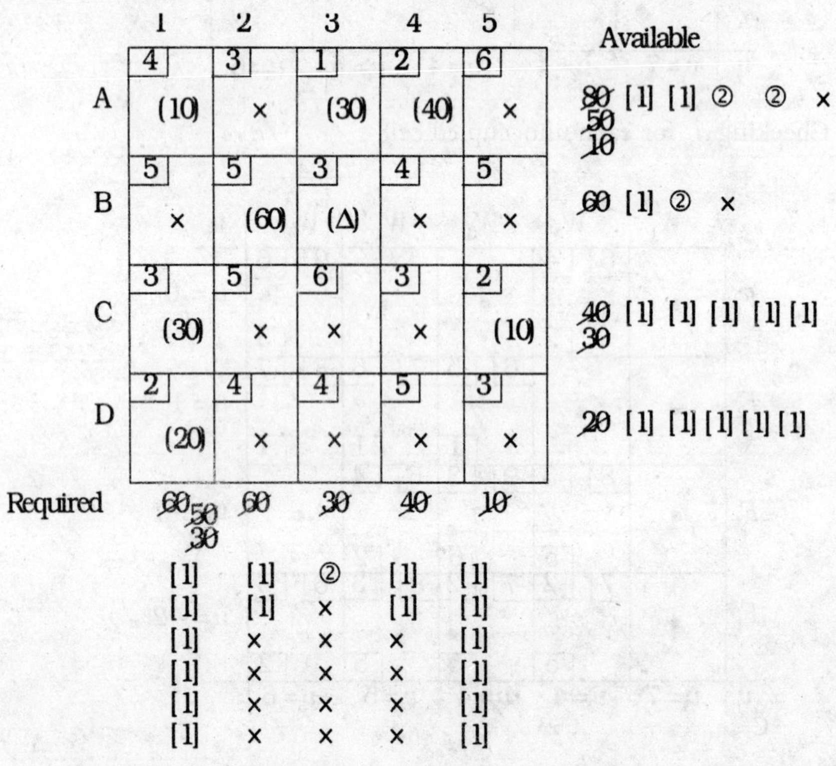

Number of occupied cell = $7 \neq (m + n - 1 \Rightarrow 8)$

Transportation cost by VAM = $(4\times10) + (1\times30) + (2\times40) + (2\times60)$
$+ (3\times\Delta) + (3\times30) + (2\times10) + (2\times20)$
$= 40 + 30 + 80 + 120 + 0 + 90 + 20 + 40$
$= 420$

Total number of occupied cells are 7, which is less than 8 (5+4−1). Now, if we allocate Δ to least cost unoccupied cell (4, 5) and (3, 4) a loop is formed. So we can allocate Δ to next minimum cost cell (1, 2) and (2, 3) which do not form a loop within occupid cells. Choosing (2, 3) arbitrarily for allocation of Δ which is very small quantity and checking for optimality using MODI's Method.

***Step II.*** Finding $u_i$ and $v_j$ $(u_i + v_j = C_{ij})$.

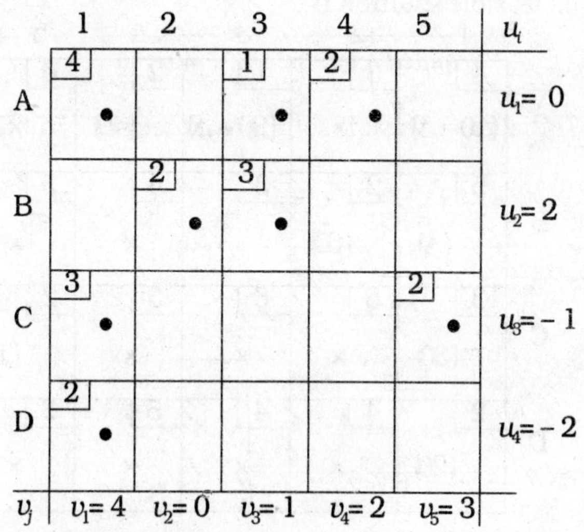


***Step III.*** Checking for $d_{ij} = C_{ij} - (u_i + v_j)$ for unoccupied cells

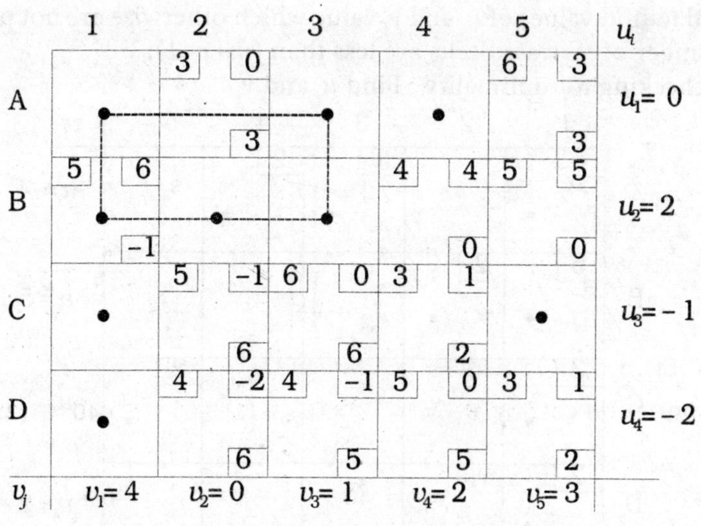


Now $\qquad d_{2,1} = -1 < 0$

Since $d_{21}$ is –ve, making an allocation in (2, 1) by adding min. {10 , Δ)} = Δ and alternatively subtracting and adding same amount in occupied cells.

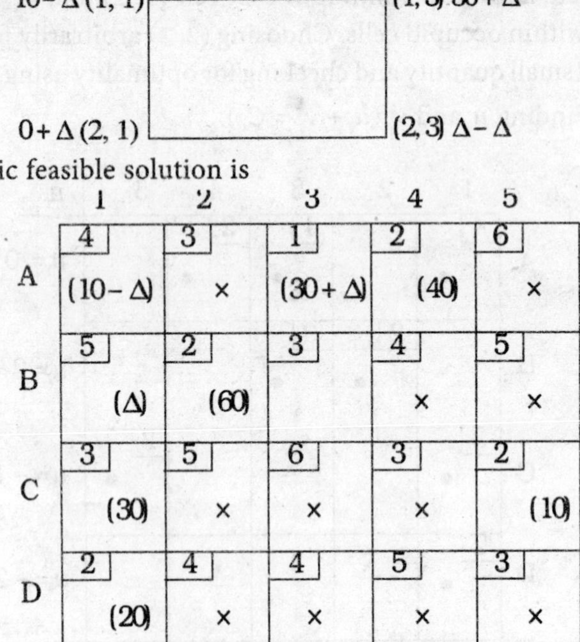


The new basic feasible solution is

| | 1 | 2 | 3 | 4 | 5 |
|---|---|---|---|---|---|
| A | [4] (10 – Δ) | [3] × | [1] (30 + Δ) | [2] (40) | [6] × |
| B | [5] (Δ) | [2] (60) | [3] | [4] × | [5] × |
| C | [3] (30) | [5] × | [6] × | [3] × | [2] (10) |
| D | [2] (20) | [4] × | [4] × | [5] × | [3] × |

The new transportation cost is 420.

**Note.** Δ plays only an auxilary role and has no contribution in final transportation cost, since it is very small quantity nearly zero. It is just introduced to find values of $u_i$ and $v_j$ value which otherwise are not possible to find as number of allocated cells are less than $(m+n-1)$.

Now checking for optimality : Find $u_i$ and $v_j$

| | 1 | 2 | 3 | 4 | 5 | $u_i$ |
|---|---|---|---|---|---|---|
| A | [4] • | | [1] • | [2] • | | $u_1 = 4$ |
| B | [5] • | [2] • | | | | $u_2 = 5$ |
| C | [3] • | | | | [2] • | $u_3 = 3$ |
| D | [2] • | | | | | $u_4 = 2$ |
| $v_j$ | $v_1 = 0$ | $v_2 = -3$ | $v_3 = -3$ | $v_4 = -2$ | $v_5 = -1$ | |

Checking for $d_{ij}$ for each unoccupied cell

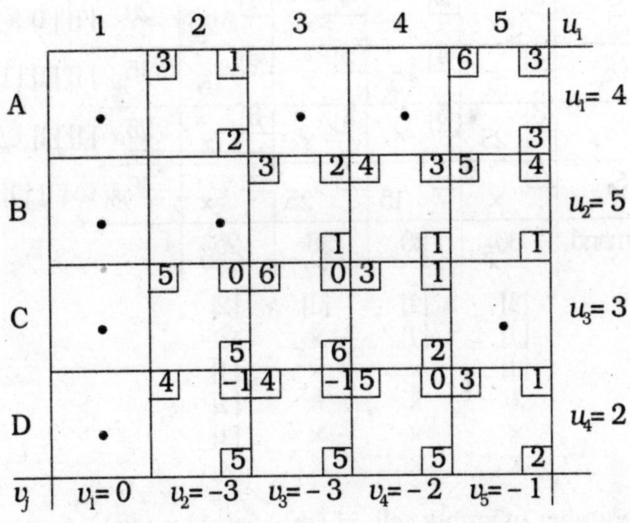


All $d_{ij} > 0$, $\therefore$ Solution is optimal and unique.

Hence current solution is the optimal solution.

**Example 5.16** *Find an initial basic feasible solution of the transportation problem. Test for its optimality. If not, find the optimal solution*

| | $D_1$ | $D_2$ | $D_3$ | $D_4$ | *Supply* |
|---|---|---|---|---|---|
| $S_1$ | 3 | 2 | 4 | 1 | 20 |
| $S_2$ | 2 | 4 | 5 | 3 | 15 |
| $S_3$ | 3 | 5 | 2 | 6 | 25 |
| $S_4$ | 4 | 3 | 1 | 4 | 40 |
| Demand | 30 | 20 | 25 | 25 | |

**[GGSIPU, End Term, 2012]**

***Solution*** ***Step I.*** Initial basics feasible solution of the given problem using VAM.

| | $D_1$ | $D_2$ | $D_3$ | $D_4$ | Supply | |
|---|---|---|---|---|---|---|
| $S_1$ | 3 × | 2 × | 4 × | 1 20 | ~~20~~ | [1] [1] × |
| $S_2$ | 2 5 | 4 5 | 5 × | 3 5 | ~~15~~ ~~10~~ ~~5~~ | [1] [1] [1] [1] [1] [1] |
| $S_3$ | 3 25 | 5 × | 2 × | 6 × | ~~25~~ 0 | [1] [1] ② × × × × |
| $S_4$ | 4 × | 3 15 | 1 25 | 4 × | ~~40~~ ~~15~~ | ② [1] [1] [1] [1] [1] |
| Demand | ~~30~~ ~~5~~ | ~~20~~ | ~~25~~ | ~~25~~ ~~5~~ | | |
| | [1] | [1] | [1] | [2] | | |
| | [1] | [1] | × | ② | | |
| | [1] | × | × | [1] | | |
| | ② | × | × | [1] | | |
| | × | × | × | [1] | | |
| | × | × | × | × | | |

Number of empty cell = $(m-1)(n-1) = (4-1)(4-1) = 9$

Number of allocated cell = $(m+n-1) = (4+4-1) = 7$

Now, transportation cost by VAM.

$$= (1\times20) + (2\times5) + (4\times5) + (3\times5) + (3\times25) + (3\times15) + (1\times25)$$

$$= 20 + 10 + 20 + 15 + 75 + 45 + 25$$

$$= 210$$

Checking for optimality using MODI's Method.

***Step II.*** To find $u_i$ and $v_j$, marking occupied cells with dots and entering associated cost in the upper left corner as follows :

| | $D_1$ | $D_2$ | $D_3$ | $D_4$ | $u_i$ |
|---|---|---|---|---|---|
| $S_1$ | | | | 1 • | $u_1 = -2$ |
| $S_2$ | 2 • | 4 • | | 3 • | $u_2 = 0$ |
| $S_3$ | 3 • | | | | $u_3 = 1$ |
| $S_4$ | | 3 • | 1 • | | $u_4 = -1$ |
| $v_j$ | $v_1 = 2$ | $v_2 = 4$ | $v_3 = 2$ | $v_4 = 3$ | |

To find $u_i$ and $v_j$, put $u_i$ corresponding to row having maximum allocations as zero. In case of tie, choose arbitrarily. Here we put $u_2 = 0$ and enter successively the value of $u_i$ and $v_j$ in the table by using $u_i + v_j = C_{ij}$ with occupied cell.

***Step III and IV.*** Putting $C_{ij}$ for unoccupied cells in top left corner and $(u_i + v_j)$ in top right corner, $d_{ij} = C_{ij} - (u_i + v_j)$ in lower right corner.

| | $D_1$ | $D_2$ | $D_3$ | $D_4$ | $u_i$ |
|---|---|---|---|---|---|
| $S_1$ | 3 | 0 ; 3 | 2 | 2 ; 0 | 4 | 0 ; 4 | • | $u_1 = -2$ |
| $S_2$ | + • | − • | 5 | 2 ; 3 | • | $u_2 = 0$ |
| $S_3$ | − • | 5 | 5 ; 0 | 2 | 3 ; −1 + | 6 | 4 ; 2 | $u_3 = 1$ |
| $S_4$ | 4 | 1 ; 3 | + • | − • | 4 | 2 ; 2 | $u_4 = 1$ |
| $v_j$ | $v_1 = 2$ | $v_2 = 4$ | $v_3 = 2$ | $v_4 = 3$ | |

***Step V.*** $d_{33} = -1 < 0$

∴ Solution is not optimal.

***Step VI.*** Since $d_{33}$ is −ve, making an allocation in (3, 3) by adding minimum of {5, 25} = 5 and alternatively subtracting and adding same amount in occupied cells

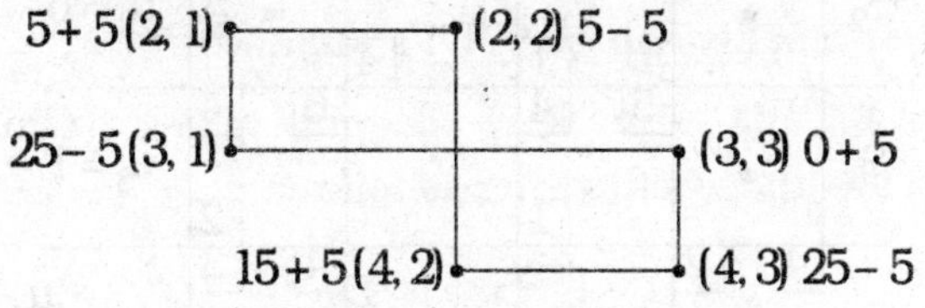

5 + 5 (2, 1) • ——— • (2, 2) 5 − 5

25 − 5 (3, 1) • ——— • (3, 3) 0 + 5

15 + 5 (4, 2) • ——— • (4, 3) 25 − 5

***Step VII.*** The new basic feasible solution is

| | $D_1$ | $D_2$ | $D_3$ | $D_4$ |
|---|---|---|---|---|
| $S_1$ | 3 | 2 | 4 | 1   20 |
| $S_2$ | 2   10 | 4 | 5 | 3   5 |
| $S_3$ | 3   20 | 5 | 2   5 | 6 |
| $S_4$ | 4 | 3   20 | 1   20 | 4 |

The new transportation cost

= (1×20) + (2×10) + (3×5) + (3×20) + (2×5) + (3×20) + (1×20)

= 20 + 20 + 15 + 60 + 10 + 60 + 20

= 205

Checking for optimality : find $u_i$ and $v_j$

| | $D_1$ | $D_2$ | $D_3$ | $D_4$ | $U_i$ |
|---|---|---|---|---|---|
| $S_1$ | | | | [1] • | $u_1 = -2$ |
| $S_2$ | [2] • | • | | [3] • | $u_2 = 0$ |
| $S_3$ | [3] • | | [2] • | | $u_3 = 1$ |
| $S_4$ | | [3] • | [1] • | | $u_4 = 0$ |
| $v_j$ | $v_1 = 2$ | $v_2 = 3$ | $v_3 = 1$ | $v_4 = 3$ | |

Checking $d_{ij}$ for each unoccupied cell

| | $D_1$ | $D_2$ | $D_3$ | $D_4$ | $u_i$ |
|---|---|---|---|---|---|
| $S_1$ | [3] [0]<br>[3] | [2] [1]<br>[1] | [4] [-1]<br>[5] | • | $u_1 = -2$ |
| $S_2$ | • | [4] [3]<br>[1] | [5] [1]<br>[4] | • | $u_2 = 0$ |
| $S_3$ | • | [5] [4]<br>[1] | | [6] [4]<br>[2] | $u_3 = 1$ |
| $S_4$ | [4] [2]<br>[2] | • | • | [4] [3]<br>[1] | $u_4 = 1$ |
| $v_j$ | $v_1 = 2$ | $v_2 = 3$ | $v_3 = 1$ | $v_4 = 3$ | |

All $d_{ij} > 0$, therefore, solution is optimal and unique.

Hence, current solution is the optimal solution.

**Example 5.17** *Use VAM method to obtain an initial basic feasible solution of the transportation problem and obtain the minimum cost.*

| | | To | | | | Supply |
|---|---|---|---|---|---|---|
| | | 1 | 2 | 3 | 4 | |
| From | 1 | 11 | 13 | 17 | 14 | 250 |
| | 2 | 16 | 18 | 14 | 10 | 300 |
| | 3 | 21 | 24 | 13 | 10 | 400 |
| Demand | | 200 | 225 | 275 | 250 | |

**[GGSIPU, End Term, 2013]**

**Solution** *Step I.* Initial basic feasible solution of the given problem using VAM.

| | $D_1$ | $D_2$ | $D_3$ | $D_4$ | Supply | |
|---|---|---|---|---|---|---|
| $S_1$ | [11] 200 | [13] 50 | [17] × | [14] × | ~~250~~ ~~50~~ | [2] [1] × |
| $S_2$ | [16] × | [18] 175 | [14] × | [10] 125 | ~~300~~ 125 | [4] [4] [4] ④ |
| $S_3$ | [21] × | [24] × | [13] 275 | [10] 125 | ~~400~~ | [3] [3] [3] [3] |
| | ~~200~~ | ~~225~~ 175 | ~~275~~ | ~~250~~ 125 | | |
| | ⑤ | [5] | [1] | [4] | | |
| | × | ⑤ | [1] | [4] | | |
| | | ⑥ | [1] | [0] | | |
| | | × | [1] | [0] | | |

Now, transportation cost by VAM

$$= (200\times11) + (50\times13) + (175\times18)$$
$$+ (125\times10) + (275\times13) + (125\times10)$$
$$= 12075$$

Checking for optimality using MODI method.

*Step II.* To find $u_i$ and $v_j$, marking occupied cells with dots and entering associated cost in the upper left corner as follows :

| | $D_1$ | $D_2$ | $D_3$ | $D_4$ | $u_i$ |
|---|---|---|---|---|---|
| $S_1$ | [11] • | [13] • | | | $u_1 = 0$ |
| $S_2$ | | [18] • | | [10] • | $u_2 = 5$ |
| $S_3$ | | | [13] • | [10] • | $u_3 = 5$ |
| $v_j$ | $v_1 = 11$ | $v_2 = 13$ | $v_3 = 8$ | $v_4 = 5$ | |

To find $u_i$ and $v_j$, put $u_i$ corresponding to row having maximum allocations as zero. In case of tie, choose arbitrarily. Here we put $u_1 = 0$, alternatively one can also take $u_3$ as zero. Hence, enter successively the value of $u_i$ and $v_j$ in the table by using $u_i + v_j = C_{ij}$ with occupied cell.

***Step III and Step IV.*** Putting $C_{ij}$ for unoccupied cells in top left corner and $u_i + v_j$ in top right corner, $d_{ij} = C_{ij} - (u_i + v_j)$ in lower right corner.

| | $D_1$ | $D_2$ | $D_3$ | $D_4$ | $u_i$ |
|---|---|---|---|---|---|
| $S_1$ | • | • | 17, 8 / 9 | 14, 5 / 9 | $u_1 = 0$ |
| $S_2$ | 16, 16 / 0 | • | 14, 13 / 1 | • | $u_2 = 5$ |
| $S_3$ | 21, 16 / 5 | 24, 18 / 6 | • | • | $u_3 = 5$ |
| $v_j$ | $v_1 = 11$ | $v_2 = 13$ | $v_3 = 8$ | $v_4 = 5$ | |

All $d_{ij} \geq 0$, therefore solution is optimal.

Hence, the current solution is the optimal solution.

**Example 5.17** *A company has three cement factories located in cities 1, 2 and 3 which supply cement to four projects located in towns I, II, III and IV. Each plant can supply 6, 1 and 10 trucks load of cement daily respectively and daily cement requirements of the projects are respectively 7, 5, 3, 2 truck loads. The transportation costs per truck load of cement from each plant to each project site are as follows :*

*Prooject sites*

| | | 1 | 2 | 3 | 4 |
|---|---|---|---|---|---|
| *Factories* | 1 | 2 | 3 | 11 | 7 |
| | 2 | 1 | 0 | 6 | 1 |
| | 3 | 5 | 8 | 15 | 9 |

*Determine the optimal distribution for the company so as to minimize the total transportation cost.* **[GGSIPU, End Term-2014]**

**Solution** *Step I.* Initial basic feasible solution of the given problem using VAM.

| | $S_1$ | $S_2$ | $S_3$ | $S_4$ | Supply | |
|---|---|---|---|---|---|---|
| $F_1$ | [2] 1 | [3] 5 | [11] | [7] | 6/1 | [1] [1] ⑤ |
| $F_2$ | [1] | [0] | [6] | [1] 1 | 1 | [1] × |
| $F_3$ | [5] 6 | [8] | [15] 3 | [9] 1 | 10/4/1 | [3] [3] [4] |
| Demand | 7/6 | 5/0 | 3 | 2/1 | | |
| | [1] | [3] | [5] | ⑥ | | |
| | [3] | ⑤ | [4] | × | | |
| | [3] | × | [4] | × | | |

Number of emply cell $= (m-1)(n-1) = (4-1)(3-1) = 6$

Number of allocated cell $= (m+n-1) = (4+3-1) = 6$

Now, transportation cost by VAM

$= (1 \times 2) + (5 \times 3) + (1 \times 1) + (6 \times 5) + (3 \times 15) + (1 \times 9)$

$= 2 + 15 + 1 + 30 + 45 + 9$

$= 102$

Checking for optimality using MODI's method.

*Step II.* To find $u_i$ and $v_j$, marking occupied cells with dots and entering associated cost in the upper left corner as follows :

| | $S_1$ | $S_2$ | $S_3$ | $S_4$ | $u_i$ |
|---|---|---|---|---|---|
| $F_1$ | [2] • | [3] • | | | $u_1 = -3$ |
| $F_2$ | | | | [1] • | $u_2 = -8$ |
| $F_3$ | [5] • | | [15] • | [9] • | $u_3 = 0$ |
| $v_j$ | $v_1 = 5$ | $v_2 = 6$ | $v_3 = 15$ | $v_4 = 9$ | |

To find $u_i$ and $v_j$, put $u_i$ corresponding to row having maximum allocations as zero. In case of tie choose arbitrarily. Here we put $u_3 = 0$, and enter successively the value of $u_i$ and $v_j$ in the table by using $u_i + v_j = C_{ij}$ with each occupied cell.

***Step III and Step IV.*** Putting $C_{ij}$ for unoccupied cells in top left corner and $(u_i + v_j)$ in top right corner, $d_{ij} = C_{ij} - (u_i + v_j)$ in lower right corner.

| | $S_1$ | $S_2$ | $S_3$ | $S_4$ | $u_i$ |
|---|---|---|---|---|---|
| $F_1$ | • | • | 11 12 / −1 | 7 6 / 1 | $u_1 = -3$ |
| $F_2$ | 1 −3 / 4 | 0 −2 / 2 | 6 7 / −1 | • | $u_2 = -8$ |
| $F_3$ | • | 8 6 / 2 | • | • | $u_3 = 0$ |
| $v_j$ | $v_1 = 5$ | $v_2 = 6$ | $v_3 = 5$ | $v_4 = 9$ | |

***Step V.*** $d_{12}, d_{23} = -1 < 0$

Therefore, solution is not optimal.

***Step VI.*** Since $d_{12}, d_{23}$ is – ve, making an allocation in (2, 3) by adding min {1, 3} = 1 and alternatively subtracting and adding same amount in occupied cells.

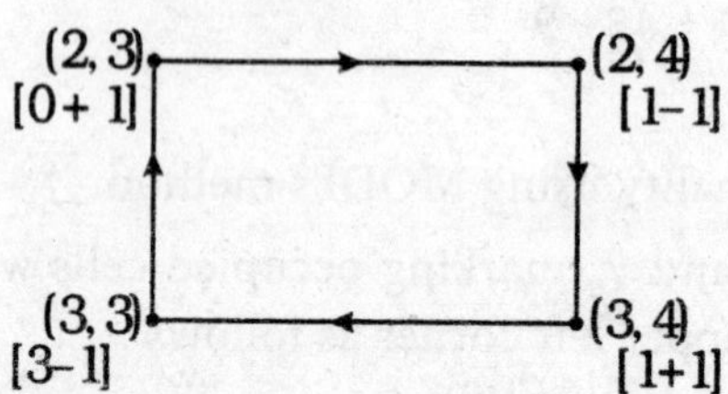


***Step VII.*** The new basic feasible solution is

| | $S_1$ | $S_2$ | $S_3$ | $S_4$ |
|---|---|---|---|---|
| $F_1$ | [2] 1 | [3] 5 | [11] | [7] |
| $F_2$ | [1] | [0] | [6] 1 | [1] |
| $F_3$ | [5] 6 | [8] | [15] 2 | [9] 2 |

$$\text{The new transportation cost} = (2\times1) + (3\times5) + (6\times1) + (15\times2) + (9\times2) + (5\times6)$$

$$= 2 + 15 + 6 + 30 + 18 + 30$$

$$= 101$$

Checking for optimality : find $u_i$ and $v_j$

| | $S_1$ | $S_2$ | $S_3$ | $S_4$ | $u_i$ |
|---|---|---|---|---|---|
| $F_1$ | 2 • | 3 • | | | $u_1 = -3$ |
| $F_2$ | | | 6 • | | $u_2 = -9$ |
| $F_3$ | 5 • | | 15 • | 9 • | $u_3 = 0$ |
| $v_j$ | $v_1 = 5$ | $v_2 = 6$ | $v_3 = 15$ | $v_4 = 9$ | |

Checking $d_{ij}$ for each unoccupied cell

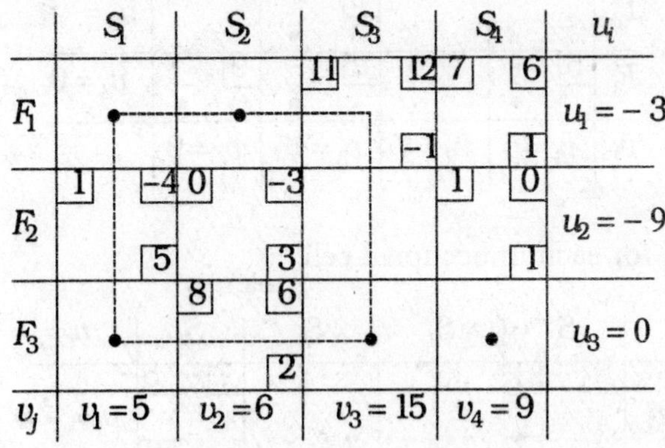

| | $S_1$ | $S_2$ | $S_3$ | $S_4$ | $u_i$ |
|---|---|---|---|---|---|
| $F_1$ | • | • | 11, 12, −1 | 7, 6, 1 | $u_1 = -3$ |
| $F_2$ | 1, −4, 5 | 0, −3, 3 | • | 1, 0, 1 | $u_2 = -9$ |
| $F_3$ | • | 8, 6, 2 | • | • | $u_3 = 0$ |
| $v_j$ | $v_1 = 5$ | $v_2 = 6$ | $v_3 = 15$ | $v_4 = 9$ | |

Now , $d_{13} = -1 < 0$

Therefore, solution is not optimal.

Now $d_{13}$ is –ve, making an allocation in (1, 3) by adding min {1, 2} = 1 and alternatively subtracting and adding same amount in occupied cells.

The new basic feasible solution is

| | $S_1$ | $S_2$ | $S_3$ | $S_4$ |
|---|---|---|---|---|
| $F_1$ | 2 | 3 (5) | 11 (1) | 7 |
| $F_2$ | 1 | 0 | 6 (1) | 1 |
| $F_3$ | 5 (7) | 8 | 15 (1) | 9 (2) |

The new transportation cost

$= (3\times5) + (11\times1) + (6\times1) + (5\times7) + (15\times1) + (9\times2)$

$= 15 + 11 + 6 + 35 + 15 + 18$

$= 100$

Checking for optimality ; find $u_i$ and $v_j$

| | $S_1$ | $S_2$ | $S_3$ | $S_4$ | $u_i$ |
|---|---|---|---|---|---|
| $F_1$ | | 3 • | 11 • | | $u_1 = -4$ |
| $F_2$ | | | 6 • | | $u_2 = -9$ |
| $F_3$ | 5 • | | 15 • | 9 • | $u_3 = 0$ |
| $v_j$ | $v_1 = 5$ | $v_2 = 7$ | $v_3 = 15$ | $v_4 = 9$ | |

Checking $d_{ij}$ for each unoccupied cell

| | $S_1$ | $S_2$ | $S_3$ | $S_4$ | $u_i$ |
|---|---|---|---|---|---|
| $F_1$ | 2 1 1 | | | 7 5 2 | $u_1 = -4$ |
| $F_2$ | 1 −4 5 | 0 −2 2 | | 1 0 1 | $u_2 = -9$ |
| $F_3$ | • | 8 7 1 | | • | $u_3 = 0$ |
| $v_j$ | $v_1 = 5$ | $v_2 = 7$ | $v_3 = 15$ | $v_4 = 9$ | |

All $d_{ij} > 0$, $\therefore$ solution is optimal and unique

## MULTIPLE CHOICE QUESTIONS

5.1. If the basic feasible solution to a transportation problem minimizes the total transportation cost then it is said to be

(a) optimal (b) unbalanced

(c) balanced (d) None

5.2. The transportation problem is

(a) an assignment problem (b) a linear programming problem

(c) a transhipment problem (d) Both a and c

5.3. What does VAM stands for

(a) Very Accurate Method (b) Vogel's Accurate Method

(c) Vogel's Advanced Method

(d) Vogel's Approximation Method

5.4. Which is the most efficient method to obtain a basic feasible solution.

(a) NWCR (b) VAM

(c) LCM (d) All the above

5.5. What is the full from of MODI?

(a) Model Distribution (b) Method of Distribution

(c) Modified Distribution (d) None

5.6. Degeneracy in a $m \times n$ transportation problem occurs when the number of occupied cell is less than

(a) $m + n$ (b) $m + n + 1$

(c) $m - n - 1$ (d) $m + n - 1$

5.7. What is the purpose of Modi method?

(a) To get an optimal solution (b) To get infeasible solution

(c) To balance the problem

(c) To get an initial basic feasible solution

5.8. If the total supply in a transportation problem is not equal to the total demand then it is said to be

(a) unbalanced (b) balanced

(c) optimal (d) infeasible

5.9. In a transportation problem the solution is optimal and alternative if [Where $d_{ij} = C_{ij} - (u_i + v_j)$]

(a) $d_{ij} > 0$ (b) $d_{ij} \geq 0$

(c) $d_{ij} < 0$ (d) None

5.10. Number of empty cell in a transportation problem can be calculated by using the relation

(a) $(m + n - 1)$ (b) $(m - 1)(n - 1)$

(c) $(m + 1)(n - 1)$ (d) $m + n + 1$

5.11. Number of allocated cell in a transportation problem can be calculated by using the relation

(a) $(m - 1)(n - 1)$ (b) $(m - n - 1)$

(c) $(m + n - 1)$ (d) $(m - n + 1)$

## ANSWERS

| | | | | | |
|---|---|---|---|---|---|
| 1. (a) | 2. (b) | 3. (d) | 4. (b) | 5. (c) | 6. (d) |
| 7. (a) | 8. (a) | 9. (b) | 10. (b) | 11. (c) | |

## SUMMARY

Transportation model is one of the most useful application of linear programming techniques. It is apparently used when we want to determine the minimum cost at which goods can be transported from given sources to specified destinations.

Let there be $m$ sources $S_i$, $i = 1, \ldots.. m$ and $n$ destination $D_j$, $j = 1, \ldots., n$. Also let $a_i$ denote availability in $i^{th}$ source $b_j$ be the requirement in $j^{th}$ destination and the cost of transportation of one unit of commodity from source $S_i$ to destination $D_j$ be $C_{ij}$.

If $x_{ij}$ denotes quality transported from origin '$S_i$' to destination $D_j$, then a general transportation problem may be formulated as

$$\text{Minimize } Z = \sum_{i=1}^{m} \sum_{j=1}^{n} C_{ij} x_{ij}$$

Subject to

$$\sum_{j=1}^{n} x_{ij} = a_i, \; i = 1, 2, \ldots., m$$

$$\sum_{i=1}^{m} x_{ij} = b_j, \; j = 1, 2, \ldots\ldots n$$

$$x_{ij} \geq 0$$

For a balanced transportation problem

$$\sum_{i=1}^{m} a_i = \sum_{j=1}^{n} b_j$$

**Finding Initial Basic Feasible Solution by**

1. North-west corner rule
2. Matrix-minima or least-cost method
3. Row-minima method
4. Column-minima method
5. Vogel's Approximation Method (VAM)

**Vogel Approximation Method (VAM) or Unit Cost Penalty Method**

**Step 1.** Write the difference of smallest and second smallest cost in each row and column on the right of corresponding row and below the corresponding column. Each difference is named as penalty.

**Step 2.** Row or column having largest penalty is allocated by maximum possible amount to the cell with lowest cost. In case of tie i.e., if two or more row/columns having equal penalities choose arbitrariiy.

**Step 3.** Recompute the row and column differences (penalties) and repeat the procedure explained in step 2. Cells in which allocations have been made or crossed cell are not taken into account.

**Step 4.** Continue the procedure till all allocations have been made.

**Finding Optimal Solution (MODI Method).** An initial basic feasible solution is obtained by using any of the methods discussed before. Most commonly used method for finding an optimal solution is MODI (Modified Distribution) method also known as UV method.

**Working Rule for Finding Optimal Solution using MODI Method. Step 1.** Find an inital basic feasible solution using any appropriate method preferably VAM

**Step 2.** Find set of numbers $u_i$ $(i = 1, 2, ...., m)$ and $v_j$ $(j = 1, 2,......, n)$ for each occupied cell $(r, s)$ such that $C_{rs} = u_r + v_s$ by putting any of $v_i$ or $v_j$ zero initially.

**Step 3.** Enter $C_{ij}$ for each unoccupied cell $(i, j)$ in the upper left corner and enter $(u_i + v_j)$ for same cell in the upper right corner.

**Step 4.** Enter $d_{ij} = C_{ij} - (u_i + v_j)$ for each unoccupied cell $(i, j)$ at the lower right corner.

**Step 5.** Examine $d_{ij}$ for all unoccupied cells.

(i) If all $d_{ij} > 0$, solution is optimal and unique.

(ii) If all $d_{ij} \geq 0$, solution is optimal and an alternative optimal solution exists.

(iii) If at least one $d_{ij} < 0$, solution is not optimal.

**Step 6.** Form a new basic feasible solution by giving an allocation to the cell for which $d_{ij}$ is most –ve by forming a loop with occupied cells and vacating one of them. Amount to be allocated is minimum allocation in occupied cells from where amount is to be subtracted.

**Step 7.** Repeat steps 2 to 5, until optimal solution is attained.

**Degeneracy in Transportation Problems.** Degeneracy in transportation problem occurs when

1. number of occupied cells is less than $m + n - 1$, where $m$ and $n$ respectively denote rows and columns in the transportation table.
2. occupied cells form a loop by vertical or horizontal movements in between them.

Degeneracy can occur in two ways

(i) Initial basic feasible solution may be degenerate

(ii) The degeneracy develops at some intermediate stage while solving for optimal solution.

**Resolution of Degeneracy During Initial Stage.** *Rule.* An extremely small quantity usually denoted by $\Delta$ or $\in$ is introduced in the least cost independent cell. If necessary two or more $\Delta s$ can be introduced in the least and second least cost cells in such a way that occupied cells do not form a loop.

CHAPTER 6

# Assignment Problems

## 6.1 INTRODUCTION

An asignment problem is a particular case of transprotation problem in which a number of operations are to be assigned to an equal number of operators, where each operator performs only one operation. The objective of assignment model is to minimize overall cost or to maximize overall profit for a given assignment schedule.

An assignment problem is a special case of transportation problem in which number of rows and columns are same, all requirements and availabilities are unity and each allocation is limited to either 0 or 1.

Mathematical formulation of a minimizing assignment problem is given as :

$$\text{Minimize } z = \sum_{i=1}^{n}\sum_{j=1}^{n} C_{ij} \cdot x_{ij}$$

Subject to

$$\sum_{i=1}^{n} x_{ij} = 1 , j = 1, 2 ....., n$$

$$\sum_{j=1}^{n} x_{ij} = 1 , j = 1, 2 ....., n$$

$$x_{ij} = 0 \text{ or } 1$$

**Note.** Each operator performs only one operation at a time

## 6.2 SOLUTION OF ASSIGNMENT PROBLEM

An assignment problem can be solved by the following methods:

(i) Enumeration method

(ii) Simplex method

(iii) Transportation method

(iv) Hungarian method

### 6.2.1 Enumeration Method

An assignment problem involves a matrix of order $n$ ($n$ jobs are to be assigned to $n$ workers). Therefore, there are $n!$ possible assignments. We evaluate best suitable of all possible assignments.This method is quite lengthy when $n$ is large. A matrix of order 4 would involve 24 possible assignments which is practically quite cumbersome to evaluate.

### 6.2.2 Simplex Method

An assignment problem of order $n$ may be formulated as an integer programming problem with $n \times n$ variables where each variable can have values 0 or 1. Also there are $n + n$ i.e. $2n$ equality constraints to be worked upon. In particular for an assignment problem of order 4, there will be 16 decision variables taking values 0 or 1 and 8 equality constraints. The procedive is quite lengthy and difficult to solve manually.

### 6.2.3 Transprotation Method

An assignment problem can also be solved as a transportation model with all requirements and availabilities as unity. However we will be getting a degenerate solution as there will be only $n$ assignments for an assignment problem of order $n$, whereas there should be $(2n - 1)$ assignments for a basic feasible solution to be non-degenerate. This will need $(n - 1)$ dummy allocations for optimality check procedure. This makes transportation method computationally inefficient for solving an assignment problem.

### 6.2.4 Hungarian Method

This is the most suitable and widely used method for solving an assignment problem.

Hungarian method (minimization case) can be summerized as follows :

*Step 1.* Prepare a square matrix if not originally. Add suitable number of dummy rows (or columns) with zero cost element for this purpose.

*Step 2.* Subtract the smallest element in each row from every element of that row.

*Step 3.* Subtract the smallest element in each column from every element of that column. Purpose is to induce a zero element in every row and column.

*Step 4.* Make assignment in rows having single zero and mark it as [0]. Cross all other zeros in the column where assignment has been made.

*Step 5.* Repeat the procedure for each column one by one starting from first column. Simultaneously cross out unassigned zeros in corresponding rows.

*Step 6.* If all $n$ assignments have been made, stop. Otherwise go to step 7.

*Step 7.* Draw horizontal and vertical lines covering all zeros by following rule :

(i) Mark '✓' to rows having no assignment.

(ii) Mark '✓' to columns having ⊗ in the marked row.

(iii) Mark '✓' to rows having [0] in marked column.

(iv) Draw horizontal and vertical lines to unmarked rows and marked columns.

*Step 8.* Select the smallest element among all uncovered elements and then subtract this value from all elements in the matrix not covered by lines and add this value to all elements that lie at the intersection of the two lines.

*Step 9.* Go back to step 4.

### 6.2.4.1 Type I. Direct solution in one interation

The problem can be solved readily in one iteration i.e. until step 6 of Hungrarian method.

**Example 6.1** *There are 4 machines and 4 operators with their respective payments (in multiples of ₹ 100) as shown below. Assign one operator to each machine, so that overall payment is minimum. Also find the minimum due payment.*

| | *I* | *II* | *III* | *IV* |
|---|---|---|---|---|
| *1* | 5 | 8 | 9 | 7 |
| *2* | 6 | 7 | 8 | 6 |
| *3* | 7 | 4 | 6 | 4 |
| *4* | 7 | 8 | 5 | 9 |

***Solution*** Select minimum element in each row and subtract from every element of that row.

| | *I* | *II* | *III* | *IV* |
|---|---|---|---|---|
| 1 | 0 | 3 | 4 | 2 |
| 2 | 0 | 1 | 2 | 0 |
| 3 | 3 | 0 | 2 | 0 |
| 4 | 2 | 3 | 0 | 4 |

Each column has 0 element, so no change after subtracting minimum element in each column. Now following Hungrarian Method for assignment schedule i.e. allocating rows with single zero with assignment and crossing out corresponding zeros in columns and there after allocating zeros with assignment column-wise.

| | *I* | *II* | *III* | *IV* |
|---|---|---|---|---|
| 1 | [0] | 3 | 4 | 2 |
| 2 | ~~0~~ | 1 | 2 | [0] |
| 3 | 3 | [0] | 2 | ~~0~~ |
| 4 | 2 | 3 | [0] | 4 |

∴ Optimal assignment schedule is :

| Operator | Machine | Payment (₹) |
|---|---|---|
| 1 | I | 500 |
| 2 | IV | 600 |
| 3 | II | 400 |
| 4 | III | 500 |

Hence, minimum due payment = ₹ 2000

### 6.2.4.2 Type II. Multi iterative solution by Hungrarian method

The given assignment problem is solved in two or more iterations.

**Example 6.2** *Find the optimal assignment schedule for assigning 4 jobs I, II, III and IV to 4 persons A, B, C and D with the following cost matrix (in multiples of thousand rupees).*

| | I | II | III | IV |
|---|---|---|---|---|
| A | 8 | 5 | 4 | 11 |
| B | 10 | 11 | 5 | 13 |
| C | 10 | 7 | 8 | 10 |
| D | 8 | 9 | 9 | 9 |

***Solution*** Subtracting minimum element in each row, so that 0 is introduced in each row :

| | I | II | III | IV |
|---|---|---|---|---|
| A | 4 | 1 | 0 | 7 |
| B | 5 | 6 | 0 | 8 |
| C | 3 | 0 | 1 | 3 |
| D | 0 | 1 | 1 | 1 |

Subtracting minimum element in each column and assigning single o's with assignments rowwise and then columnwise.

| | I | II | III | IV |
|---|---|---|---|---|
| A | 4 | 1 | [0] | 6 |
| B | 5 | 6 | ~~0~~ | 7 |
| C | 3 | [0] | 1 | 2 |
| D | [0] | 1 | 1 | ~~0~~ |

Marking '✓'s as per step 7 as all the assignments have not been made and drawing horizontal and vertical lines to unmarked rows and marked columns :

| | I | II | III ✓(ii) | IV | |
|---|---|---|---|---|---|
| A | 4 | 1 | [0] | 6 | ✓ (iii) |
| B | 5 | 6 | ~~0~~ | 7 | ✓ (i) |
| C | 3 | [0] | 1 | 2 | |
| D | [0] | 1 | 1 | 0 | |

Selecting the minimum among the uncovered elements which is min{4, 1, 6, 5, 6, 7} = 1 and subtracting this from all the elements, not covered by lines. Also adding 1 at intersection of horizontal and vertical lines.

| | I | II | III | IV |
|---|---|---|---|---|
| A | 3 | 0 | 0 | 5 |
| B | 4 | 5 | 0 | 6 |
| C | 3 | 0 | 2 | 2 |
| D | 0 | 1 | 2 | 0 |

Now repeating the procedure by assigning single 0's with an assignment and marking '✓'s if assignments are not made fully.

| | I | II ✓ (ii) | III ✓ (ii) | IV | |
|---|---|---|---|---|---|
| A | 3 | ~~0~~ | ~~0~~ | 5 | ✓ (i) |
| B | 4 | 5 | [0] | 6 | ✓ (iii) |
| C | 3 | [0] | 2 | 2 | ✓ (iii) |
| D | [0] | 1 | 2 | ~~0~~ | |

Taking minimum of uncovered elements i.e. min {3, 5, 4, 6, 3, 2} = 2, subtracting from uncovered elements and adding at junction of horizontal and vertical lines and making assignments to single o's row-wise and then columnwise :

| | I | II | III | IV |
|---|---|---|---|---|
| A | 1 | [0] | ~~0~~ | 3 |
| B | 2 | 5 | [0] | 4 |
| C | 1 | ~~0~~ | 2 | [0] |
| D | [0] | 3 | 4 | ~~0~~ |

∴ Optimum assignment schedule is :

| Person | Job | Cost |
|---|---|---|
| A | II | ₹5,000 |
| B | III | ₹ 5,000 |
| C | IV | ₹ 10,000 |
| D | I | ₹ 8,000 |

Minimum assignment cost is ₹ 28,000.

**Example 6.3** *Assign 4 trucks for 3 destinations, so that distance travelled is minimized. The matrix shows distance in Kilometres.*

| | A | B | C |
|---|---|---|---|
| $T_1$ | 20 | 11 | 12 |
| $T_2$ | 26 | 15 | 17 |
| $T_3$ | 30 | 19 | 21 |
| $T_4$ | 34 | 21 | 24 |

***Solution*** Adding a dummy destination '*D*' its make a square matrix and subtracting minimum element in each row and column.

| | A | B | C | D |
|---|---|---|---|---|
| $T_1$ | 0 | 0 | 0 | 0 |
| $T_2$ | 6 | 4 | 5 | 0 |
| $T_3$ | 10 | 8 | 9 | 0 |
| $T_4$ | 14 | 10 | 12 | 0 |

Marking single 0's with assignment row-wise and then column-wise, there by putting '✓'s if assignment are not made.

| | A | B | C | D ✓ (ii) | |
|---|---|---|---|---|---|
| $T_1$ | [0] | ~~0~~ | ~~0~~ | ~~0~~ | |
| $T_2$ | 6 | 4 | 5 | [0] | ✓ (iii) |
| $T_3$ | 10 | 8 | 9 | ~~0~~ | ✓ (i) |
| $T_4$ | 14 | 10 | 12 | ~~0~~ | ✓ (i) |

Taking minimum of uncovered elements which is min{6, 4, 5, 10, 8, 9, 14, 10, 12} = 4, and subtracting from uncovered elements, also adding at junctions of horizontal and vertical lines and making assignments :

| | $A$ | $B$ | $C$ | $D$ ✓(ii) | |
|---|---|---|---|---|---|
| $T_1$ | [0] | ~~0~~ | ~~0~~ | 4 | |
| $T_2$ | 2 | [0] | 1 | ~~0~~ | |
| $T_3$ | 6 | 4 | 5 | [0] | ✓ (iii) |
| $T_4$ | 10 | 6 | 8 | ~~0~~ | ✓ (i) |

Taking minimum of uncovered elements, which is min{6, 4, 5, 10, 6, 8} = 4 and subtracting from uncovered elements, also adding at junctions of horizontal and vertical lines and making assignments.

| | $A$ | $B$ | $C$ | $D$ ✓(ii) | |
|---|---|---|---|---|---|
| $T_1$ | [0] | ~~0~~ | ~~0~~ | 8 | |
| $T_2$ | 2 | [0] | 1 | 4 | |
| $T_3$ | 2 | ~~0~~ | 1 | [0] | ✓ (iii) |
| $T_4$ | 6 | 2 | 4 | ~~0~~ | ✓ (i) |

Taking min{2, 1, 6, 2, 4} = 1 and subtracting from uncovered elements, adding at junctions of intersecting lines and making assignments :

| | $A$ | $B$ | $C$ | $D$ |
|---|---|---|---|---|
| $T_1$ | [0] | ~~0~~ | ~~0~~ | 9 |
| $T_2$ | 2 | [0] | 1 | 5 |
| $T_3$ | 1 | ~~0~~ | [0] | ~~0~~ |
| $T_4$ | 5 | 1 | 3 | [0] |

All the assignments have been made.

Optimal assignment schedule is :

| Truck | Destination | Distance |
|---|---|---|
| $T_1$ | A | 20 |
| $T_2$ | B | 15 |
| $T_3$ | C | 21 |

Total minimum distance = 20 km.

#### 6.2.4.3 Type III. Problems having alternate optimal solution

The assignment problem having more than one optimal solution is specifically distinguished by a set of 4, 6, 8 ...... unmarked zeros, which are neither assigned nor crossed out. To solve further, we arbitrarily make an assignment and cross out other zeros in that particular row and column.

**Example 6.4** *A multinational company has four expert programmes and four application programs to be developed. The estimated computer time in days by the four programmers are given as follows :*

| | $A_1$ | $A_2$ | $A_3$ | $A_4$ |
|---|---|---|---|---|
| $P_1$ | 20 | 30 | 40 | 45 |
| $P_2$ | 25 | 33 | 41 | 43 |
| $P_3$ | 15 | 23 | 31 | 33 |
| $P_4$ | 30 | 36 | 42 | 50 |

*Assign the programmers $P_1$, $P_2$, $P_3$, $P_4$ with application so that total computer time is minimized.*

***Solution*** Subracting minimum element in each row from every element of the row :

| | $A_1$ | $A_2$ | $A_3$ | $A_4$ |
|---|---|---|---|---|
| $P_1$ | 0 | 10 | 20 | 25 |
| $P_2$ | 0 | 8 | 16 | 18 |
| $P_3$ | 0 | 8 | 16 | 18 |
| $P_4$ | 0 | 6 | 12 | 20 |

Subtracting minimum element in each column from every element of the column so that a zero is induced in every column and making assignments.

| | ✓(ii) $A_1$ | $A_2$ | $A_3$ | ✓(ii) $A_4$ | |
|---|---|---|---|---|---|
| $P_1$ | [0] | 4 | 8 | 7 | ✓(iii) |
| $P_2$ | ~~0~~ | 2 | 4 | [0] | ✓(iii) |
| $P_3$ | ~~0~~ | 2 | 4 | ~~0~~ | ✓(i) |
| $P_4$ | ~~0~~ | [0] | ~~0~~ | 2 | |

Subtracting minimum of uncovered elements, i.e. min{4, 8, 2, 4, 2, 4} = 2 from all uncovered elements, adding it at junctions of horizontal and vertical lines and making assignments

| | $A_1$ | $A_2$ | $A_3$ | $A_4$ |
|---|---|---|---|---|
| $P_1$ | [0] | 2 | 6 | 7 |
| $P_2$ | ~~0~~ | 0 | 2 | 0 |
| $P_3$ | ~~0~~ | 0 | 2 | 0 |
| $P_4$ | 2 | ~~0~~ | [0] | 4 |

Clearly there are 4 unassigned zeros at $(P_2, A_2)$, $(P_2, A_4)$, $(P_3, A_2)$ and $(P_3, A_4)$ which indicate an alternate optimal solution. Making arbitrary assignments we get two solutions as shown below :

**Solution I**

| | $A_1$ | $A_2$ | $A_3$ | $A_4$ |
|---|---|---|---|---|
| $P_1$ | [0] | 2 | 6 | 7 |
| $P_2$ | ~~0~~ | [0] | 2 | ~~0~~ |
| $P_3$ | ~~0~~ | ~~0~~ | 2 | [0] |
| $P_4$ | 2 | ~~0~~ | [0] | 4 |

Optimal assignment schedule

| Programmer | Application | Times (Days) |
|---|---|---|
| $P_1$ | $A_1$ | 20 |
| $P_2$ | $A_2$ | 33 |
| $P_3$ | $A_4$ | 33 |
| $P_4$ | $A_3$ | 42 |

Minimum computer time = 128 days

**Solution II**

| | $A_1$ | $A_2$ | $A_3$ | $A_4$ |
|---|---|---|---|---|
| $P_1$ | [0] | 2 | 6 | 7 |
| $P_2$ | ~~0~~ | ~~0~~ | 2 | [0] |
| $P_3$ | ~~0~~ | [0] | 2 | ~~0~~ |
| $P_4$ | 2 | ~~0~~ | [0] | 4 |

Optimal assignment schedule

| Programmer | Application | Times (Days) |
|---|---|---|
| $P_1$ | $A_1$ | 20 |
| $P_2$ | $A_4$ | 43 |
| $P_3$ | $A_2$ | 23 |
| $P_4$ | $A_3$ | 42 |

Minimum computer time = 128 days

### 6.2.4.4 Type IV. Restrictions on assignments

In some special type of assignment problems, a specific job can not be assigned to a particular person due to some reason, may it be his incompetence for the specified job or refusal or any other peculiar boundation. To solve such type of assignment problems, we assign very high cost (say ∞) against that particular entry, so that it is never assigned.

**Examples 6.5** *There are 5 jobs A, B, C, D and E to be assigned to 4 persons $P_1$, $P_2$, $P_3$ and $P_4$. Approximate time (in Days) for each person for a particular job is given in the following table.*

| | A | B | C | D | E |
|---|---|---|---|---|---|
| $P_1$ | 3 | 5 | 9 | 4 | 5 |
| $P_2$ | 6 | 3 | – | 4 | 3 |
| $P_3$ | – | 5 | 8 | 5 | 2 |
| $P_4$ | 8 | 2 | 7 | – | 2 |

*$P_2$, $P_3$ and $P_4$ can not be assigned with jobs C, A and D respectively due to incapabilities in performing them. Give an optimal assignment schedule for minimal time.*

***Solution*** Assigning a substantially high time for restricted assignments, and adding a dummy row $P_5$.

| | A | B | C | D | E |
|---|---|---|---|---|---|
| $P_1$ | 3 | 5 | 9 | 4 | 5 |
| $P_2$ | 6 | 3 | ∞ | 4 | 3 |
| $P_3$ | ∞ | 5 | 8 | 5 | 2 |
| $P_4$ | 8 | 2 | 7 | ∞ | 2 |
| $P_5$ | 0 | 0 | 0 | 0 | 0 |

Subtracting minimum element in each row from every element of that row so that a zero element is induced in every row.

| | A | B | C | D | E |
|---|---|---|---|---|---|
| $P_1$ | 0 | 2 | 6 | 1 | 2 |
| $P_2$ | 3 | 0 | ∞ | 1 | 0 |
| $P_3$ | ∞ | 3 | 6 | 3 | 0 |
| $P_4$ | 6 | 0 | 5 | ∞ | 0 |
| $P_5$ | 0 | 0 | 0 | 0 | 0 |

Every column already has zero element. so starting assignment at single zeros first row-wise and then column-wise.

| | A | B ✓(ii) | C | D | E ✓(ii) | |
|---|---|---|---|---|---|---|
| $P_1$ | $\boxed{0}$ | 2 | 6 | 1 | 2 | |
| $P_2$ | 3 | $\boxed{0}$ | ∞ | 1 | $\not{0}$ | ✓(iii) |
| $P_3$ | ∞ | 3 | 6 | 3 | $\boxed{0}$ | ✓(iii) |
| $P_4$ | 6 | $\not{0}$ | 5 | ∞ | $\not{0}$ | ✓ (i) |
| $P_5$ | $\not{0}$ | $\not{0}$ | $\boxed{0}$ | $\not{0}$ | $\not{0}$ | |

Subtracting min{3, 1, 6, 3, 6, 5} = 1 from uncovered elements, adding it at junction of horizontal and vertical lines and making assignments.

| | A | B | C | D | E |
|---|---|---|---|---|---|
| $P_1$ | $\boxed{0}$ | 3 | 6 | 1 | 3 |
| $P_2$ | 2 | $\not{0}$ | ∞ | $\boxed{0}$ | $\not{0}$ |
| $P_3$ | ∞ | 3 | 5 | 2 | $\boxed{0}$ |
| $P_4$ | 5 | $\boxed{0}$ | 4 | ∞ | $\not{0}$ |
| $P_5$ | $\not{0}$ | 1 | $\boxed{0}$ | $\not{0}$ | 1 |

All assignments have been made.

Optimal assignment schedule is given below :

| Person | Job | Time (Days) |
|---|---|---|
| $P_1$ | A | 3 |
| $P_2$ | D | 4 |
| $P_3$ | E | 2 |
| $P_4$ | B | 2 |

Minimum assignment time = 11 days

### 6.2.4.5 Type V. Maximization problem

In case of maximization assignment, it can be converted into minimization problem to be solved by Hungrarian Method. This can be done in either of two ways.

(i) Subtracting all the elements of the matrix from maximum value of the assignment table.

(ii) Putting a –ve sign before every element of the assignment table

**Example 6.6** *A company has a team of 4 salesmen and there are 4 districts. The company estimates the profit per day in rupees (multiples of 100) that each salesman can yield is as given in following table :*

| | $D_1$ | $D_2$ | $D_3$ | $D_4$ |
|---|---|---|---|---|
| A | 20 | 13 | 18 | 14 |
| B | 17 | 15 | 18 | 18 |
| C | 18 | 18 | 16 | 15 |
| D | 16 | 14 | 17 | 16 |

*Find the assignmentof salesmen to various districts which will yield maximum profit.*

***Solution*** Subtracting each element from maximum element (which is 20) in the given table

| | $D_1$ | $D_2$ | $D_3$ | $D_4$ |
|---|---|---|---|---|
| A | 0 | 7 | 2 | 6 |
| B | 3 | 5 | 2 | 2 |
| C | 2 | 2 | 4 | 5 |
| D | 4 | 6 | 3 | 4 |

Now, Hungrarian method can be applied as usual. Subtracting minimum element in each row from each element of the row, so that a zero element is induced in each row.

| | $D_1$ | $D_2$ | $D_3$ | $D_4$ |
|---|---|---|---|---|
| A | 0 | 7 | 2 | 6 |
| B | 1 | 3 | 0 | 0 |
| C | 0 | 0 | 2 | 3 |
| D | 1 | 3 | 0 | 1 |

Each column is having a zero element, so starting with assignments first row-wise and then column-wise.

| | $D_1$ | $D_2$ | $D_3$ | $D_4$ |
|---|---|---|---|---|
| A | [0] | 7 | 2 | 6 |
| B | 1 | 3 | ~~0~~ | [0] |
| C | ~~0~~ | [0] | 2 | 3 |
| D | 1 | 3 | [0] | 1 |

Optimal assignment schedule is as given below :

| Salesmen | District | Profit |
|---|---|---|
| A | $D_1$ | 2000 |
| B | $D_4$ | 1800 |
| C | $D_2$ | 1800 |
| D | $D_4$ | 1600 |

Maximum profit per day for the given assignment schedule is ₹ 7200.

**Example 6.7** *Following table shows sales (in lacs rupees) of 4 brands A, B, C and D in 4 outlets $O_1$, $O_2$, $O_3$ and $O_1$ per month. Give an assignment schedule to maximize the sale.*

| | $O_1$ | $O_2$ | $O_3$ | $O_4$ |
|---|---|---|---|---|
| A | 6 | 5 | 10 | 3 |
| B | 3 | 2 | 10 | 5 |
| C | 4 | 6 | 5 | 3 |
| D | 6 | 3 | 3 | 2 |

***Solution*** Since objective of assignment is to maximize the sales, taking –ve of each value in the table.

| | $O_1$ | $O_2$ | $O_3$ | $O_4$ |
|---|---|---|---|---|
| A | –6 | –5 | –10 | –3 |
| B | –3 | –2 | –10 | –5 |
| C | –4 | –6 | –5 | –3 |
| D | –6 | –3 | –3 | –2 |

Now following Hungrarian method as usual i.e. subtracting minimum element in each row from every element of the row.

| | $O_1$ | $O_2$ | $O_3$ | $O_4$ |
|---|---|---|---|---|
| A | 4 | 5 | 0 | 7 |
| B | 7 | 8 | 0 | 5 |
| C | 2 | 0 | 1 | 3 |
| D | 0 | 3 | 3 | 4 |

Subtracting minimum element in each column and making assignments.

| | $O_1$ | $O_2$ | $O_3$ ✓(ii) | $O_4$ | |
|---|---|---|---|---|---|
| A | 4 | 5 | [0] | 4 | ✓(iii) |
| B | 7 | 8 | ~~0~~ | 2 | ✓(i) |
| C | 2 | [0] | 1 | ~~0~~ | |
| D | [0] | 3 | 3 | 1 | |

Subracting min{4, 5, 4, 7, 8, 2} = 2 from uncovered elements in the matrix, adding it at junction of horizontal and vertical lines and making assignments.

| | $O_1$ | $O_2$ | $O_3$ | $O_4$ |
|---|---|---|---|---|
| A | 2 | 3 | [0] | 2 |
| B | 5 | 6 | ~~0~~ | [0] |
| C | 2 | [0] | 3 | ~~0~~ |
| D | [0] | 3 | 5 | 1 |

All assignments have been made.

Optimal assignment schedule to maximize the sales target.

| Brand | Outlet | Sale (₹) |
|---|---|---|
| A | $O_3$ | 10 lacs |
| B | $O_4$ | 5 lacs |
| C | $O_2$ | 6 lacs |
| D | $O_1$ | 6 lacs |

Total maximum sale = ₹ 27 lacs/per month.

## 6.3 MISCELLANEOUS ASSIGNMENT PROBLEMS

Assignment model has various real life applications. Some of them are being discussed here.

### 6.3.1 Travelling Salesman Problem or Routing Problem

A salesman is required to visit certain number of cities alloted to him. He knows the distance (or cost or time) of journey between every pair of cities. His problem is to select such a route that starts from his home city, passes through each city once and only once and returns to his home city in shortest possible distance (or at least cost or in least time). Problem can be solved using assignment model. Solution procedure consists of following steps :

*Step 1.* Solve the problem as usual assignment model.

*Step 2.* If the solution found is cyclic in nature, starting from home-town and not repeating any city, it is the final solution.

*Step 3.* If solution is not cyclic, choose the next lowest entry in the matrix, excluding zeros and make assignment deleting other assignments in that particular row or column .

*Step 4.* Repeat the procedure until cyclic solution is obtained.

**Example 6.8** *A salesman wants to visit cities A, B, C, D. He does not want to visit any city twice. Distances in km between various cities is given in table below. The home city is A. Use assignment method to determine the route that will minimize the total distance of visiting all cities and returning home.*

| | A | B | C | D |
|---|---|---|---|---|
| A | – | 4 | 8 | 4 |
| B | 6 | – | 4 | 7 |
| C | 4 | 6 | – | 8 |
| D | 5 | 3 | 8 | – |

***Solution*** Since the salesman can not travel from $A$ to $A$, $B$ to $B$ etc., assigning a very high distance say $\infty$ to these positions and subtracting minimum element in each row from all the elements in that row.

| | A | B | C | D |
|---|---|---|---|---|
| **A** | ∞ | 0 | 4 | 0 |
| **B** | 2 | ∞ | 0 | 3 |
| **C** | 0 | 2 | ∞ | 4 |
| **D** | 2 | 0 | 5 | ∞ |

Each column is having a zero element, so start making assignments

| | A | B | C | D |
|---|---|---|---|---|
| **A** | ∞ | ~~0~~ | 4 | [0] |
| **B** | 2 | ∞ | [0] | 3 |
| **C** | [0] | 2 | ∞ | 4 |
| **D** | 2 | [0] | 5 | ∞ |

The optimal assignment is

$A \to D \to B \to C \to A$ which is cyclic.

Hence, the minimum distance travelled is

$$4 + 3 + 4 + 4 = 15 \text{ km}$$

**Example 6.9** *A salesman needs to visit 5 cities. He wishes to start from a particular city, visit each city once and only once and return to its starting point. The travelling cost (in hundred rupees) is given in the table below :*

| | A | B | C | D | E |
|---|---|---|---|---|---|
| **A** | – | 3 | 7 | 2 | 3 |
| **B** | 3 | – | 5 | 2 | 3 |
| **C** | 7 | 5 | – | 7 | 4 |
| **D** | 2 | 2 | 7 | – | 7 |
| **E** | 3 | 3 | 4 | 7 | – |

*Find the least cost route.*

***Solution*** Assigning an extremely high cost in between self and subtracting minimum element in each row from all the elements of the row:

| | *A* | *B* | *C* | *D* | *E* |
|---|---|---|---|---|---|
| *A* | ∞ | 1 | 5 | 0 | 1 |
| *B* | 1 | ∞ | 3 | 0 | 1 |
| *C* | 3 | 1 | ∞ | 3 | 0 |
| *D* | 0 | 0 | 5 | ∞ | 5 |
| *E* | 0 | 0 | 1 | 4 | ∞ |

Subtracting minimum element in each column and making assignments

| | *A* | *B* | *C* | *D* ✓(ii) | *E* | |
|---|---|---|---|---|---|---|
| *A* | ∞ | 1 | 4 | [0] | 1 | ✓(iii) |
| *B* | 1 | ∞ | 2 | ~~0~~ | 1 | ✓(i) |
| *C* | 3 | 1 | ∞ | 3 | [0] | |
| *D* | [0] | ~~0~~ | 4 | ∞ | 5 | |
| *E* | ~~0~~ | ~~0~~ | [0] | 4 | ∞ | |

Subracting min{1, 4, 1, 1, 2, 1} = 1, from uncovered elements, adding at junctions and making assignments again.

| | *A* | *B* | *C* | *D* | *E* |
|---|---|---|---|---|---|
| *A* | ∞ | 0 | 3 | 0 | ~~0~~ |
| *B* | 0 | ∞ | 1 | 0 | ~~0~~ |
| *C* | 3 | 1 | ∞ | 4 | [0] |
| *D* | 0 | 0 | 4 | ∞ | 5 |
| *E* | ~~0~~ | ~~0~~ | [0] | 5 | ∞ |

Multiple optimal solution exists as shown :

**Solution I**

| | $A$ | $B$ | $C$ | $D$ | $E$ |
|---|---|---|---|---|---|
| $A$ | ∞ | [0] | 3 | ~~0~~ | ~~0~~ |
| $B$ | 0 | ∞ | 1 | [0] | ~~0~~ |
| $C$ | 3 | 1 | ∞ | 4 | [0] |
| $D$ | [0] | ~~0~~ | 4 | ∞ | 5 |
| $E$ | ~~0~~ | ~~0~~ | [0] | 5 | ∞ |

$A \to B \to D \to A$

**Solution II**

| | $A$ | $B$ | $C$ | $D$ | $E$ |
|---|---|---|---|---|---|
| $A$ | ∞ | ~~0~~ | 3 | [0] | ~~0~~ |
| $B$ | [0] | ∞ | 1 | ~~0~~ | ~~0~~ |
| $C$ | 3 | 1 | ∞ | 4 | [0] |
| $D$ | ~~0~~ | [0] | 4 | ∞ | 5 |
| $E$ | ~~0~~ | ~~0~~ | [0] | 5 | ∞ |

$A \to D \to B \to A$

Both the solutions violate the condition that salesman should visit each city once and only once.

∴ Looking for the next best solution by giving assignments to minimum value in the table, other than zero in solutions I and II.

*Case 1.* Making assignments at 1 ($B$, $C$) in solution I, discarding other assignments in its row and column, following two solutions arise :

**Solution III**

| | $A$ | $B$ | $C$ | $D$ | $E$ |
|---|---|---|---|---|---|
| $A$ | ∞ | ~~0~~ | 3 | [0] | ~~0~~ |
| $B$ | ~~0~~ | ∞ | [1] | ~~0~~ | ~~0~~ |
| $C$ | 3 | 1 | ∞ | 4 | [0] |
| $D$ | [0] | ~~0~~ | 4 | ∞ | 5 |
| $E$ | ~~0~~ | [0] | ~~0~~ | 5 | ∞ |

$A \to D \to A$

**Solution IV**

| | $A$ | $B$ | $C$ | $D$ | $E$ |
|---|---|---|---|---|---|
| $A$ | ∞ | ~~0~~ | 3 | [0] | ~~0~~ |
| $B$ | ~~0~~ | ∞ | [1] | ~~0~~ | ~~0~~ |
| $C$ | 3 | 1 | ∞ | 4 | [0] |
| $D$ | ~~0~~ | [0] | 4 | ∞ | 5 |
| $E$ | [0] | ~~0~~ | ~~0~~ | 5 | ∞ |

$A \to D \to B \to C \to E \to A$

*Case 2.* Making assignment at 1 ($C$, $B$) in solution I and discarding the other assignments in its row and column, following two solutions arise.

**Solution V**

| | A | B | C | D | E |
|---|---|---|---|---|---|
| A | ∞ | Ø | 3 | [0] | Ø |
| B | Ø | ∞ | 1 | Ø | [0] |
| C | 3 | [1] | ∞ | 4 | Ø |
| D | [0] | Ø | 4 | ∞ | 5 |
| E | Ø | Ø | [0] | 5 | ∞ |

$A \rightarrow D \rightarrow A$

**Solution VI**

| | A | B | C | D | E |
|---|---|---|---|---|---|
| A | ∞ | Ø | 3 | Ø | [0] |
| B | Ø | ∞ | 1 | [0] | Ø |
| C | 3 | [1] | ∞ | 4 | Ø |
| D | [0] | Ø | 4 | ∞ | 5 |
| E | Ø | Ø | [0] | 5 | ∞ |

$A \rightarrow E \rightarrow C \rightarrow B \rightarrow D \rightarrow A$

Solution II would yield similar set of solutions if assignments are given at next minimum value other than zero. Of these only two solution viz solution IV and solution VI satisfy the condition that each city should be visited only once. Cost of travelling in solution IV

$$A \rightarrow D \rightarrow B \rightarrow C \rightarrow E \rightarrow A$$

$$= 200 + 200 + 500 + 400 + 300$$

$$= ₹\ 1600$$

Cost of travelling in solution VI.

$$A \rightarrow E \rightarrow C \rightarrow B \rightarrow D \rightarrow A$$

$$= 300 + 400 + 500 + 200 + 200$$

$$= ₹\ 1600$$

∴ Both solutions IV and VI are optimal solutions.

### 6.3.2 Scheduling Problem

Scheduling is required to manage transport system in order to minimize lay over time. This may be required to schedule aircrafts, trains, buses and other commercial transport systems.

**Example 6.10** *An airline operates seven days a week in between Delhi and Mumbai and has the time table as shown below. Crews must have a minimum layover time of four hours between the flights. Suggest an allocation for the crew to be based at the city that results in minimum layover. For each pair also mention the city where the crew should be based.*

| *Flight No.* | *Delhi-Mumbai* | | *Flight No.* | *Mumbai-Delhi* | |
|---|---|---|---|---|---|
| | *Dep.* | *Arrival* | | *Dep.* | *Arrival* |
| *101* | *6 AM* | *8 PM* | *201* | *7 AM* | *9 PM* |
| *102* | *8 AM* | *10 PM* | *202* | *10 AM* | *12 PM* |
| *103* | *2 PM* | *4 PM* | *203* | *4 PM* | *6 PM* |
| *104* | *6.30 AM* | *8.30 PM* | *204* | *8.30 AM* | *10.30 PM* |

***Solution*** For smooth operation of flights, some of the crews should be based at Delhi and some at Mumbai. Following table shows the layover time for different flights when all crew members are based at Delhi, keeping in view minimum layover time of 4 hours.

Layover time : Crew at Delhi (Table 1)

**Table 1.**

| | 201 | 202 | 203 | 204 |
|---|---|---|---|---|
| 101 | 23 | 26 | 8 | 12.5 |
| 102 | 21 | 24 | 6 | 10.5 |
| 103 | 15 | 18 | 24 | 4.5 |
| 104 | 10.5 | 13.5 | 19.5 | 24 |

Also layover time for different flights when all crew members are based at Mumbai, keeping in view minimum layover time of 4 hours are given in following table :

Layover time : Crew at Mumbai (Table 2)

**Table 2.**

| | 201 | 202 | 203 | 204 |
|---|---|---|---|---|
| 101 | 21 | 18 | 12 | 7.5 |
| 102 | 23 | 24 | 14 | 9.5 |
| 103 | 5 | 26 | 20 | 15.5 |
| 104 | 9.5 | 6.5 | 24.5 | 20 |

Since crew can be based at either of two bases Delhi and Mumbai, taking minimum layover time for the two crews, based at Delhi and Mumbai.

Minimum layover time : Crew at Delhi/Mumbai

| | 201 | 202 | 203 | 204 |
|---|---|---|---|---|
| 101 | 21 | 18 | 8* | 7.5 |
| 102 | 21* | 20 | 6* | 9.5 |
| 103 | 5 | 18* | 20 | 4.5* |
| 104 | 9.5 | 6.5 | 19.5* | 20 |

* Crew based at Delhi

Now apply Hungrarian method to obtain optimal pairing. Subtracting mininum element in each row from every element of the row/column and making assignments.

| | 201 | 202 | 203 | 204 ✓(ii) | |
|---|---|---|---|---|---|
| 101 | 13.5 | 10.5 | 0.5 | [0] | ✓(iii) |
| 102 | 15 | 14 | [0] | 3.5 | |
| 103 | 0.5 | 13.5 | 15.5 | ~~0~~ | ✓ (i) |
| 104 | 3 | [0] | 13 | 13.5 | |

Subtracting min{13.5, 10.5, 0.5, 13.5, 15.5} = 0.5 from uncovered elements, adding at junctions and making assignments.

| | 201 | 202 | 203 | 204 |
|---|---|---|---|---|
| 101 | 13 | 10 | ~~0~~ | [0] |
| 102 | 15 | 14 | [0] | 4 |
| 103 | [0] | 13 | 15 | ~~0~~ |
| 104 | 3 | [0] | 13 | 14 |

∴ Optimal schedule minimizing the layover time is as given below :

| Flight No. | Flight No. | Layover time (hours) | Crew based at |
|---|---|---|---|
| 101 | 204 | 7.5 | Mumbai |
| 102 | 203 | 6 | Delhi |
| 103 | 201 | 5 | Mumbai |
| 104 | 202 | 6.5 | Mumbai |

Minimum layover time = 25 hours

**Note.** Consider Table 1 and Table 2

*layover time at (1, 1) in Table 1

+ layover time at (1, 1) in Table 2 = 44

= (48 – 4 )

= 2 days – to and fro travel time.

*layover time at (4, 1) in Table 1

+ layover time at (4, 1) in Table 2 = 20

= (24 – 4 )

= 1 day – to and from travel time.

Hence, sum of layover time at corresponding positions is either 44 or 20. This insight may help in self-checking the solution.

## PREVIOUS YEARS QUESTIONS

**Example 6.11** *Find the optimal assignment for the assignment problem with the following Cost Matrix.*

Jobs

| Persons | I | II | III | IV | V |
|---|---|---|---|---|---|
| A | 5 | 3 | 4 | 7 | 1 |
| B | 2 | 3 | 7 | 6 | 5 |
| C | 4 | 1 | 5 | 2 | 4 |
| D | 6 | 8 | 1 | 2 | 3 |
| E | 4 | 2 | 5 | 7 | 1 |

**[GGSIPU, Feb-2012]**

***Solution*** Encircle the minimum element in each row.

| | | *Jobs* | | | |
|---|---|---|---|---|---|
| | *I* | *II* | *III* | *IV* | *V* |
| *A* | 5 | 3 | 4 | 7 | ① |
| *B* | ② | 3 | 7 | 6 | 5 |
| *Persons C* | 4 | ① | 5 | ② | 4 |
| *D* | 6 | 8 | 1 | 2 | 3 |
| *E* | 4 | ② | 5 | 7 | ① |

Rowwise subtraction (Min value)

| | I | II | III | IV | V |
|---|---|---|---|---|---|
| A | 4 | 2 | 3 | 6 | 0 |
| B | 0 | 1 | 5 | 4 | 3 |
| C | 3 | 0 | 4 | 1 | 3 |
| D | 5 | 7 | 0 | 1 | 2 |
| E | 3 | 1 | 4 | 6 | 0 |

Coloumn wise subtraction (Min value) and making assigments

| | *I* | *II* | *III* | *IV* | *V* ✓(ii) | |
|---|---|---|---|---|---|---|
| *A* | 4 | 2 | 3 | 5 | [0] | ✓ (iii) |
| *B* | [0] | 1 | 5 | 3 | 3 | |
| *C* | 3 | [0] | 4 | ~~0~~ | 3 | |
| *D* | 5 | 7 | [0] | ~~0~~ | 2 | |
| *E* | 3 | 1 | 4 | 5 | ~~0~~ | ✓ (i) |

Now subtracting min {4, 2, 3, 5, 3, 1, 4, 5} = 1 from uncovered elements, adding at junctions and making assignments.

| | I | II | III | IV | V |
|---|---|---|---|---|---|
| A | 3 | 1 | 2 | 4 | [0] |
| B | [0] | 1 | 5 | 3 | 4 |
| C | 3 | ~~0~~ | 4 | [0] | 4 |
| D | 5 | 7 | [0] | ~~0~~ | 3 |
| E | 2 | [0] | 3 | 4 | ~~0~~ |

Optimal assignment for Cost Matrix

*A* – V, *B* – I, *C* – IV, *D* – III, *E* – II. **Ans.**

**Example 6.12** *A department has four subordinates and four tasks are to be performed. The estimates of the profit in rupees each man would earn is given in the matrix. How should the tasks be allocated, one to each man, so as to maximize the total earning?*

| | | Tasks | | | |
|---|---|---|---|---|---|
| | | 1 | 2 | 3 | 4 |
| | A | 5 | 40 | 20 | 5 |
| Persons | B | 25 | 35 | 30 | 25 |
| | C | 15 | 25 | 20 | 10 |
| | D | 15 | 5 | 30 | 15 |

**[GGSIPU, Feb-2012 (Reappear)]**

***Solution*** Encircle the minimum element in each row,

| | | Tasks | | | |
|---|---|---|---|---|---|
| | | 1 | 2 | 3 | 4 |
| | A | (5) | 40 | 20 | (5) |
| Persons | B | (25) | 35 | 30 | (25) |
| | C | 15 | 25 | 20 | (10) |
| | D | 15 | (5) | 30 | 15 |

Row-wise subtraction (min value in each row)

| | 1 | 2 | 3 | 4 |
|---|---|---|---|---|
| A | 0 | 35 | 15 | 0 |
| B | 0 | 10 | 5 | 0 |
| C | 5 | 15 | 10 | 0 |
| D | 10 | 0 | 25 | 10 |

Column-wise subraction of minimum element and making assignments

| | 1 | 2 | 3 | 4 |
|---|---|---|---|---|
| A | [0] | 35 | 10 | ~~0~~ |
| B | ~~0~~ | 10 | [0] | ~~0~~ |
| C | 5 | 15 | 5 | [0] |
| D | 10 | [0] | 20 | 10 |

To maximize the total earning the tasks should be allocated as

A – 1, B – 3, C – 4, D – 2. **Ans.**

**Example 6.13** *An engineer wants to assign 3 jobs $J_1$, $J_2$, $J_3$ to three machines $M_1$, $M_2$, $M_3$ in such a way that each job is assigned to some machine and not machine works on more than one job. The cost of assigning job $J_i$ to machine $M_i$ is given below :*

| | $M_1$ | $M_2$ | $M_3$ |
|---|---|---|---|
| $J_1$ | 15 | 10 | 9 |
| $J_2$ | 9 | 15 | 10 |
| $J_3$ | 10 | 12 | 8 |

*Find the optimal assignment using Hungarian method.*

**[GGSIPU, End Term 2013]**

***Solution*** Encircle minimum elements in each row.

| | $M_1$ | $M_2$ | $M_3$ |
|---|---|---|---|
| $J_1$ | 15 | 10 | (9) |
| $J_2$ | (9) | 15 | 10 |
| $J_3$ | 10 | 12 | (8) |

Row-wise minimum subtraction from each element

| | $M_1$ | $M_2$ | $M_3$ |
|---|---|---|---|
| $J_1$ | 6 | 1 | 0 |
| $J_2$ | 0 | 6 | 1 |
| $J_3$ | 2 | 4 | 0 |

Colomn-wise minimum subtraction and making assignments

| | $M_1$ | $M_2$ | $M_3$ |
|---|---|---|---|
| $J_1$ | 6 | [0] | ~~0~~ |
| $J_2$ | [0] | 5 | 1 |
| $J_3$ | 2 | 3 | [0] |

Now,

| Jobs | M/c's | Cost |
|---|---|---|
| $J_1$ | $M_2$ | 10 |
| $J_2$ | $M_1$ | 9 |
| $J_3$ | $M_3$ | 8 |
| | | 27 |

**Example 6.14** *A company is to assign 4 jobs to 5 trainees. The expected revenue earnings for each person on each job are as follows:*

jobs

| | | $J_1$ | $J_2$ | $J_3$ | $J_4$ |
|---|---|---|---|---|---|
| | $T_1$ | 86 | 78 | 62 | 81 |
| | $T_2$ | 55 | 79 | 65 | 60 |
| *Trainees* | $T_3$ | 72 | 65 | 63 | 80 |
| | $T_4$ | 86 | 70 | 65 | 71 |
| | $T_5$ | 72 | 70 | 71 | 60 |

*(i) Formulate it as LPP*

*(ii) Find the optimal assignment using a method of your choice.*

**[GGSIPU, End Term 2013]**

***Solution*** (i) LPP formation is as given below :

$$\text{Maximize } Z = 86\,x_{11} + 78\,x_{12} + \ldots\ldots + 60\,x_{54}$$

Subject to

$$86\,x_{11} + 78\,x_{12} + 62\,x_{13} + 81\,x_{14} = 1$$
$$55\,x_{21} + 79\,x_{22} + 65\,x_{23} + 60\,x_{24} = 1$$
$$72\,x_{31} + 65\,x_{32} + 63\,x_{33} + 80\,x_{34} = 1$$
$$86\,x_{41} + 70\,x_{42} + 65\,x_{43} + 71\,x_{44} = 1$$
$$72\,x_{51} + 70\,x_{52} + 71\,x_{53} + 60\,x_{54} = 1$$
$$86\,x_{11} + 79\,x_{21} + 72\,x_{31} + 86\,x_{41} + 72\,x_{51} = 1$$
$$78\,x_{12} + 79\,x_{22} + 65\,x_{32} + 70\,x_{42} + 70\,x_{52} = 1$$
$$62\,x_{13} + 65\,x_{23} + 63\,x_{33} + 65\,x_{43} + 71\,x_{53} = 1$$
$$81\,x_{14} + 60\,x_{24} + 80\,x_{34} + 71\,x_{44} + 60\,x_{54} = 1$$
$$x_{11}, x_{12}, \ldots\ldots, x_{54} = 0 \text{ or } 1$$

(ii) Solving the given problem using Hungarain Method :

Adding a dummy job $J_5$ to make a square matrix.

| | $J_1$ | $J_2$ | $J_3$ | $J_4$ | $J_5$ |
|---|---|---|---|---|---|
| $T_1$ | 86 | 78 | 62 | 81 | 0 |
| $T_2$ | 55 | 79 | 65 | 60 | 0 |
| $T_3$ | 72 | 65 | 63 | 80 | 0 |
| $T_4$ | 86 | 70 | 65 | 71 | 0 |
| $T_5$ | 72 | 70 | 71 | 60 | 0 |

Since it is a maximization problem, subtracting each element from maximum element in the table i.e. from 86, new table is

| | $J_1$ | $J_2$ | $J_3$ | $J_4$ | $J_5$ |
|---|---|---|---|---|---|
| $T_1$ | 0 | 8 | 24 | 5 | 86 |
| $T_2$ | 31 | 7 | 21 | 26 | 86 |
| $T_3$ | 14 | 21 | 23 | 6 | 86 |
| $T_4$ | 0 | 16 | 21 | 15 | 86 |
| $T_5$ | 14 | 16 | 15 | 26 | 86 |

Now subtracting minimum element from each row

| | $J_1$ | $J_2$ | $J_3$ | $J_4$ | $J_5$ |
|---|---|---|---|---|---|
| $T_1$ | 0 | 8 | 24 | 5 | 86 |
| $T_2$ | 24 | 0 | 14 | 19 | 79 |
| $T_3$ | 8 | 15 | 17 | 0 | 80 |
| $T_4$ | 0 | 6 | 21 | 15 | 86 |
| $T_5$ | 0 | 2 | 1 | 12 | 72 |

Subtracting minimum element from each column and making assignments

| | $J_1$ ✓(ii) | $J_2$ | $J_3$ | $J_4$ | $J_5$ | |
|---|---|---|---|---|---|---|
| $T_1$ | [0] | 8 | 23 | 5 | 14 | ✓(iii) |
| $T_2$ | 24 | [0] | 13 | 19 | 7 | |
| $T_3$ | 8 | 15 | 16 | [0] | 8 | |
| $T_4$ | ~~0~~ | 16 | 20 | 15 | 14 | ✓(i) |
| $T_5$ | ~~0~~ | 2 | [0] | 12 | ~~0~~ | |

Min of uncovered elements is 5, subtracting it from each uncovered element and adding at intersections of two lines. Also making assigments.

| | $J_1$ ✓(ii) | $J_2$ | $J_3$ | $J_4$ ✓(iii) | $J_5$ | |
|---|---|---|---|---|---|---|
| $T_1$ | ~~0~~ | 3 | 18 | ~~0~~ | 9 | ✓(i) |
| $T_2$ | 29 | [0] | 13 | 19 | 7 | |
| $T_3$ | 13 | 15 | 16 | [0] | 8 | ✓(ii) |
| $T_4$ | [0] | 11 | 15 | 10 | 9 | ✓(iii) |
| $T_5$ | 5 | 2 | [0] | 12 | ~~0~~ | |

Subtracting min {3, 18, 9, 15, 16, 8, 199, 15, 9} = 3 from uncovered elements and adding at intersaction of two lines, making assignments thereafter.

|  | ✓(ii) $J_1$ | ✓(iii) $J_2$ | $J_3$ | ✓(iii) $J_4$ | $J_5$ |  |
|---|---|---|---|---|---|---|
| $T_1$ | 0̸ | 0̸ | 15 | 0̸ | 6 | ✓(i) |
| $T_2$ | 32 | [0] | 13 | 22 | 7 | ✓(iii) |
| $T_3$ | 13 | 12 | 13 | [0] | 5 | ✓(iii) |
| $T_4$ | [0] | 8 | 12 | 10 | 6 | ✓(iii) |
| $T_5$ | 8 | 2 | [0] | 15 | 0̸ |  |

Subtracting min {15, 6, 13, 7, 13, 5, 12, 6} = 5 from uncovered elements and adding at intersection of two lines, making assignments thereafter

|  | $J_1$ | $J_2$ | $J_3$ | $J_4$ | $J_5$ |
|---|---|---|---|---|---|
| $T_1$ | 0̸ | 0̸ | 10 | [0] | 1 |
| $T_2$ | 32 | [0] | 8 | 22 | 2 |
| $T_3$ | 13 | 12 | 8 | 0̸ | [0] |
| $T_4$ | [0] | 8 | 7 | 10 | 1 |
| $T_5$ | 13 | 7 | [0] | 20 | 0̸ |

∴ Optimal assignments for maximum revenue are $T_1 \to J_4$, $T_2 \to J_2$, $T_4 \to J_1$, $T_5 \to J_3$ and maximum revenue is 81 + 79 + 86 + 71 = 317

**Example 6.15** *Salesman wants to visit cities A.B.C.D and E. He does not want to visit any city twice before completing his tour to all the cities and wishes to return to point of starting journey. Cost of going from one city to another is shown in the table given below. Find the least cost route.*

|  | A | B | C | D | E |
|---|---|---|---|---|---|
| A | 0 | 2 | 5 | 7 | 1 |
| B | 6 | 0 | 3 | 8 | 2 |
| C | 8 | 7 | 0 | 4 | 7 |
| D | 12 | 4 | 6 | 0 | 5 |
| E | 1 | 3 | 2 | 8 | 0 |

**[GGSIPU, End Term 2013]**

***Solution*** Solving the given travelling salesman problem as an assignment problem, by Hungraian method.Subtracting minimum element from each row and thereafter each column, making restricted assignments ∞ :

To city

From city

| | A | B | C | D | E |
|---|---|---|---|---|---|
| A | ∞ | 1 | 3 | 6 | [0] |
| B | 4 | ∞ | [0] | 6 | ~~0~~ |
| C | 4 | 3 | ∞ | [0] | 3 |
| D | 8 | [0] | 1 | ∞ | 1 |
| E | [0] | 2 | ~~0~~ | 7 | ∞ |

However this solution is not the solution to the travelling salesman problem as it gives the sequence *A-E-A*. This violates the condition that salesman can visit each city only once.

The 'next best' solution can be obtained by bringing the next (non-zero) minimum element, i.e. 1 into the solution.The cost 1 occurs at three different places.

∴ Consider all three different places.

*Case 1.* Make the unit assignment in the cell (*A*, *B*) instead of zero assignment in the cell (*A*, *E*) and delete row *A* and column *B* so as to eliminate the possibility of any other assignment in row *A* and column *B*. Now make the assignments in usual manner.

To city

From city

| | A | B | C | D | E |
|---|---|---|---|---|---|
| A | ∞ | [1] | 3 | 6 | 0 |
| B | 4 | ∞ | [0] | 6 | 0 |
| C | 4 | 3 | ∞ | [0] | 3 |
| D | 8 | 0 | 1 | ∞ | [1] |
| E | [0] | 2 | 0 | 7 | ∞ |

**The sequence :** $A \to B, B \to C, C \to D, D \to E, E \to A$

The cost corresponding to this feasible solution is ₹ 15.

*Case 2.* If we make the assignment in the cell $(D, C)$ instead of $(D, E)$ then no feasible solution is obtained in terms of zeros or in forms of one which may give cost less than ₹ 15. **Ans.**

Hence, the best solution is $A—B—C—D—E—A$ and the total cost associated with this solution is ₹ 15.

**Example 6.16** *A company has 5 operators and 5 machines. The assignment costs are given in the table. Operator A cannot be assigned to machine $M_3$ and operator C cannot be assigned to machine $M_4$. Find the optimum assignment schedule.*

*Machine*

| *Operators* | $M_1$ | $M_2$ | $M_3$ | $M_4$ | $M_5$ |
|---|---|---|---|---|---|
| A | 7 | 7 | – | 4 | 8 |
| B | 9 | 6 | 4 | 5 | 6 |
| C | 11 | 5 | 7 | – | 5 |
| D | 9 | 4 | 8 | 9 | 4 |
| E | 8 | 7 | 9 | 11 | 3 |

**[GGSIPU, Feb-2013]**

***Solution*** Assigning cost $\propto$ to $(A, M_3)$ and $(C, M_4)$

Machine

| Operators | $M_1$ | $M_2$ | $M_3$ | $M_4$ | $M_5$ |
|---|---|---|---|---|---|
| A | 7 | 7 | $\propto$ | 4 | 8 |
| B | 9 | 6 | 4 | 5 | 6 |
| C | 11 | 5 | 7 | $\propto$ | 5 |
| D | 9 | 4 | 8 | 9 | 4 |
| E | 8 | 7 | 9 | 11 | 3 |

Row-wise minimum value subtraction :

| | $M_1$ | $M_2$ | $M_3$ | $M_4$ | $M_5$ |
|---|---|---|---|---|---|
| A | 3 | 3 | ∝ | 0 | 4 |
| B | 5 | 2 | 0 | 1 | 2 |
| C | 6 | 0 | 2 | ∝ | 0 |
| D | 5 | 0 | 4 | 5 | 0 |
| E | 5 | 4 | 6 | 8 | 0 |

Coloum-wise minimum value subtraction and making assignments :

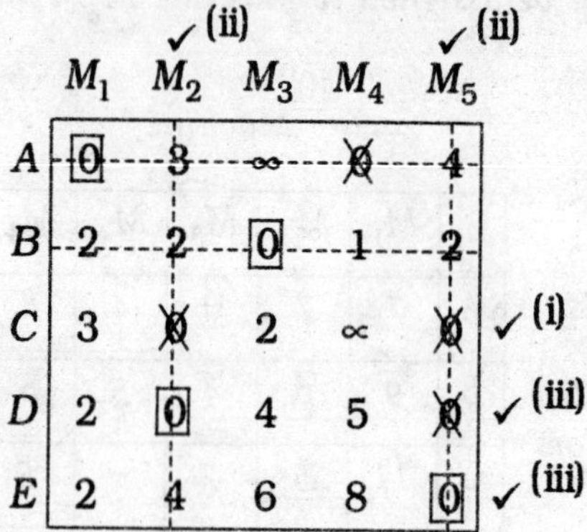

Note. ($D$, $M_2$) has been assignment arbitrarily

Min = 2, Subtract 2 from uncovered element and add 2 at intersection points; and making assignments thereafter

| | $M_1$ | $M_2$ | $M_3$ | $M_4$ | $M_5$ |
|---|---|---|---|---|---|
| A | ~~0~~ | 5 | ∝ | [0] | 6 |
| B | 2 | 4 | [0] | 1 | 4 |
| C | 1 | [0] | ~~0~~ | ∝ | 0 |
| D | [0] | ~~0~~ | 2 | 3 | ~~0~~ |
| E | ~~0~~ | 4 | 4 | 6 | [0] |

Assigning cost $\Rightarrow 4+4+5+9+3 = 25$

*Or*

| | $M_1$ | $M_2$ | $M_3$ | $M_4$ | $M_5$ |
|---|---|---|---|---|---|
| A | ~~0~~ | 5 | ∝ | [0] | 6 |
| B | 2 | 4 | [0] | 1 | 4 |
| C | 1 | [0] | ~~0~~ | ∝ | ~~0~~ |
| D | ~~0~~ | ~~0~~ | 2 | 3 | [0] |
| E | [0] | 4 | 4 | 6 | ~~0~~ |

Assigning cost $\Rightarrow 4+4+5+4+8=25$

*Or*

| | $M_1$ | $M_2$ | $M_3$ | $M_4$ | $M_5$ |
|---|---|---|---|---|---|
| A | ~~0~~ | 5 | ∝ | [0] | 6 |
| B | 2 | 4 | [0] | 1 | 4 |
| C | 1 | ~~0~~ | ~~0~~ | ∝ | [0] |
| D | ~~0~~ | [0] | 2 | 3 | ~~0~~ |
| E | [0] | 4 | 4 | 6 | ~~0~~ |

Assigning cost $\Rightarrow 4+4+5+4+8=25$

**Example 6.17** *A machine tool company decides to make four sub-assemblies through four contractors. Each contractor is to receive only one sub-assembly. The cost of each sub-asssembly is determined by the bids submitted by each contractor and is shown in table below. Assign different assemblies to contractors so as to minimize the total cost.*

| Sub-assemblies \ Contractor | A | B | C | D |
|---|---|---|---|---|
| I | 15 | 13 | 14 | 17 |
| II | 11 | 12 | 15 | 13 |
| III | 18 | 12 | 10 | 11 |
| IV | 15 | 17 | 14 | 16 |

**[GGSIPU, End Term-2014]**

***Solution*** Select minimum element in each row and subtract from every element of that row.

| | A | B | C | D |
|---|---|---|---|---|
| I | 2 | 0 | 1 | 4 |
| II | 0 | 1 | 4 | 2 |
| III | 8 | 2 | 0 | 1 |
| IV | 1 | 3 | 0 | 2 |

Subtract the smallest element in each column from every element of that column.

| | A | B | C | D |
|---|---|---|---|---|
| I | 2 | 0 | 1 | 3 |
| II | 0 | 1 | 4 | 1 |
| III | 8 | 2 | 0 | 0 |
| IV | 1 | 3 | 0 | 1 |

Now following Hungrarian method for assignment schedule *i.e.* allocating rows having single zero with assignment and crossing out corresponding zeros in columns and thereafter allocating zeros with assignment column-wise.

| | A | B | C | D |
|---|---|---|---|---|
| I | 2 | [0] (*i*) | 1 | 3 |
| II | [0] (*ii*) | 1 | 4 | 1 |
| III | 8 | 2 | ~~0~~ | [0] (*iv*) |
| IV | 1 | 3 | [0] (*iii*) | 1 |

∴ Optimal assignment schedule is :

| *Sub-assembly* | *Contractor* | *Cost* |
|---|---|---|
| I | *B* | 13 |
| II | *A* | 11 |
| III | *D* | 11 |
| IV | *C* | 14 |
| | | 49 |

Hence, total minimum cost = 49. **Ans.**

**Example 618** *A machine operator processes five types of items on his machine each week and must choose a sequence for them. The set up cost per change depends on the item presently on the machine and the set up to be made according to the following table :*

| *From item* | *To item* A | B | C | D | E |
|---|---|---|---|---|---|
| A | – | 4 | 7 | 3 | 4 |
| B | 4 | – | 6 | 3 | 4 |
| C | 7 | 6 | – | 7 | 5 |
| D | 3 | 3 | 7 | – | 7 |
| E | 4 | 4 | 5 | 7 | – |

*If the operator processes each type of item once and only once each week, how should he sequence the items on his machine in order to minimize the total set up cost?* **[GGSIPU, Feb-2014]**

***Solution*** Assigning cost ∞ to restricted assignments :

| | A | B | C | D | E |
|---|---|---|---|---|---|
| A | ∞ | 4 | 7 | 3 | 4 |
| B | 4 | ∞ | 6 | 3 | 4 |
| C | 7 | 6 | ∞ | 7 | 5 |
| D | 3 | 3 | 7 | ∞ | 7 |
| E | 4 | 4 | 5 | 7 | ∞ |

Applying Hungarian method (Row-wise minimum value subtraction)

| | A | B | C | D | E |
|---|---|---|---|---|---|
| A | ∞ | 1 | 4 | 0 | 1 |
| B | 1 | ∞ | 3 | 0 | 1 |
| C | 2 | 1 | ∞ | 2 | 0 |
| D | 0 | 0 | 4 | ∞ | 4 |
| E | 0 | 0 | 1 | 3 | ∞ |

Column-wise minimum value subtraction and making assignments

| | A | B | C | D ✓ | E | |
|---|---|---|---|---|---|---|
| A | ∞ | 1 | 3 | [0] | 1 | ✓ |
| B | 1 | ∞ | 2 | ~~0~~ | 1 | ✓ |
| C | 2 | 1 | ∞ | 2 | [0] | |
| D | ~~0~~ | [0] | 3 | ∞ | 4 | |
| E | ~~0~~ | ~~0~~ | [0] | 3 | ∞ | |

Minimum of uncovered elements is 1, subtracting it from each uncovered element and adding at intersections of lines and making assignments.

| | *A* | *B* | *C* | *D* | *E* |
|---|---|---|---|---|---|
| *A* | ∞ | ~~0~~ | 2 | [0] III | ~~0~~ |
| *B* | [0] IV | ∞ | 1 | ~~0~~ | ~~0~~ |
| *C* | 2 | 1 | ∞ | 3 | [0] I |
| *D* | ~~0~~ | [0] V | 3 | ∞ | 4 |
| *E* | ~~0~~ | ~~0~~ | [0] II | 4 | ∞ |

**Sequence :** $A \rightarrow D, D \rightarrow B, B \rightarrow A, C \rightarrow E, E \rightarrow C$

It violates given condition.

***Case I.*** Now next least element is 1, try to bring 1 in to the solution.

| | *A* | *B* | *C* | *D* | *E* |
|---|---|---|---|---|---|
| *A* | ∞ | ~~0~~ | 2 | [0] | ~~0~~ |
| *B* | ~~0~~ | ∞ | [1] | ~~0~~ | ~~0~~ |
| *C* | 2 | 1 | ∞ | 3 | [0] |
| *D* | ~~0~~ | [0] | 3 | ∞ | 4 |
| *E* | [0] | ~~0~~ | ~~0~~ | 4 | ∞ |

**Sequence :** $A \rightarrow D, D \rightarrow B, B \rightarrow C, C \rightarrow E, E \rightarrow A$

Setup cost = 21

***Case II.***

| | *A* | *B* | *C* | *D* | *E* |
|---|---|---|---|---|---|
| *A* | ∞ | ~~0~~ | 2 | ~~0~~ | [0] |
| *B* | ~~0~~ | ∞ | 1 | [0] | ~~0~~ |
| *C* | 2 | [1] | ∞ | 3 | ~~0~~ |
| *D* | [0] | ~~0~~ | 3 | ∞ | 4 |
| *E* | ~~0~~ | ~~0~~ | [0] | 4 | ∞ |

**Sequence :** $A \rightarrow E, E \rightarrow C, C \rightarrow B, B \rightarrow D, D \rightarrow A$

Setup cost = 21

## MULTIPLE CHOICE QUESTIONS

6.1. The assignment problem is a special case of

(a) an LPP (b) a transportation problem

(c) an duality (d) None

6.2. Which method is used to get an optimal assignment?

(a) Simplex method (b) Transprotation method

(c) Enumeration method (d) Hungarian method

6.3. In A.P. the Cost Matrix is

(a) square matrix (b) diagonal matrix

(c) row matrix (d) column matrix

6.4. In an unbalanced problem cost matrix is not a square matrix, how it can be balanced?

(a) Adding row with all the entries zero

(b) Adding column with all the entries zero

(c) Either (a) or (b)

(d) Both (a) and (b)

6.5. The transportation method cannot be used to solve the A.P. because of

(a) Dummy allocation (b) Degeneracy

(c) Non degeneracy (d) Both (a) and (b)

## ANSWERS

1. (b) 2. (d) 3.(a) 4.(c) 5.( b)

## SUMMARY

An assignment problem is a particular case of transprotation problem in which a number of operations are to be assigned to an equal number of operators, where each operator performs onlye one operation. The objective of assignment model is to minimize overall cost or to maximize overall profit for a given assignment schedule.

An assignment problem in which number of rows and column are same, all requirements and availabilities are unity and each allocation is limited to either 0 or 1.

Mathematical formulation of a minimizing assignment problem is given as :

$$\text{Minimize } z = \sum_{i=1}^{n}\sum_{j=1}^{n} C_{ij} \cdot x_{ij}$$

Subject to

$$\sum_{i=1}^{n} x_{ij} = 1, \; j = 1, 2 \ldots\ldots, n$$

$$\sum_{j=1}^{n} x_{ij} = 1 , j = 1, 2 ....., n$$

$$x_{ij} = 0 \text{ or } 1$$

Solution of Assignment Problem

An assignment problem can be solved by the following methods.

(i) Enumeration method
(ii) Simplex method
(iii) Transportation method
(iv) Hengarian method

***Hungarian method*** (minimization case) can be summerized as follows :

*Step 1.* Prepare a square matrix if not originally. Add suitable number of dummy rows (or columns) with zero cost element for this purpose.

*Step 2.* Subtract the smallest element in each row from every element of that row.

*Step 3.* Subtract the smallest element in each column from every element of that column. Purpose is to induce a zero element in every row and column.

*Step 4.* Make assignment in rows having single zero and mark it as $\boxed{0}$. Cross all other zeros in the column where assignment has been made.

*Step 5.* Repeat the procedure for each column one by one starting from first column. Simultaneouly cross out unassigned zeros in corresponding rows.

*Step 6.* If all $n$ assignments have been made, stop otherwise go to step 7.

*Step 7.* Draw horizontal and vertical lines covering all zeros by following rule :

(i) Mark '✓' to rows having no assignment.
(ii) Mark '✓' to columns having $\otimes$ in the marked row.
(iii) Mark '✓' to rows having $\boxed{0}$' in marked column.
(iv) Draw horizontal and vertical lines to unmarked rows and marked columns.

*Step 8.* Select the smallest element among all uncovered elements and then subtract this value from all elements in the matrix not covered by lines and add this value to all elements that lie at the intersection of the two lines.

*Step 9.* Go back to step 4.

CHAPTER 7

# Queueing Theory

## 7.1 INTRODUCTION

A group of items or persons waiting to receive service including those receiving the service is known as a waiting line or a queue.

In everyday life queueing or waiting in a line is a very common situation. Waiting lines are generally seen at ticket booths, grocery shops, bus stops, bank counters, bill pay counters, post offices and so on. Queueing is a very common civilized life phenomenon. Generally, queues are formed when either customers requiring service exceeds the number of service facilities, or service facilities take more time to serve a customer. The study of queues has been applied to variety of situation to reduce the customer's waiting time and to increase the rate of service facilities.

## 7.2 QUEUEING COST STRUCTURE

Queuing cost structure is a graphical representation of the cost of service provider and waiting time of the customers. The queueing models suggest ways to improve the effeciency of service and also indicates the effective uses of existing man power and other resources. Queuing theory determine the balance between cost of offering service and cost incured due to delay in offering service. The relationship between queueing and service rate can be diagrammatically explained using the cost curve as shown in Fig. 7.1.

The Figure 7.1 indicates that, an increase in service facilities would reduced customer waiting time. Conversely and decrease in the service facilities would result in long queues.

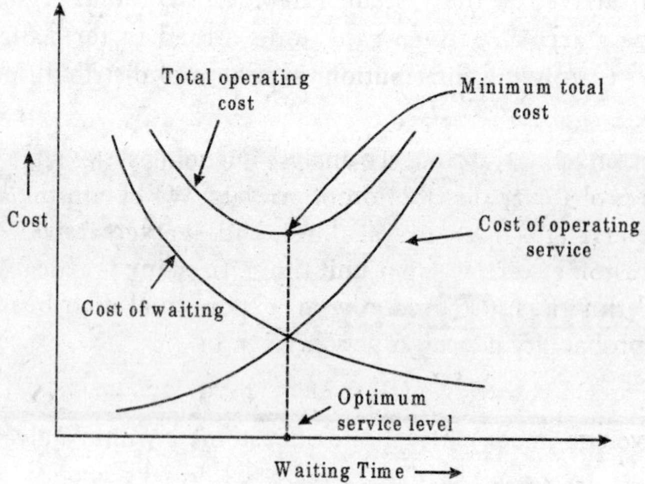


**Fig. 7.1**

The optimum service level is the one that minimize the operating cost of the entire system.

## 7.3 QUEUEING SYSTEM

A queueing system is composed of the following components :

(i) Input source (Calling population)

(ii) Queueing process

(iii) Queueing discipline (FIFO, LIFO, SIRO, SIP)

(iv) Service process

(v) Customer behaviour

Figure 7.2 will summerised the five features of Queue.

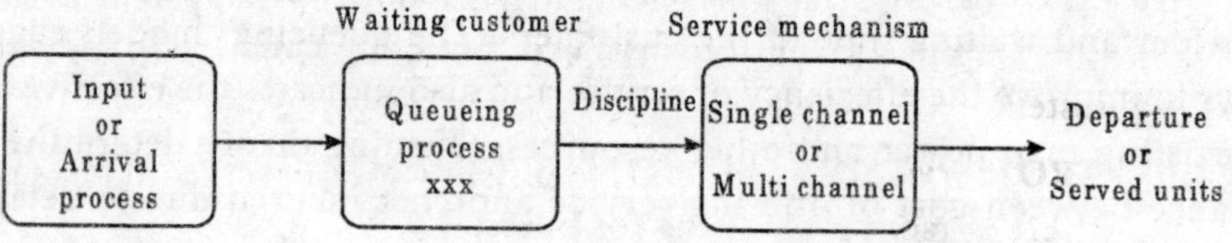


**Fig. 7.2**

### 7.3.1 Input Sources

The input process is the arrival pattern in which the number of customers (population) arrived at the system. They don't come at regular intervals therefore their arrival pattern can be described in term of probability distribution i.e. Poission distribution, exponential distribution and Erlang distribution.

**Distribution of arrival time.** To analyse queueing system it is necessary to describe the probability distribution of arrivals. We obtain the average time between the arrivals (Inter arrival time), and average arrival rate (i.e. the number of customers arriving per unit time). In many practical situation the inter arrival time is approximate by an exponential distribution with the continuous probability density function given by

$$f(t) = \lambda e^{-\lambda t} \ (t \geq 0)$$

where $\lambda$ is average (mean) arrival rate of customers within a given time ($t$).

(i) $\lambda$ — Average arrival rate

(ii) $\lambda t$ — Expected number of customers arriving in time $t$.

(iii) $t$ — Time between arrival

### 7.3.2 Queueing Process

It is also known as queue configuration. This means that arrangement of number of queue and their respective lengths. The number of queue depends upon the layout of the service system, while the length of queue depends upon operational situation.

### 7.3.3 Queue Discipline

It is basically a rule or a manner in which customers are selected from the queue for services the most common ways are :

(i) **FIFO (First In First Out).** Most commonly observed system, e.g. waiting lines at railway booths.

(ii) **LIFO (Last In First Out).** Generally, this does not happen in a system where human beings are involved. But quite common in inventory system.

(iii) **SIRO (Service In Random Order).** This can be observed at bus stops, where persons are waiting for buses.

(iv) **SIP (Service In Priority).** Generally in case of some emergency in hospitals etc.

### 7.3.4 Service Mechanism

It is a manner in which customers are served and after that they leave the service system. The service to the customer depend upon the server behaviour and distribution of service time. The time taken by the server to complete the service for a customer is known as service time. We again use exponential distribution with density function as

$$f(t) = \mu e^{-\mu t} \ (t \geq 0)$$

(i) $\mu$ — Average (mean) service rate.

(ii) $\mu t$ — Average number of customers served in time ($t$)

(iii) $t$ — Time of service

(iv) $1/\mu$ — Average service completion time.

### 7.3.5 Customer Behaviour

The customers generally behave in different manner while waiting ina queue, some of the important ways are :

(i) *Balking.* When a customer doesn't enter the queue because the queue is too long to suit him and decided not to enter in it, such customers are called balked (who leave the queue before getting into the system).

(ii) *Reneging.* These types of customers have very little patience they may enter the queue but quickly looses patience and leave the queue after spending some time in the system. Such customers are called Reneged.

(iii) *Jocking.* Whenever there are two or more serving queues, some customers keep switching from one queue to another to get advantage of being served earlier that is called Jockey for position.

## 7.4 STATE OF THE SYSTEM

(i) **Transient state.** When its operating characteristics are dependent on time.

(ii) **Steady state.** When its operating characteristics become indepenent of time

$$\lim_{t \to \infty} P_n(t) = P_n \text{ (Independent of time } t)$$

$$\frac{d}{dt} P_n(t) = 0 \text{ (In steady state)}$$

(iii) ***Explosive state.*** If the arrival rate of the system is more than its service rate, the length of queue will go on increasing with time and will tend to infinity as $t \to \infty$.

## 7.5 KENDALL'S NOTATION FOR REPRESENTING QUEUEING MODELS

The useful notation for queueing model introduced by D.G. Kendall and A.M. Lee. and the standard representation is given below :

$$(A/B/C) \ : \ (d/e)$$

where,

- $A$ : Arrival pattern of the units which is given by probability distribution of inter arrival time unit.
- $B$ : Probability distribution of servicing time of the unit.
- $C$ : Number of parallel servicing channels in the system.
- $d$ : Capacity of the system (in service + waiting).
- $e$ : Queue discipline.

### 7.5.1 Symbols and notations

- $T$ : The inter arrival time between two successive customers.
- $n$ : Number of customers in the queueing system.
- $E_n$ : The state in which there are $n$ calling units in the system.
- $P_n(t)$ : Transient state–probability that there are exactly $n$ units in the system at any time $t$.
- $P_n$ : Steady state probability of having $n$ units in the systems.
- $\lambda_n$ : Mean arrival rate (Expected number of arrivals per unit time), when $n$ units are present in the system.
- $\lambda$ : Mean arrival rate of customers (Independent of $n$).
- $\mu_n$ : Mean service rate (When $n$ units are present in the system).
- $\mu$ : Mean service rate (Independent of $n$).
- $C$ : Number of parallel service stations.
- $\rho$ : $\lambda/\mu$ = Traffic intensity (Utilization factor)
- $\phi(w)$ : Probability density function of waiting time in the system.
- $\phi_T(n)$ : Probability of $n$ services in time $T$.
- $E(L_s)$ : Expected number of customers in the system, i.e. expected queue length.
- $E(L_q)$ : Expected number of customers in the queue, i.e. expected queue length.
- $E(W_s)$ : Expected waiting time per customer in the system (Included service time).

$E(W_q)$ : Expected waiting time per customer in the queue (excluded service time).

$E(L/L > 0)$ : Expected length of the non-empty queue.

$E(W/W > 0)$ : Expected waiting time of a customer, who has to wait.

## 7.6 CLASSIFICATION OF QUEUEING MODELS

**Model I.** [(M/M/C) : (∝/FCFS)]. Birth and death process model. Where arrival is in Poisson distribution and service time is in exponential distribution. The letter M represents the Markovian property of exponential process.

**Model II.** Multiserver Model [(M/M/C) : (∝/FCFS)]. In this Model C represents the number of service channels.

### Distribution of Arrival and Service Time

The common waiting line models have been developed on the assumption that arrival rate follows the Possion distribution (Random arrival) and service time follows the exponential distribution unless mentioned otherwise follow solution steps.

**Solution of Queueing Model Consists of following Steps**

**Step 1.** First we shall obtain the system of steady state equations governing the queue.

**Step 2.** We solve the equations obtained in step 1 and find the probability distribution of queue length.

**Step 3.** Obtain various probability density function and derive formulae for $E(L_s)$, $E(L_q)$, $E(W_s)$, $E(W_q)$, $E(L/L > 0)$ and $E(W/W > 0)$.

### 7.6.1 Solution of model–I [ (M/M/1):(∝/FCFS) ] (Birth and death process)

M - Poisson arrival time
M - Exponential service time } Both are denoted by 'M' due to Markorian property of exponential process

1 – Single service

∝ – Infinite capacity

FCFS – First Come, First Served

Let $\lambda$ = Mean arrival rate of units

$\mu$ = Mean service rate.

**Step 1.** *To find the system of steady state equations,* If there are $n > 0$ units in the system at time $(t + \Delta t)$.

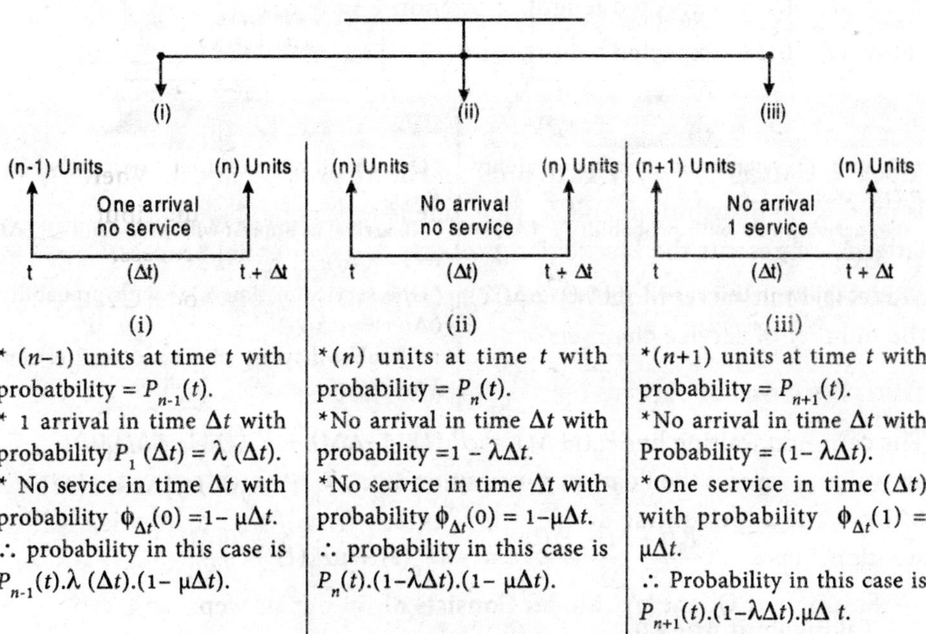


(i)

* $(n-1)$ units at time $t$ with probatbility $= P_{n-1}(t)$.
* 1 arrival in time $\Delta t$ with probability $P_1(\Delta t) = \lambda(\Delta t)$.
* No sevice in time $\Delta t$ with probability $\phi_{\Delta t}(0) = 1 - \mu\Delta t$.

$\therefore$ probability in this case is $P_{n-1}(t).\lambda(\Delta t).(1-\mu\Delta t)$.

(ii)

* $(n)$ units at time $t$ with probability $= P_n(t)$.
* No arrival in time $\Delta t$ with probability $= 1 - \lambda\Delta t$.
* No service in time $\Delta t$ with probability $\phi_{\Delta t}(0) = 1 - \mu\Delta t$.

$\therefore$ probability in this case is $P_n(t).(1-\lambda\Delta t).(1-\mu\Delta t)$.

(iii)

* $(n+1)$ units at time $t$ with probability $= P_{n+1}(t)$.
* No arrival in time $\Delta t$ with Probability $= (1 - \lambda\Delta t)$.
* One service in time $(\Delta t)$ with probability $\phi_{\Delta t}(1) = \mu\Delta t$.

$\therefore$ Probability in this case is $P_{n+1}(t).(1-\lambda\Delta t).\mu\Delta t$.

Hence all the above three cases are mutually exclusive;

$$P_n(t+\Delta t) = P_{n-1}(t).\lambda\Delta t.(1-\mu\Delta t) + P_n(t).(1-\lambda\Delta t).(1-\mu\Delta t) + P_{n+1}(t).(1-\lambda\Delta t).\mu\Delta t$$

$$\Rightarrow \quad P_n(t+\Delta t) = [\lambda P_{n-1}(t) - (\lambda+\mu)P_n(t) + \mu P_{n+1}(t)]\Delta t + P_n(t) + 0(\Delta t)$$

$\{\because 0(\Delta t)$ indicates terms having $(\Delta t)^2$ and higher, also $\underset{\Delta t - 0}{\text{Lt}} \dfrac{0(\Delta t)}{(\Delta t)} = 0\}$

$$\therefore \quad \frac{P_n(t+\Delta t) - P_n(t)}{\Delta t} = [\lambda P_{n-1}(t) - (\lambda+\mu)P_n(t) + \mu P_{n+1}(t)] + \frac{0(\Delta t)}{(\Delta t)}$$

Taking limit $\Delta t \to 0$

$$\underset{\Delta t \to 0}{\text{Lt}} \frac{P_n(t+\Delta t) - P_n(t)}{\Delta t} = \lambda P_{n-1}(t) - (\lambda+\mu)P_n(t) + \mu P_{n+1}(t)$$

$$\Rightarrow \quad \frac{d}{dt}P_n(t) = \lambda P_{n-1}(t) - (\lambda+\mu)P_n(t) + \mu P_{n+1}(t) \quad \dots (1)$$

Again for $n = 0$

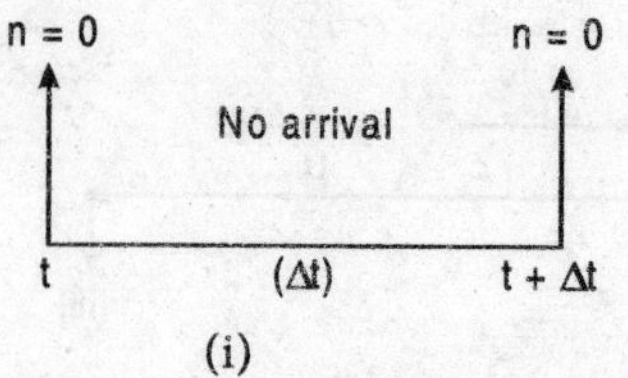


(i)

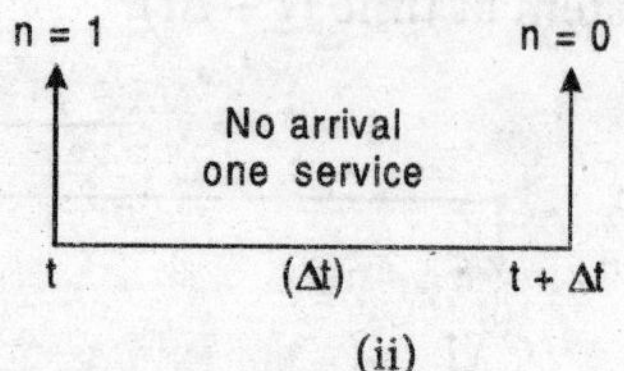


(ii)

* No units in system at time $t$ with probability $P_0(t)$.
* No arrival in $\Delta t$ with probability $P_0(\Delta t) = (1-\lambda\Delta t)$.

$\therefore$ Probability in this case is $P_0(t).(1-\lambda\Delta t)$.

*One unit in system at time $t$ with probability $P_1(t)$.

*No arrival in time $\Delta t$ with probability $P_0(\Delta t) = (1-\lambda\,\Delta t)$.

*One service in time $(\Delta t)$ with probability $\phi\Delta t(1) = \mu\Delta t$.

$\therefore$ Probability in this case is $P_1(t).(1-\lambda\Delta t).(\mu\Delta t)$

$$\therefore \quad P_0(t+\Delta t) = P_0(t).(1-\lambda\Delta t) + P_1(t).(1-\lambda\Delta t).\mu\Delta t$$

$$\Rightarrow \quad P_0(t+\Delta t) = P_0(t) + [-\lambda P_0(t) + \mu P_1(t)]\Delta t + 0(\Delta t)$$

$$\therefore \quad \frac{P_0(t+\Delta t)-P_0(t)}{\Delta t} = -\lambda P_0(t) + \mu P_1(t) + \frac{0(\Delta t)}{(\Delta t)}$$

Taking limit $\Delta t \to 0$.

$$\underset{\Delta t\to 0}{\text{Lt}} \frac{P_0(t+\Delta t)-P_0(t)}{\Delta t} = -\lambda P_0(t) + \mu P_1(t)$$

$$\Rightarrow \quad \frac{d}{dt}P_0(t) = -\lambda P_0(t) + \mu P_1(t) \qquad \text{... (2)}$$

**When system reaches steady state,**

$$\frac{d}{dt}P_n(t) = 0 \text{ and } \frac{d}{dt}P_0(t) = 0$$

$\therefore$ under steady state the system of equations (1) and (2)

$$\lambda P_{n-1} - (\lambda+\mu)P_n + \mu P_{n+1} = 0 \qquad \text{...(A)}$$

$$-\lambda P_0 + \mu P_1 = 0 \qquad \text{...(B)}$$

Equations (A) and (B) are the steady state equations of the system

**Note.** The system will settle to steady state only if $\lambda<\mu$, i.e. $\rho = \frac{\lambda}{\mu} < 1$

**Step 2.** To solve the steady state Eqs (A) and (B) obtained in step 1.

From B,

$$P_1 = \frac{\lambda}{\mu}P_0 = \rho P_0 \quad \left(\because \frac{\lambda}{\mu} = \rho\right) \qquad \text{... (3)}$$

Putting $n = 1$ in Eq. (A),

$$\lambda P_0 - (\lambda+\mu)P_1 + \mu P_2 = 0$$

$$\Rightarrow \quad P_2 = -\frac{\lambda}{\mu}P_0 + \left(\frac{\lambda}{\mu}+1\right)P_1$$

$$P_2 = -\frac{\lambda}{\mu}P_0 + \frac{\lambda}{\mu}P_1 + P_1$$

$$P_2 = -\rho P_0 + \rho P_1 + P_1 \qquad \left(\because \frac{\lambda}{\mu} = \rho\right)$$

$$P_2 = -P_1 + \rho P_1 + P_1 \qquad \text{(Using Eq. 3)}$$

$$P_2 = \rho P_1 \Rightarrow \rho^2 P_0$$

$$\Rightarrow \quad P_2 = \rho^2 P_0$$

$$\vdots \quad \vdots \quad \vdots$$

$$P_n = \rho^n P_0 \text{ (for } n \geq 0) \qquad \text{... (4)}$$

But $\quad \sum_{n=0}^{\infty} P_n = 1 \text{ or } P_0 + P_1 + P_2 + \text{....} P_{\infty} = 1$

$$\Rightarrow \quad P_0 + \rho P_0 + \rho^2 P_0 + \text{.....} = 1$$

$$\Rightarrow \quad P_0(1 + \rho + \rho^2 + \text{.....}) = 1$$

$$\Rightarrow \quad P_0\left(\frac{1}{1-\rho}\right) = 1 \qquad \left(\text{Infinite G.P. with } a = 1, r = \rho, S_{\infty} = \frac{a}{1-r}\right)$$

$$\Rightarrow \quad P_0 = (1-\rho) \qquad \text{... (5)}$$

$$\Rightarrow \quad P_n = \rho^n(1-\rho) \qquad \text{Using (4) and (5)}$$

Which gives the probability that there are $n$ units in the system at any time.

**Step 3** (a)**To find probability (queue size $\geq N$)**

$$\text{Probability (queue size} \geq N) = \sum_{n=N}^{\infty} P_n = \sum_{n=0}^{\infty} P_n - \sum_{n=0}^{N-1} P_n$$

$$= 1 - [P_0 + P_1 + P_2 + .... + P_{N-1}]$$

$$= 1 - P_0[1 + \rho + \rho^2 + ....\rho^{N-1}] \text{[using eq. (4)]}$$

$$= 1 - (1-\rho) \cdot \frac{1(1-\rho^N)}{(1-\rho)}$$

$$\left( \because P_0 = 1-\rho \quad S_n \text{ of G.P.} = \frac{a(1-r^n)}{(1-r)} \right)$$

$$= 1 - 1 + \rho^N$$

$$P \text{ (queue size} \geq N) = \rho^N \Rightarrow \left(\frac{\lambda}{\mu}\right)^N$$

(b) **To find $E(L_s)$.** Average number of customers in system.

$$E(n) = E(L_s) = \sum_{n=0}^{\infty} np_n = \sum_{n=0}^{\infty} n(1-\rho)\rho^n$$

$$[\because P_n = \rho^n(1-\rho)]$$

$$= (1-\rho) \cdot \rho \sum_{n=0}^{\infty} n \cdot \rho^{n-1}$$

$$= (1-\rho) \cdot \rho \; [0 + 1 + 2\rho + 3\rho^2 + ......]$$

$$= \rho\,(1-\rho) \cdot (1-\rho)^{-2}$$

$$\left[\because 1 + nx + \frac{n(n-1)}{\lfloor 2}x^2 + .... = (1-x)^{-n}\right]$$

$$E(L_s) = \frac{\rho}{(1-\rho)} = \frac{\lambda}{\mu - \lambda}$$

(c) **To find $E(L_q)$.** Expected number of customers queue or average length of waiting line.

$$E(L_q) = \sum_{n=1}^{\infty} (n-1)P_n$$

(Since there are $(n-1)$ units in the queue excluded one in service)

$$E(L_q) = \sum_{n=1}^{\infty} nP_n - \sum_{n=1}^{\infty} P_n$$

$$E(L_q) = E(L_s) - [P_1 + P_2 + \ldots\ldots]$$

$$E(L_q) = \frac{\rho}{(1-\rho)} - [\rho + \rho^2 + \ldots.]P_0$$

$\therefore$

$$[P_n = \rho^2 P_0 \text{ and } E(L_s) = \frac{\rho}{(1-\rho)}]$$

$$E(L_q) = \frac{\rho}{(1-\rho)} - \left[\frac{\rho}{(1-\rho)}\right](1-\rho) \; [\because P_0 = (1-\rho)]$$

$$E(L_q) = \frac{\rho}{(1-\rho)} - \rho = \frac{\rho^2}{(1-\rho)}$$

(d) **To find $E(L/L>0)$.** Expected length of non-empty queue.

$$E(L/L>0) = \frac{E(L_q)}{\text{Probability [On arrival has to wait, L>0]}}$$

$$= \frac{E(L_q)}{P((n-1)>0)} = \frac{E(L_q)}{P(n>1)}$$

$$E(L/L>0) = \frac{E(L_q)}{\sum_{n=2}^{\infty} P_n} = \frac{E(L_q)}{P_2 + P_3 + P_4 + \ldots.}$$

$$E(L/L>0) = \frac{E(L_q)}{(\rho^2 + \rho^3 + \rho^4 + \ldots.)P_0}$$

$$E(L/L>0) = \frac{\rho^2}{(1-\rho)} \Big/ \frac{\rho^2}{(1-\rho)} \cdot (1-\rho)$$

$$E(L/L>0) = \frac{1}{1-\rho}$$

$\therefore$

$$E(L/L>0) = \frac{1}{1-\rho} = \frac{\mu}{(\mu-\lambda)}$$

(e) **To find $E(W/W > 0)$.** Expected waiting time of a customer who has to wait $(W/W>0)$

$$E(W/W>0) = \frac{1}{\mu(1-\rho)}$$

(f) **To find fluctuation (variance) of queue length.**

By definition,

$$\text{Var}(n) = E(n^2) - \{E(n^2)\}^2$$

$$= \sum_{n=1}^{\infty} n^2 P_n - \left[\sum_{n=1}^{\infty} nP_n\right]^2$$

$$= \sum_{n=1}^{\infty} n^2(1-\rho)\rho^n - \left(\frac{\rho}{1-\rho}\right)^2$$

[From part (b), $P_n = (1 - \rho)\rho^n$]

$$\text{Var}(n) = \rho(1-\rho)\sum_{n=1}^{\infty} n^2\rho^{n-1} - \left(\frac{\rho}{1-\rho}\right)^2$$

$$\text{Var}(n) = \rho(1-\rho)[1 + 2^2\rho + 3^2\rho^2 + ....] - \frac{\rho^2}{(1-\rho)^2} \quad ... (1)$$

Let $S = 1 + 2^2\rho + 3^2\rho^2 + ....$

Integrating both sides w.r.t. ρ form 0 to ρ.

$$\int_0^{\rho} S d\rho = \left[\rho + 2\rho^2 + 3\rho^3 + ....\right]_0^{\rho}$$

$$= \rho + 2\rho^2 + 3\rho^3 + ....$$

$$= \rho(1-\rho)^{-2} \Rightarrow \frac{\rho}{(1-\rho)^2}$$

Differentiate both sides w.r.t. ρ, we have

$$S = \frac{1+\rho}{(1-\rho)^3} \quad ... (2)$$

Using (2) in (1)

$$\text{Var}(n) = \rho(1-\rho)\cdot\frac{(1+\rho)}{(1-\rho)^3} - \frac{\rho^2}{(1-\rho)^2}$$

$$\text{Var}(n) = \frac{\rho}{(1-\rho)^2}$$

**Model I. (*M*/*M*/1) : (∝/*FCFC*) Birth and death model (C=1)**

(i) Probability that the system will be idle ($P_0$)

$$P_0 = (1-\rho) \qquad (\because \rho = \frac{\lambda}{\mu})$$

(ii) Probability of $n$ units in the system ($P_n$)

$$P_n = \rho^n(1-\rho)$$

(iii) Expected (Average) number of units in the system ($L_s$)

$$L_s = \frac{\lambda}{\mu-\lambda} = \frac{\rho}{(1-\rho)}$$

(iv) Average queue length ($L_q$)

$$L_q = \frac{\lambda^2}{\mu(\mu-\lambda)} = \frac{\rho^2}{(1-\rho)}$$

(v) Expected waiting time in the queue

$$W_q = \frac{L_q}{\lambda} = \frac{\lambda}{\mu(\mu-\lambda)}$$

(vi) Expected waiting time in the system

$$W_s = \frac{L_s}{\lambda} = \frac{1}{\mu-\lambda}$$

(vii) Probability (queue size ≥ $N$),

$$P(n \geq N) = \rho^N = \left(\frac{\lambda}{\mu}\right)^N$$

(viii) Expected length of non-empty queue

$$E(L/L>0) = \frac{1}{(1-\rho)} = \frac{\mu}{(\mu-\lambda)}$$

(ix) Expected waiting time of a customer who has to wait

$$E(W/W>0) = \frac{1}{\mu(1-\rho)}$$

(x) To find fluctuation (Variance) of queue length

$$\text{Var}(n) = \frac{\rho}{(1-\rho)^2}$$

**Examples 7.1** *At a one-man barber shop, customers arrive according to Poisson distribution with a mean arrival rate of 5 per hour and his hair cutting time was exponentially distributed with an average hair cutting taking 10 minutes. Find*

*(i) Average number of customers in the shop.*

*(ii) Average number of customers waiting for a hair cut.*

*(iii) The % of customers who have to wait before getting into barber's chair.*

*(iv) The person of no time arrival and walked without having to wait.*

***Solution***

Average arrival rate — $\lambda = 5$ per hour

Average service rate — $\mu = \frac{1}{10}$ per minute

$$= \frac{60}{10} \text{ per hour}$$

$$= 6 \text{ per hour}$$

(i) Average number of customers in the shop

$$L_s = \frac{\lambda}{\mu-\lambda} = \frac{5}{6-5} = 5 \text{ customers}$$

(ii) Average number of customers waiting for a hair cut = $\frac{\lambda^2}{\mu(\mu-\lambda)}$

$$\text{(Average queue length) } L_q = \frac{5^2}{6(6-5)}$$

$$= \frac{25}{6} = 4.16 = 4 \text{ approx.}$$

(iii) The % of customers who have to wait before getting into barber's chair (Traffic Intensity)

$$\rho_{n>0} = \frac{\lambda}{\mu} = 0.83$$

∴ 83% customers will have to wait.

(iv)
$$P_0 = 1 - \frac{\lambda}{\mu} = 1 - \frac{5}{6} = \frac{1}{6} = 0.167 = 0.17$$

⇒ 17% customers may walked without having hair cut.

**Example** 7.2 *An air-conditioner repairman finds that the time spent on his job has an exponential distribution with mean 45 minutes. If he repairs A/C in the order in which they come in, and if the arrival of A/C is approx Poisson distribution with a average rate of 12 per 10 hour day, what is repairman's expected idle time each day? How many jobs are in the queue?*

***Solution*** Given,

$$\text{Arrival rate } (\lambda) = \frac{12}{10\times 60} = \frac{1}{50}$$

$$\text{Service rate } (\mu) = \frac{1}{45}$$

$$\rho = \text{Traffic intensity}$$

$$= \frac{\lambda}{\mu} = \frac{45}{50} = \frac{9}{10}$$

∴ Number of hours for which the repairman remain busy in an 10 hours shift is busy time.

$$\text{Busy time} = 10 \times \frac{9}{10} = 9 \text{ hours}$$

∴
$$\text{Idle time} = 10 - 9 = 1 \text{ hour}$$

And jobs are in the queue (Expected)

$$E(L_s) = \frac{\rho}{1-\rho} = \frac{\lambda}{\mu-\lambda}$$

$$= \frac{9/10}{1-9/10}$$

$$= \frac{9}{10} \times \frac{10}{(10-9)}$$

$$= 9 \text{ Jobs.}$$

**Ans.**

**Example 7.3** *In a bus depot, buses arrive at a route of 32 buses per day. Assuming that inter arrival time follow an exponential distribution and the service time distribution is also exponential distribution with an average of 40 minutes. Calculate the following :*

(*i*) *Average number of buses in the system.*

(*ii*) *The probability that queue size exceeds 10.*

*If the input of buses increase to an average 35 per day what will be the change in (i) and (ii) ?*

***Solution*** Given,

$$\text{Arrival rate} = \frac{32}{60 \times 24} = \frac{1}{45} \text{ buses per minute}$$

$$\text{Service rate} = \frac{1}{40} \text{ buses per minute}$$

$$\rho = \text{Traffic Intensity}$$

$$= \frac{\lambda}{\mu} = \frac{40}{45} = 0.88$$

(i) Average number of buses in the system

$$E(L_s) = \frac{\rho}{1-\rho} = \frac{0.88}{1-0.88}$$

$$= 7 \text{ buses}$$

(ii) Probability that queue size exceeds (10) $\{P(\geq 10)\}$

$$= \rho^{10}$$

$$= (0.88)^{10}$$

$$= \mathbf{0.27}$$

If the input of buses increases to an average of 35 per day

$$\therefore \quad \lambda = \frac{35}{60 \times 24} = \frac{7}{288}$$

$$\mu = \frac{1}{40}$$

$$\rho = \frac{7 \times 40}{288} = 0.97$$

(i) $$E(L_s) = \frac{\rho}{1-\rho} = \frac{0.97}{(1-0.97)} = 32 \text{ buses}$$

(ii) $$P(\geq 10) = \rho^{10} = (0.97)^{10} = 0.74$$ **Ans.**

**Example 7.4** *In a grocery shop, customers arrive according to poisson distribution with a mean arrival rate of 5 per hour and the service time distribution is exponential distribution with an average of 10 minutes. If the condition of single channel queueing model is applied to this problem. Determine:*

*(i) The average number of customers in the shop and the average number of customer waiting for their turn.*

*(ii) The percentage of customers who have to eait before getting service.*

***Solution*** Given,

$$\text{Arrival rate } (\lambda) = \frac{5}{60} = \frac{1}{12} \text{ per minute}$$

$$\text{Service rate } (\mu) = \frac{1}{10} \text{ per minute}$$

$$\rho = \text{Traffic intensity}$$

$$= \frac{\lambda}{\mu} = \frac{10}{12} = 0.83$$

(i) Average number of customer in the shop (system)

$$E(L_s) = \frac{\rho}{1-\rho}$$

$$= \frac{0.83}{1-0.83} = 5$$

Average number of customers waiting (Average queue length)

$$E(L_q) = \frac{\rho^2}{1-\rho}$$

$$= \frac{(0.83)^2}{1-0.83} = 4$$

(ii) Percentage of customer who have to wait ($p \geq 1$)

$$\rho = 0.83 = 83\%$$ **Ans.**

**Example 7.5** *A telephone bill receiving window has only one casheir and the cashier takes on an average 6 minutes per customer. The persons coming for bill* *payment with an average of 8 per hour. The authority receive so many complaints regarding customers long queue, so for solving this problem, determine:*

(i) *Average length of queue.*

(ii) *Average time for which a person has to wait to pay his telaphone bill.*

(iii) *The time on an average, the cashier is idle.*

***Solution*** Given,

$$\text{Arrival rate } (\lambda) = \frac{8}{60} \text{ per minute}$$

$$\text{Service rate } (\mu) = \frac{1}{6} \text{ per minute}$$

$$\rho = \text{Traffic intensity}$$

$$= \frac{8 \times 6}{60} = 0.8$$

(i) $$\text{Average length of queue } (L_q) = \frac{\lambda^2}{\mu(\mu - \lambda)}$$

$$= \frac{8 \times 8 \times 60}{60 \times 60\left(\frac{1}{6} - \frac{8}{60}\right)}$$

$$= \frac{8 \times 8 \times 60}{10 \times 60(1-8)}$$

$$= \frac{64}{20} = 3$$

$$\Rightarrow \qquad = \textbf{3 persons}$$

(ii) $$W_q = \frac{L_q}{\lambda}$$

$$= \frac{\lambda}{\mu(\mu - \lambda)}$$

$$= \frac{8 \times 6 \times 60}{60(10-8)}$$

$$= \frac{48}{2}$$

$$= \textbf{24 minutes}$$

(iii) $$P_0 = 1-\rho = 1 - 0.8 = 0.2$$

Probability that the cashier will be idle is 20%. **Ans.**

**Example 7.6** *A Vodafone company plans to open a single server customer care facility, where customers arrive at a rate of 25 customers per hour and the average service time of the customer care centre is 2 minutes. Calculate the following :*

(*i*) *The probability of time that the system will be idle.*

(*ii*) *Average time a customer is waiting in the queue.*

(*ii*) *Average number of customers waiting in a queue.*

***Solution*** Given,

$$\text{Arrival rate } (\lambda) = \frac{25}{60} \text{ per minute}$$

$$\text{Service rate } (\mu) = \frac{1}{2} \text{ per minute}$$

$$\rho = \text{Traffic intensity}$$

$$= \frac{\lambda}{\mu} = \frac{25}{60} \times \frac{2}{1} = 0.83$$

(i) $$P_0 = 1 - \rho = 1 - 0.83 = 0.17$$

$$\Rightarrow 17\%$$

Probability that the system will be idle is 17%

(ii) $$W_q = \frac{L_q}{\lambda} = \frac{\lambda}{\mu(\mu-\lambda)}$$

$$= \frac{25\times 2}{60\left(\frac{1}{2}-\frac{25}{60}\right)}$$

$$= \frac{25\times 2\times \cancel{60}}{\cancel{60}(30-25)}$$

$$= \frac{25\times 2}{5} \Rightarrow 10 \text{ minutes}$$

(iii)
$$L_q = \frac{\lambda^2}{\mu(\mu-\lambda)}$$

$$= \frac{25\times25\times60\times2}{60\times60(30-25)}$$

$$= \frac{25\times25}{30\times5}$$

$$= \frac{125}{30}$$

$= 4.16 \approx$ **4 customers.** **Ans.**

**Example** 7.7 *On an average, 96 patients per 24-hour day require the service of an emergency clinic. Also, on an average patient requires 10 minutes of active attention. Assume that the facility can handle only one emergency at a time. Suppose that it costs the clinic ₹ 100 per patient treated, to obtain an average servicing time of 10 minutes and thus, each minute of decrease in this average time would cost ₹ 10 per patient treated. How much would have to be budgeted by the clinic to decrease the average size of the queue from $1\frac{1}{3}$ patients to $\frac{1}{2}$ patient?* **[GGSIPU, B.Tech 2010]**

***Solution*** Given,

$$\lambda = \frac{96}{24\times60} = \frac{1}{15} \text{ patient/minute}$$

$$\mu = \frac{1}{10} \text{ patient/minute}$$

Average number of patients in the queue,

$$L_q = \frac{\lambda}{\mu} \cdot \frac{\lambda}{\lambda-\mu} = \frac{\left(\frac{\lambda}{\mu}\right)^2}{1-\frac{\lambda}{\mu}}$$

$$= \frac{\left(\frac{1}{15}\right)^2}{\left(\frac{1}{10}-\frac{1}{15}\right)\frac{1}{10}} = 1\frac{1}{3} \text{ patients}$$

But, $L_q = 1\frac{1}{3}$ is reduced to $L_q' = \frac{1}{2}$

Substituting $L_q' = \frac{1}{2}, \lambda' = \lambda = \frac{1}{15}$ in the formula

$$L_q' = \frac{\lambda'^2}{\mu'(\mu'-\lambda')}$$

$$\frac{1}{2} = \frac{\left(\frac{1}{15}\right)^2}{\mu'\left(\mu'-\frac{1}{15}\right)}$$

$$\mu' = \frac{2}{15} \text{ patient/minute}$$

Hence, the average rate of treatment required is ,

$$\frac{1}{\mu} = 7.5 \text{ minute.}$$

Decrease in time required by each patient $= 10 - \frac{15}{2} = \frac{5}{2}$ min

∴ The budget required for each patient

$$= 100 + \frac{5}{2} \times 10 = ₹\ 125$$

So, in order to get the required size of the queue, the budget should be increased from ₹ 100 to ₹ 125 per patient.

### 7.6.2 Model II. (*M/M/2*) : (∞/*FCFC*) Multiservice model (C=2)

(i) Traffice Intensity ($\rho$) $= \frac{\lambda}{C\mu}$ or $\frac{\lambda}{2C}$

(ii) Probability of zero customer in the system ($P_0$)

$$P_0 = \left[1+\left(\frac{\lambda}{\mu}\right)+\frac{\left(\frac{\lambda}{\mu}\right)^2}{2}\times\left(\frac{1}{1-\rho}\right)\right]^{-1}, \left(\rho=\frac{\lambda}{2\mu}\right)$$

$$P_0 = \left[1 + 2\rho + \frac{2\rho^2}{(1-\rho)}\right]^{-1}$$

(iii) Probability that an arrival has to wait (*i.e.* number of customers are more than number to service channels)

$$P(w > 0) = \frac{2\rho^2}{(1-\rho)} P_0$$

(iv) Average number of customers in the queue

$$L_q = \frac{2\rho^3}{(1-\rho)^2} P_0 = \frac{\lambda\mu\left(\frac{\lambda}{\mu}\right)^2}{(2\mu-\lambda)^2} P_0$$

(v) Average number of customers in the system

$$L_s = L_q + \rho = L_q + \frac{\lambda}{2\mu}$$

(vi) Waiting time in queue

$$W_q = \frac{L_q}{\lambda}$$

(vii) Waiting time in system

$$W_s = \frac{L_s}{\lambda}$$

(viii) Expected arerage length of non-empty queue

$$E(L/L>0) = \frac{1}{(1-\rho)} = \frac{\mu}{\mu-\lambda}$$

(ix) Expected average waiting time of a customer who has to wait

$$E(W/W>0) = \frac{1}{C.\mu(1-\rho)} = \frac{1}{2\mu(1-\rho)}$$

**Example 7.8** *A telephone exchange has two long distance operators. The company finds that during the peak load, long distance calls arrived in Poisson fashion at an average rate of 15 per hour. The length of service on these calls is approximately exponentially distributed with mean length 5 minutes.*

*(i) What is the probability that a subscriber will have to wait for his long distance call during the peak hours of the day?*

*(ii) If the subscriber will wait and are serviced in turn, what is the expected waiting time?*

***Solution*** Given,

$$C = 2, \lambda = \frac{15}{60} = \frac{1}{4} \text{ per minute}$$

$$\mu = \frac{1}{5} \text{ per minute}$$

$$\text{Traffic intensity } (\rho) = \frac{\lambda}{C\mu} = \frac{5}{8}$$

$$P_0 = \left[1 + 2\rho + \frac{2\rho^2}{(1-\rho)}\right]^{-1}$$

$$= \left[1 + \frac{10}{8} + \frac{2\left(\frac{5}{8}\right)^2}{\left(1 - \frac{5}{8}\right)}\right]^{-1}$$

$$= \frac{3}{13}$$

(i)
$$P(w > 0) = \frac{2\rho^2}{(1-\rho)} \cdot P_0$$

$$= \frac{2\left(\frac{5}{8}\right)^2}{1 - \left(\frac{5}{8}\right)} \times \left(\frac{3}{13}\right)$$

$$= 0.48$$

(ii)
$$W_q = \frac{L_q}{\lambda}$$

$$L_q = \frac{2\rho^3}{(1-\rho)^2} \cdot P_0$$

$$= \frac{2\left(\frac{5}{8}\right)^3}{\left(1-\frac{5}{8}\right)^2} \times \left(\frac{3}{13}\right)$$

$$= \frac{250}{312} = \frac{125}{156}$$

$$\therefore \quad W_q = \frac{125}{156} \times \frac{4}{1} = \textbf{3.2 minutes.}$$

**Example 7.9** *A car washing station has two Hydraulic lift system. If the service time follows the exponential distribution with mean 5 minutes and car arrived in a Poisson fashion at the rate of 10 cars per hour, then calculate :*

(*i*) *The expected idle time of car washing system.*

(*ii*) *The probability that a customer has to wait for the service.*

(*iii*) *Average number of customers in the queue.*

***Solution*** Given,

$$C = 2, \lambda = 10 \text{ per hour} = \frac{10}{60} = \frac{1}{6} \text{ per minute}$$

$$\mu = \frac{1}{5} \text{ per minute}$$

$$\text{Traffic intensity } \rho = \frac{\lambda}{C\mu} = \frac{1 \times 5}{2 \times 6 \times 1} = \frac{5}{12}$$

(i) Idle time for car washing station $(1 - \rho) = 1 - \frac{5}{12} \Rightarrow \frac{7}{12}$

$$= 0.58$$

$$= \textbf{58\% (approx)}$$

(ii)

$$P_0 = \left[1 + 2\rho + \frac{2\rho^2}{(1-\rho)}\right]^{-1}$$

$$= \left[1 + 2 \times \frac{5}{12} + \frac{2 \times 5 \times 5 \times 12}{12 \times 12(12-5)}\right]^{-1}$$

$$= \left[1+\frac{10}{12}+\frac{50}{12\times 7}\right]^{-1}$$

$$= \left[\frac{84+70+50}{12\times 7}\right]^{-1}$$

$$= \left[\frac{204}{84}\right]^{-1} \Rightarrow \frac{7}{17}$$

Probability that an arrival has to wait

$$P(w>0) = \frac{2\rho^2}{(1-\rho)}\times P_0$$

$$= \frac{2\times 5\times 5\times 12}{12\times 12\times 7}\times\frac{7}{17}$$

$$= \frac{25}{102} = 0.245 \text{ Approx.}$$

(iii) Average number of customer in the queue

$$(L_q) = \frac{2\rho^3}{(1-\rho)^2}\times P_0$$

$$= \frac{2\times 5\times 5\times 5\times 12\times 12}{12\times 12\times 12\times (12-5)^2}\times\frac{7}{17}$$

$$= \frac{2\times 5\times 5\times 5\times 7}{12\times 7\times 7\times 17}$$

$$= \frac{125}{714} = 0.175$$

**Ans.**

**Example 7.10** *A beauty parlour has two girls for handling the customers. If customer arrive in a Poisson fashion at the rate of 8 per hour and the service time for each customer is exponential with mean 6 minutes, then find out :*

(*i*) *Percetage of idle time for each girl.*

(*ii*) *The probability of customers waiting time for service.*

(*iii*) *Expected length of waiting time of a customer.*

(*iv*) *Expected average number of customers who actually wait.*

***Solution*** Given,

$$C = 2, \lambda = \frac{8}{60} = \frac{2}{15} \text{ per minute}$$

$$\mu = \frac{1}{6} \text{ per minute}$$

$$\rho = \text{Traffic indensity} = \frac{\lambda}{2\mu} = \frac{2\times 6}{15\times 2} = \frac{2}{5}$$

$$P_0 = \left[1+2\times\frac{2}{5}+\frac{2\times 4\times 5}{5\times 5\times(5-2)}\right]^{-1}$$

$$= \left[1+\frac{4}{5}+\frac{8}{15}\right]^{-1}$$

$$= \left[\frac{15+12+8}{15}\right]^{-1}$$

$$= \left[\frac{35}{15}\right]^{-1} = \frac{3}{7}$$

(i) Percentage of idle time $= (1 - \rho)$

$$= \left(1-\frac{2}{5}\right)$$

$$= \frac{3}{5} = 0.6 \Rightarrow \mathbf{60\%}$$

(ii) $$P(w>0) = \frac{2\rho^2}{(1-\rho)}\times P_0$$

$$= \frac{2\times 2\times 2\times 5}{5\times 5\times 3}\times\frac{3}{7}$$

$$= \frac{8}{35} \Rightarrow \mathbf{0.228}$$

(iii) $$E(w/w>0) = \frac{1}{C.\mu(1-\rho)}$$

$$= \frac{1 \times 6 \times 5}{2 \times 1 \times (3)}$$

$$= 5 \text{ minutes}$$

(iv) $$E(L/L>0) = \frac{1}{(1-\rho)}$$

$$= \frac{1 \times 5}{3} = 1.666$$

## PREVIOUS YEARS QUESTIONS

**Example 7.11** *Arrival rate of telephone calls at a telephone booth are according to Poisson distribution with an average time of 9 minutes between two consecutive arrivals. The length of telephone call is assumed to be exponentially distributed with mean 3 minutes*

*(a) Determine the probability that a person arriving at the booth will have to wait.*

*(b) Find the average queue length that is formed from time to time.*

*(c) Telephone company will install a second booth when convinced that an arrival would expect to have to wait at least 4 minutes for the phone, Find the increase in flow of arrivals which will justify a second booth.*

*(d) Find the fraction of the day that the phone will be in use.*

**[GGSIPU, 2014]**

***Solution*** Here $\lambda = 1/9$ per minute and $\mu = 1/3$ per minute

(a) Probability that a person will have to wait

$$\rho = \frac{\lambda}{\mu} = \frac{1/9}{1/3} = 0.33 = \frac{1}{3}$$

(b) Average queue length (Lq) $= \dfrac{\lambda^2}{\mu(\mu-\lambda)}$

$$= \frac{(1/9)^2}{1/3(1/3-1/9)}$$

$$= \frac{1}{6} \text{ person}$$

(c) If the arrival rate is greater than waiting time, that justifies the provision of a second booth

$$(W_q) = \frac{\lambda}{\mu(\mu-\lambda)} = \frac{Lq}{\lambda}$$

$$\text{Here } (W_q) = 4 \text{ minutes}, \mu = 1/3 \text{ per minute}$$

To find out, mean arrival rate = $\lambda'$

$$\therefore \quad 4 = \frac{\lambda'}{1/3(1/3-\lambda')}$$

$$\text{or} \quad \lambda' = \frac{4}{21} \text{ arrival/minute} = 0.19/\text{minute}$$

The arrival rate should become 0.19 person/minute to justify the second booth.

$$\therefore \text{ Increase in the flow of arrivals} = \frac{4}{21} - \frac{1}{9} = \frac{5}{63} \text{ per minute}$$

(d) The expected fraction of a day that the phone will be in use

$$= \frac{\lambda}{\mu} = 0.33 = \text{traffic intensity}$$

**Example 7.12** *Workers come to tool store room to receive special tools for accomplishing a particular project assigned to them. The average time between two arrivals is 60 seconds and the arrivals are assumed to be in Poisson distribution. The average service time of the tool room attendant is 40 seconds.*

*(i) Average queue length.*

*(ii) Average length of non-empty queues.*

*(iii) Average number of workers in system including the worker being attended.*

*(iv) Mean waiting time of an arrival.*

*(v) Average waiting time of an arrival who waits and*

*(vi) Probability that the tool room attendant remains, idle.*

**[GGSIPU, B.Tech-2013]**

***Solution***

$$\lambda = 1/60 \text{ per second} = 1 \text{ per minute}$$

$$\mu = 1/40 \text{ per second} = 1.5 \text{ per minute}$$

(i) Average queue length,

$$L_q = \frac{\lambda}{\mu} \times \left(\frac{\lambda}{\mu - \lambda}\right)$$

$$= \frac{1}{5} \times \left(\frac{1}{1.5 - 1}\right)$$

$$= \frac{1}{0.75} = \frac{4}{3} \text{ workers}$$

(ii) Average length of non empty queues,

$$L_n = \frac{\mu}{\mu - \lambda}$$

$$= \frac{1.5}{1.5 - 1} = 3 \text{ workers}$$

(iii) Average number of workers in the system,

$$L_s = \frac{\lambda}{\mu - \lambda}$$

$$= \frac{1}{1.5 - 1} = 2 \text{ workers}$$

(iv) Mean waiting time of an arrival,

$$W_q = \frac{1}{\mu} \times \frac{\lambda}{\mu - \lambda}$$

$$= \frac{1}{1.5 - 1} \times \frac{1}{1.5}$$

$$= \frac{4}{3} \text{ min}$$

(v) Average waiting time of an arrival who waits,

$$W_n = \frac{1}{\mu - \lambda}$$

$$= \frac{1}{1.5 - 1} = 2 \text{ min}$$

(vi) Probability that the tool room attendent remains idle,

$$P_0 = \left(1 - \frac{\lambda}{\mu}\right)$$

$$= 1-\frac{1}{1.5} = \frac{1}{3}$$ **Ans.**

**Example 7.13** *A shipping company has a single unloading dock with ships arriving in a Poisson fashion at an average rate of 3 ships per day. The unloading time distribution for a ship with n unloading crews is found to be exponential with average unloading time l/2n days. The company has a large labour supply without regular working hours and to avoid long waiting times, the company has a policy of using as many loading crews as there are ships waiting in line or being unloaded. Find,*

*(i) The average number of unloading crews working at any time.*

*(ii) The probability that more than 4 crews will be needed.*

**[GGSIPU, B.Tech-2013]**

***Solution*** Here the mean service rate is not constant as in Model 1.

We have $\lambda = 3$ ships/day and mean service rate with one unloading crew = $\mu = 2$ ships/day.

Now,
$$P_n = \frac{\rho^n}{\underline{|n}} e^{-\rho}$$

(i) Expected number of crews at any specified instant

$$E(L_s) = \sum_{n=0}^{\infty} nP_n$$

$$= \sum_{n=0}^{\infty} n\frac{\rho^n}{\underline{|n}} e^{-\rho}$$

$$= \rho e^{-\rho} \sum_{n=1}^{\infty} \frac{\rho^{n-1}}{\underline{|n-1}}$$

$$= \rho e^{-\rho}\left[1+\rho+\frac{\rho^2}{\underline{|2}}+\frac{\rho^3}{\underline{|3}}+....\right]$$

$$= \rho e^{-\rho}(e^{\rho})$$

$$= \rho = \frac{\lambda}{\mu}$$

$$= \frac{3}{2} = 1.5 \text{ crew}$$

(b) Probability that a ship entering service will require more than 4 crews = Probability that there are at least 5 ships in the system at any instant

$$= \sum_{n=5}^{\infty} P_n$$

$$= \sum_{n=0}^{\infty} P_n - \sum_{n=0}^{4} P_n$$

$$= 1 - \left[P_0 + P_1 + P_2 + P_3 + P_4\right]$$

$$= 1 - \left[1 + \rho + \frac{\rho^2}{\lfloor 2} + \frac{\rho^3}{\lfloor 3} + \frac{\rho^4}{\lfloor 4}\right] e^{-\rho}$$

$$= 1 - \left[1 + 1.5 + \frac{(1.5)^2}{\lfloor 2} + \frac{(1.5)^3}{\lfloor 3} + \frac{(1.5)^4}{\lfloor 4}\right] e^{-1.5}$$

$$= 0.019$$ **Ans.**

**Example 7.14** *In a railway yard, goods trains arrive at a rate of 30 trains per day. Assuming that the inter-arrival time follows an exponential distribution and the service time distribution is also exponential with an average 36 minutes. Calculate the following :*

*(a) The mean queue size*

*(b) The probability that the queue size exceeds 10. In case the input of trains increases to an average 33 per day. What will be the change in (a) and (b)?* **[GGSIPU, B.Tech-2012]**

**Solution** $\lambda = 30$ trains per day

$$= \frac{30}{24 \times 60} \text{ trains per minute}$$

$$\lambda = \frac{1}{48} \text{ trains per minute}$$

$$\mu = \frac{1}{36} \text{ trains per minute}$$

$$\rho = \frac{\lambda}{\mu}$$

$$= \frac{1}{48} \times \frac{36}{1} = \frac{6}{8} = \frac{3}{4} = 0.75$$

(a) Mean queue size $E(L_s) = \frac{\rho}{1-\rho}$

$$= \frac{\frac{3}{4}}{1-\frac{3}{4}}$$

$$= \frac{3}{4} \times \frac{4}{1} = 3 \text{ trains}$$

(b) $P(m \geq 10) = \rho^{10} = \left(\frac{3}{4}\right)^{10} = (0.75)^{10} = 0.06$

When the input of trains increases to 33 trains per day

We have, $\lambda = \frac{33}{60 \times 24}$

$$= \frac{11}{480} \text{ trains per minute}$$

$$\mu = \frac{1}{36} \text{ trains per minute}$$

$$\rho = \frac{\lambda}{\mu} = \frac{11}{480} \times \frac{36}{1}$$

$$= \frac{11 \times 6}{80} = \frac{33}{40} = 0.83$$

(a) $E(L_s) = \frac{\rho}{1-\rho}$

$$= \frac{0.83}{1-0.83} = \frac{0.83}{0.17}$$

$$= 4.8 \Rightarrow 5 \text{ trains}$$

(b) $P(m \geq 10) = \rho^{10} = (0.83)^{10} = 0.2$ **Ans.**

**Example** 7.15 *A duplicating machine maintained for office use is used and operated by people in office the need to make copies, mostly secretaries. Since the work to be copied varies in length (number of pages of the original) and copies required, the service rate is randomly distributed, but it does approximate a Poisson having a mean service rate of 10 jobs per hour. Generally the requirements for use are random over the entire 8-hoiir working but arrive at a rate of 5/hour. Several people have noted that a waiting line develops occasionally and have questioned the policy of maintaining only one unit. If the time of a secretary is valued at ₹3.50 per hour, then determine:*

*(a) Equipment utilization*

*(b) The percent time an arrival has to wait*

*(c) Average waiting time of an arrival in the system*

*(d) The average cost due to waiting and operating the machine.*

***Solution***

$$\lambda = S \text{ arrivals per hour}$$

$$\mu = 10 \text{ jobs per hour}$$

(a) Utilization factor $\rho$(rho)= $\frac{\lambda}{\mu} = \frac{5}{10} = \frac{1}{2}$

Thus equipment is used $= \frac{1}{2}$ hr × 100% = 50% of time.

(b) The percentage of time an arrival has to wait is busy i.e. 0.50

(c) Average waiting time of (arrival)

$$E(W_s) = \frac{1}{\mu - \lambda}$$

$$= \frac{1}{10-5} = \frac{1}{5} \text{ hr}$$

$$= 0.20 \text{ hours}$$

(d) Average, cost per job = average time per job × ₹ per hour

$$= \frac{1}{5} \text{hr} \times ₹\ 3.50 = 0.20 \times 3.5 = ₹\ 70$$

= No. of jobs processed per day
+ average cost per job

$$\text{Cost per day} = ₹\, 8 \times 5 \times 0.70$$
$$= ₹\, 40 \times 0.70$$
$$= ₹\, 28 \text{ per day.}$$

**Example 7.16** *A self service store employee's one cashier at its counter. An average of nine customer arrives every 5 minutes while the cashier can serve 10 customers in 5 minutes. Assuming poisson distribution for arrival rate and exponential distribution for service rate, find :*

(i) *Average number of customer in the system.*

(ii) *Average number of customers in the queue.*

**[GGSIPU, B.Tech-2014]**

*Solution*

$$\text{Arrival rate } \lambda = 9/5 = 1.8 \text{ customers/minute}$$

$$\text{Service rate } \mu = 10/5 = 2 \text{ customers/minute}$$

$$\text{Utilization factor} \quad \rho = 1.8/2 = 0.9$$

(a) Average no. of customers in the system

$$L_s = \frac{\lambda}{\mu-\lambda}$$

$$= \frac{1.8}{2-1.8}$$

$$= \frac{1.8}{0.2} = 9$$

(b) Average no. of customeres in the queue

$$L_q = \frac{\lambda^2}{\mu(\mu-\lambda)}$$

$$= \frac{(1.8)^2}{2(2-1.8)}$$

$$= \frac{3.24}{2\times(0.2)}$$

$$= 8.1$$

**Ans.**

## MULTIPLE CHOICE QUESTIONS

7.1. The passenger and the bus in queueing system describes

(a) Customer and server (b) Server and customer

(c) Both server (d) None of these

7.2. If the inter arrival time is exponential with mean 1/l the number of arrival in (0, *t*) is

(a) Poisson distribution (b) Exponential distribution

(c) Binomial distribution (d) None of these

7.3. If the rate of birth is 12 per minute, then the average number of birth per year is

(a) 43,800 (b) 45,000

(c) 40,000 (d) None of these

7.4. The arrival of customer (with no departure) in system, in queueing theory can be stated as

(a) The pure birth process (b) The pure death process

(d) The birth death process (d) None of these

7.5. The arrival and departure of customers in queueing system can be stated as

(a) The pure birth process (b) The pure death process

(d) The birth death process (d) None of these

7.6. The departure and arrival in queueing system are normally cousidered respectively *M*/*M*/1 the arrival and departure distribution are

(a) Both Markovian (b) Binonial

(c) General (d) None of these

7.7. In queue description *M*/*M*/2, the number of servers are

(a) 1 (b) M

(c) 2 (d) None of these

7.8. For *M*/*M*/1 model the expected number of busy servers are equal to

(a) Traffic intensity ($\rho$) (b) Arrival rate ($\lambda$)

(c) Service rate ($\mu$) (d) None of these

7.9. The utilization factor for *M*/*M*/1 system is given by

(a) $\lambda/\mu$ (b) $\mu/\lambda$

(c) $\lambda\mu$ (d) None of these

7.10. For *M*/*M*/1 system, the expected waiting time in the queue is

(a) $\lambda/\mu$ (b) $\lambda/\mu\ (\mu - \lambda)$

(c) $\lambda/(\mu - \lambda)$ (d) None of these

7.11. For *M*/*M*/1 queueing system, the expected number of customers in the system are

(a) $\rho/(1-\rho)$ (b) $(1-\rho)/\rho$

(c) $(1-\rho)$ (d) None of these

7.12. In Kendall's Notation $\{(a/b/c):(d/e)\}$, $d$ indicates

(a) Arrival patterns of unit (b) Service time of unit

(c) Queue discipline (d) Queueing capacity

7.13. For $M/M/1$ queueing system if arrival rate is 10 customers per day and service rate is 30 customers per day then average number of customers in the queue

(a) 1/3 (b) 1/6

(c) 6 (d) None of these

7.14. In a departmental store customers arrive at a rate of 20 customers per hour. The average number of customers that can be handled by cashier is 24 per hour. What is service rate in the problem?

(a) 20 (b) 3

(c) 24 (d) 10

7.15. What is arrival rate in the above mentioned problem?

(a) 20 (b) 3

(c) 24 (d) 10

7.16. In Kendall's Notation $\{(a/b/c):(d/e)\}$, $e$ indicates

(a) Number of paralles service channel

(b) Queueing capacity

(c) Queue discipline

(d) Arrival pattern of units

7.17. In inventory system the queue discipline is

(a) FIFO (b) SIRO

(c) SIP (d) LIFO

7.18. Customers are selected for service at random. This is known as the ____ discipline.

(a) FIFO (b) SIRO

(c) LIFO (d) SIP

7.19. ____ refers to the number of customers in the waiting line.

(a) Queue length (b) Service order

(c) Waiting time in queue (d) None of the above

7.20. Mean and average number of arrivals per unit time period is called ____.

(a) Mean service rate (b) Mean arrival rate

(c) Avarage rate service (d) None of the above

## ANSWERS

| | | | | | | |
|---|---|---|---|---|---|---|
| 1. (a) | 2. (b) | 3. (a) | 4. (a) | 5. (c) | 6. (a) | 7. (c) |
| 8. (a) | 9. (a) | 10. (b) | 11. (a) | 12. (d) | 13. (b) | 14. (c) |
| 15. (a) | 16. (c) | 17. (d) | 18. (b) | 19. (a) | 20. (b) | |

## SUMMARY

**Queueing system**

(i) Input source (Calling popullation)

(ii) Queueing process

(iii) Queueing descipline (FIFO, LIFO, SIRO, SIP)

(iv) Service process

(v) Customer behavior

**State of the system**

**Transient state and Steady state.** A queueing system will be influenced by initial state, when it has just started operation. The affecting fectors may include the number of customers in the system and the elapsed time. In this state, the system is said to be in transient state

In a subsequent time, the system becomes independent of the initial state and elapsed time. This state is known as steady state

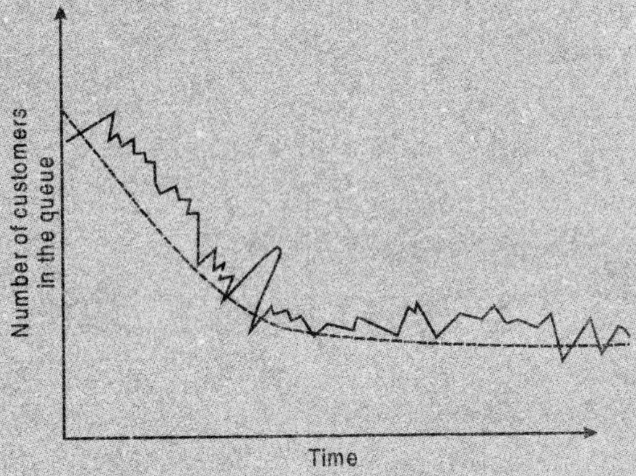


**Waiting lines (Queueing Theory)**

A group of items/persons waiting to receive service including those receiving the service is known as a waiting line or a queue.

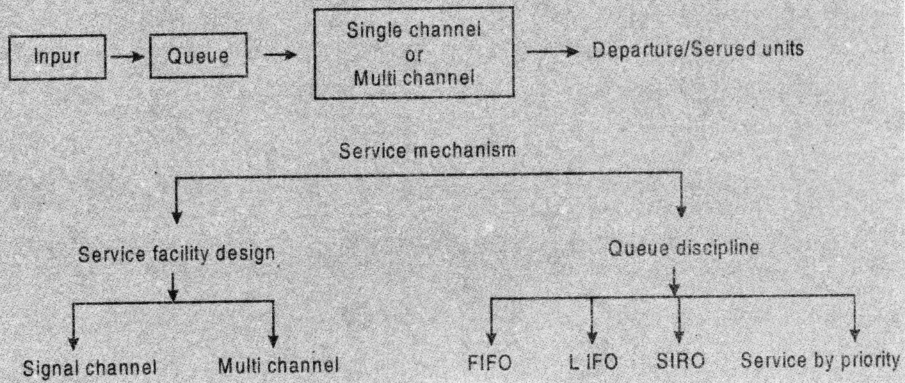

**Kendall's notation for representign queueing models**

$(A/B/C) : (d/e)$ where,

$A$ : Arrival pattern of the units

$B$ : Probability distribution of service time.

$C$ : Number of || parallel service channels in the system.

$d$ : Capacity of the system (in service + waiting).

$e$ : Queue discipline.

(M/M/C) : (∞/FCFS)

I (M/M/1) : (∞/FCFS)

II (M/M/2) : (∞/FCFS)

**Model I.** [(M/M/C) : (∞/FCFS)].

**Notations**

$\lambda$ = Mean arrival rate of units

$\mu$ = Mean service rate of units

$C$ = Number of parallel service channel

$$\rho \text{ (traffic intensity)} = \frac{\lambda}{C\mu}$$

$P_0$ =(No. of customer in the queue)

$$P_0 = \left[\sum_{n=0}^{C-1} \frac{1}{\lfloor n}\left[\frac{\lambda}{\mu}\right]^n + \frac{1}{\lfloor c}\left[\frac{\lambda}{\mu}\right]^C \left[\frac{C\mu}{C\mu-\lambda}\right]\right]^{-1}$$

C = 1.

$$P_0 = \left[1+\left(\frac{\lambda}{\mu}\right)\left(\frac{\mu}{\mu-\lambda}\right)\right]^{-1}$$

$$= \left[1+\frac{\lambda}{\mu-\lambda}\right]^{-1}$$

$$= \left[\frac{\mu}{\mu-\lambda}\right]^{-1}$$

$$= \frac{\mu-\lambda}{\mu}$$

$$= 1-\rho$$

C = 2

$$P_0 = \left[1+\frac{\lambda}{\mu}+\frac{\left(\frac{\lambda}{\mu}\right)^2}{2}\times\left(\frac{1}{1-\rho}\right)\right]^{-1},$$

$$\rho = \frac{\lambda}{2\mu}$$

$L_q$ : Average no. of customers in the queue.

$L_s$ : Average no. of customers in the system.

$W_q$ : Waiting time in the queue.

$W_s$ : Waiting time in the system.

| | (M/M/C) : (∝/FCFS) | Model I, C=1 | Model II, C=2 |
|---|---|---|---|
| (i) Probability that an arrival has to wait (No. of customers are more than no. of service no. of channels.) | $P(n \geq C) = \sum_{n=c}^{\infty} P_n$ <br> $\frac{\mu\left[\frac{\lambda}{\mu}\right]^C}{\underline{C-1}(C\mu-\lambda)} \rho_0$ | $P_0 = (1-\rho)$ <br> $\rho = \left(\frac{\lambda}{\mu}\right)$ | $\frac{\mu\left(\frac{\lambda}{\mu}\right)^2}{2\mu-\lambda}\left[1+\frac{\lambda}{\mu}+\frac{\left(\frac{\lambda}{\mu}\right)^2}{2}\times\frac{1}{1-\rho}\right]$ <br> $1-p(n \geq 2)$ |
| (ii) No waiting time. | $1-p\ (n \geq C)$ | $(1-\rho)$ | $[1-p\ (n \geq 2)]$ |
| (iii) Average no. of customers in the queue ($L_q$). | $\frac{\lambda\mu\left[\frac{\lambda}{\mu}\right]^C}{\underline{C-1}(C\mu-\lambda)^2} \rho_0$ | $\frac{\rho^2}{1-\rho} = \frac{\lambda^2}{\mu(\mu-\lambda)}$ | $\frac{\lambda\mu\left(\frac{\lambda}{\mu}\right)^2}{(2\mu-\lambda)^2} \rho_0$ |
| (iv) Average no. of customers in the system ($L_s$). | $L_q+\rho = L_q+\frac{\lambda}{c\mu}$ | $L_q+\rho = L_q+\frac{\lambda}{\mu}$ <br> $= \frac{\rho}{1-\rho}$ | $L_q+\rho = L_q+\frac{\lambda}{2\mu}$ |
| (v) Waiting time in queue ($W_q$). | $\frac{L_q}{\lambda}$ | $\frac{L_q}{\lambda}$ | $\frac{L_q}{\lambda}$ |
| (vi) Waiting time in system ($W_s$). | $\frac{L_s}{\lambda}$ | $\frac{L_s}{\lambda}$ | $\frac{L_s}{\lambda}$ |

CHAPTER 8

# Sequencing Problems

## 8.1 INTRODUCTION

Sequencing is an organised activity of producing useful products from raw material. The selection of a path or order which is to be followed for converting raw material into finished product is called 'Sequencing'. A sequence is the order in which jobs are to be performed and sequencing is the selection of appropriate order in which Waiting Customer or Jobs are serviced. The problem of sequencing arise whenever there is a need for determining an optimum order of perfoming a number of jobs by number of facilities.

Sequencing is using the available facilities in some pre assigned order so as to minimise the cost (time) involved.

### Situations for Sequencing Arised

(i) Customers in banks.

(ii) In hospitals (patient waiting for different types of test).

(iii) In production units (factories) where variety of equipments are used.

### Basic Assumptions

Basic assumptions of sequencing are as following :

1. Only one job can be processed on a given machine at a particular time.
2. Once a job operation started, it must be performed till completion.
3. Only one machine of each type is available.
4. The time required for shifting a job from one machine to other is negligible.

5. Capability and capacity of all machines are different.
6. A job is processed as soon as possible, but in specified order.
7. Each operation must be completed before its succeeding operation can start.

## 8.2 TYPES OF SEQUENCING PROBLEMS

(i) $n$ jobs on two machines (Johnson's Algorithm)

(ii) $n$ jobs on $m$ machines.

(iii) 2 jobs on $m$ machines.

### 8.2.1 Sequencing of *n* Jobs on two Machines

The sequencing algorithm which minimizes the time to complete $n$ jobs through two machines was developed by Johnson and is called Johnson's Algorithm. Steps for Johnson's Algorithm are

**Step 1 :** Examine the $A_i$'s and $B_i$'s for $i$ = 1, 2, .......$n$ and find out minimum [$Ai$, $Bi$].

**Step 2 :** (i) If this minimum be $A_k$ For some $i = k$, process the $k^{th}$ job first of all.

(ii) If this minimum be $B_r$ for some $i = r$, process the $r^{th}$ job last of all.

**Step 3 :**(i) If there is a tie for minima, $A_k = B_r$, process the $k^{th}$ job first of all and $r^{th}$ job in the last.

(iii) If tie occurs among $A_i$'s , select the job corresponding to the greater of $B_i$'s and process it first of all.

(iv) If tie for minimum occurs among $B_i$'s select the job corresponding to the greater of $A_i$'s and process it in the last.

**Step 4 :** Cross-out the jobs already assigned and repeat steps 1 to 3, arranging the jobs, until all the jobs have been assigned.

**Example 8.1** *The are 5 jobs, each of which go through the two machines A and B and the processing times are given below.*

| *Jobs* → | *a* | *b* | *c* | *d* | *e* |
|---|---|---|---|---|---|
| *Machine A* | 6 | 2 | 10 | 4 | 11 |
| *Machine B* | 3 | 7 | 8 | 9 | 5 |

*Determine a sequence for 5 jobs that will minimize the total elapsed time T.*

**Solution** Follwoing the proscedure as given in algorithm.

**Step 1.** Since the minimum time is '2' for job '*b*' against machine *A*, put in 1st slot

**Step 2.** Next minimum time is '3' for job '*a*' against machine '*B*' so puting it in last

Remaining blocks are filled in similar manner. Optimal sequence obtained is

| *b* | *d* | *c* | *e* | *a* |
|---|---|---|---|---|

Now Computing total elapsed time and idle time for machines :

| *Job sequence* | *Machine A* | | | *Machine B* | | | *Idle Time* |
|---|---|---|---|---|---|---|---|
| | *Time in* | *P.T.* | *Time out* | *Time in* | *P.T.* | *Time out* | |
| *b* | 0 | 2 | 2 | 2 | 7 | 9 | 2 |
| *d* | 2 | 4 | 6 | 9 | 9 | 18 | — |
| *c* | 6 | 10 | 16 | 18 | 8 | 26 | — |
| *e* | 16 | 11 | 27 | 27 | 5 | 32 | 1 |
| *a* | 27 | 6 | 33 | 33 | 3 | 36 | 1 |

∴ Total elapsed time = 36 hours

Idle time for Machine *A* = 36 – 33

= 3 hours

Idle time for Machine *B* = 4 hours **Ans.**

**Example 8.2** *Determine a sequence of these jobs that will minimize the total elapsed time (T). Also find the total elapsed time and idle time for each machine.*

| *Jobs →* | *A* | *B* | *C* | *D* | *E* | *F* | *G* |
|---|---|---|---|---|---|---|---|
| *M1* | *3* | *12* | *15* | *6* | *10* | *11* | *9* |
| *M2* | *8* | *10* | *10* | *7* | *12* | *1* | *3* |

**Solution** The optimial sequence obtained using Johnson's algorithm is :

| A | D | E | B | C | G | F |
|---|---|---|---|---|---|---|

Compulation of total elapsed time and idle time for machines :

| Job sequence | Machine M1 | | | Machine M2 | | |
|---|---|---|---|---|---|---|
| | Time in | P.T. | Time Out | Time In | P.T. | Time Out |
| A | 0 | 3 | 3 | 3 | 8 | 11 |
| D | 3 | 6 | 9 | 11 | 7 | 18 |
| E | 9 | 10 | 19 | 19 | 12 | 31 |
| B | 19 | 12 | 31 | 31 | 10 | 41 |
| C | 31 | 15 | 46 | 46 | 10 | 56 |
| G | 46 | 9 | 55 | 56 | 3 | 59 |
| F | 55 | 11 | 66 | 66 | 1 | 67 |

Minimum elapsed time = 67 hrs

Idle time for M1 = (67 – 66) = 1 hr

Idle time for M2 = (3 – 0) + (19 – 18) + (46 – 41) + (66 – 59)

= 3 + 1 + 5 + 7 = 16 hr.

**Example 8.3** *In a factory there are 6 jobs to perform, each of which should go through machine A and B. The processing time (in hours) for the jobs are given in table. Determine the sequence for performing the jobs that would minimised the total elapsed time (T). What is the value of T? Also draw Gantt chart showing idle time of 2 machines.*

| *Jobs* → | $J_1$ | $J_2$ | $J_3$ | $J_4$ | $J_5$ | $J_6$ |
|---|---|---|---|---|---|---|
| *Machine A* | 1 | 3 | 8 | 5 | 6 | 3 |
| *Machine B* | 5 | 6 | 3 | 2 | 2 | 10 |

***Solution*** The optimal sequence is obtained as

| $J_1$ | $J_6$ | $J_2$ | $J_3$ | $J_4$ | $J_5$ |
|---|---|---|---|---|---|

Compulation of total elapsed time and Idle time for machines :

| Job sequence | Machine A | | | Machine B | | | Idle Time |
|---|---|---|---|---|---|---|---|
| | In | PT | Out | In | PT | Out | Mahcine B |
| $J_1$ | 0 | 1 | 1 | 1 | 5 | 6 | 1 |
| $J_6$ | 1 | 3 | 4 | 6 | 10 | 16 | – |
| $J_2$ | 4 | 3 | 7 | 16 | 6 | 22 | – |
| $J_3$ | 7 | 8 | 15 | 22 | 3 | 25 | – |
| $J_4$ | 15 | 5 | 20 | 25 | 2 | 27 | – |
| $J_5$ | 20 | 6 | 26 | 27 | 2 | 29 | – |

∴ Total elapsed time = 29 hrs

Idle time for machine $A$ = 29 – 26 = 3 hrs

Idle time for machine $B$ = 1 hr

The Gantt chart for the above problem can be illustrated as follows :

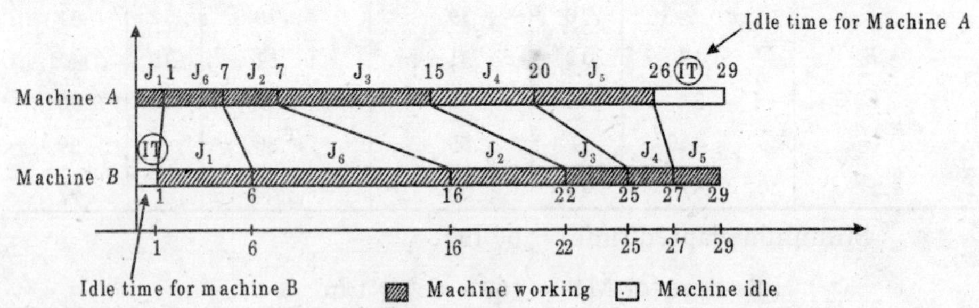


**Example 8.4** *Find the sequence that minimizes the total elapsed time (T) required to complete the following task on two machines. The processing time (in hours) for the tasks are given in table below.*

| Jobs → Machines | A | B | C | D | E | F | G | H | I |
|---|---|---|---|---|---|---|---|---|---|
| $M_1$ | 2 | 5 | 4 | 9 | 6 | 8 | 7 | 5 | 4 |
| $M_2$ | 6 | 8 | 7 | 4 | 3 | 9 | 3 | 8 | 11 |

***Solution*** The optimal sequence following Johnson's algorithm is obtained as :

| A | C | I | B | H | F | D | G | E |
|---|---|---|---|---|---|---|---|---|

Computation of total elapsed time and idle time for machines.

| *Job sequence* | *Machine $M_1$* | | | *Machine $M_2$* | | | *Idle Time* |
|---|---|---|---|---|---|---|---|
| | *In* | *PT* | *Out* | *In* | *PT* | *Out* | *Mahcine $M_2$* |
| *A* | 0 | 2 | 2 | 2 | 6 | 8 | 2 |
| *C* | 2 | 4 | 6 | 8 | 7 | 15 | – |
| *I* | 6 | 4 | 10 | 15 | 11 | 26 | – |
| *B* | 10 | 5 | 15 | 26 | 8 | 34 | – |
| *H* | 15 | 5 | 20 | 34 | 8 | 42 | – |
| *F* | 20 | 8 | 28 | 42 | 9 | 51 | – |
| *D* | 28 | 9 | 37 | 51 | 4 | 55 | – |
| *G* | 37 | 7 | 44 | 55 | 3 | 58 | – |
| *E* | 44 | 6 | 50 | 58 | 3 | 61 | – |

We find, Total elapsed time = 61 hrs

Idle time for machine $M_1$ = 61 – 50 = 11 hrs

Idle time for machine $M_2$ = 2 hrs **Ans.**

**Example 8.5** *A book binder has one printing press, one binding machine and manuscrips of a number of books. Time required to perform the printing and binding operations on each book are shown below. The binder wishes to determine order in which the books should be processed, so that the total time required to process all books is minimized.*

| *Book* | *A* | *B* | *C* | *D* | *E* | *F* |
|---|---|---|---|---|---|---|
| *Printing time ( hours)* | *30* | *120* | *50* | *20* | *90* | *110* |
| *Binding time (hours)* | *80* | *100* | *90* | *60* | *30* | *10* |

***Solution*** The optimal sequence following Johnson's algorithm is obtained as :

| D | A | C | B | E | F |
|---|---|---|---|---|---|

Computation of total elapsed time and idle time for Machines.

| *Job sequence (Books)* | *Printing machine* | | | *Binding machine* | | | *Idle Time* |
|---|---|---|---|---|---|---|---|
| | *In* | *PT* | *Out* | *In* | *PT* | *Out* | *binding mahcine* |
| D | 0 | 20 | 20 | 20 | 60 | 80 | 20 |
| A | 20 | 30 | 50 | 80 | 80 | 160 | – |
| C | 50 | 50 | 100 | 160 | 90 | 250 | – |
| B | 100 | 120 | 220 | 250 | 100 | 350 | – |
| E | 220 | 90 | 310 | 350 | 30 | 380 | – |
| F | 310 | 110 | 420 | 420 | 10 | 430 | 40 |

Total elapsed time = 430 hrs

Idle time for printing machine = 430 – 420 = 10 hrs

Idle time for binding machine = 60 hrs **Ans.**

**Example 8.6** *We have seven jobs each of which has to go through the machine $M_1$ and $M_2$ in the order $M_1$, $M_2$. Processing times (in hours) are gives as:*

| *Job* | *1* | *2* | *3* | *4* | *5* | *6* | *7* |
|---|---|---|---|---|---|---|---|
| Machine $M_1$ | 3 | 12 | 15 | 6 | 10 | 11 | 9 |
| Machine $M_2$ | 8 | 10 | 10 | 6 | 12 | 1 | 3 |

Determine a sequence of these jobs that will minimize the total elapsed time. Also compute idle time for each machine.

***Solution***

| *Job* | *1* | *2* | *3* | *4* | *5* | *6* | *7* |
|---|---|---|---|---|---|---|---|
| $M_1$ | 3 | 12 | 15 | 6 | 10 | 11 | 9 |
| $M_2$ | 8 | 10 | 10 | 6 | 12 | 1 | 3 |

The optimal sequence following Johnson's algorithm are :

| 1 | 4 | 5 | 2 | 3 | 7 | 6 |
|---|---|---|---|---|---|---|

or

| 1 | 5 | 2 | 3 | 4 | 7 | 6 |
|---|---|---|---|---|---|---|

Case 1 : Computation of total elapsed time and idle time for machines.

| *Job* | $M_1$ | | | $M_2$ | | | |
|---|---|---|---|---|---|---|---|
| | *In* | *P.T.* | *Out* | *In* | *P.T.* | *Out* | *I.T.* |
| 1 | 0 | 3 | 3 | 3 | 8 | 11 | 3 |
| 4 | 3 | 6 | 9 | 11 | 6 | 17 | – |
| 5 | 9 | 10 | 19 | 19 | 12 | 31 | 2 |
| 2 | 19 | 12 | 31 | 31 | 10 | 41 | – |
| 3 | 31 | 15 | 46 | 46 | 10 | 56 | 5 |
| 7 | 46 | 9 | 55 | 56 | 3 | 59 | – |
| 6 | 55 | 11 | 66 | 66 | 1 | 67 | 7 |

Total elapsed time = 67 hrs

Idle time for Machine $M_1$ = 67 – 66 = 1 hrs

Idle time for Machine $M_2$ = 17 hrs

**Same for Case 2**

### 8.2.2 Sequencing of *n* Jobs on *m* Machines

Consider *n* jobs (1, 2, 3, .... *n*) processing through *m* machines ($M_1$, $M_2$, ....$M_m$) as shown in table below.

| *Machines* | *Jobs* | | | |
|---|---|---|---|---|
| | 1 | 2 | 3 ........................ | *n* |
| $M_1$ | $t_{11}$ | $t_{12}$ | $t_{13}$ ........................ | $t_{1n}$ |
| $M_2$ | $t_{21}$ | $t_{22}$ | $t_{23}$ ........................ | $t_{2n}$ |
| ⋮ | ⋮ | ⋮ | ⋮ | ⋮ |
| $M_{m-1}$ | | | | |
| $M_m$ | $t_{m1}$ | $t_{m2}$ | $t_{m3}$ ........................ | $t_{mn}$ |

An optimum solution to this problem can be obtained if either or both of the following conditions are satisfied.

(i) The minimum time on machine $M_1 \geq$ the maximum time on machines $M_2$ ......... $M_{m-1}$.

(ii) The minimum time on Machine $M_m \geq$ the maximum time on machines $M_2$ ....... $M_{m-1}$.

Otherwise method fails.

**Procedure do determine optimal sequence.**

**Step 1 :** Convert the *m* machine problem into two machine problem by indroducing two fictitious machines say *G* and *H*.

Such that,

$$t_{Gj} = t_{1j} + t_{2j} + \text{........} + t_{(m-1)j}$$
$$t_{Hj} = t_{2j} + t_{3j} + \text{........} + t_{mj}$$

**Step 2 :** Determine the optimal sequence for *n* jobs and 2 machines by following usual procedure.

**Example 8.6** *5 jobs are to be processed on 4 machines. The processing time (in hours) for each job on each machine is given as below. Determine the sequence of jobs that minimizes total elapsed time to complete the jobs.*

| *Machines* | *Jobs* | | | | |
|---|---|---|---|---|---|
| | *A* | *B* | *C* | *D* | *E* |
| $M_1$ | 9 | 11 | 7 | 14 | 15 |
| $M_2$ | 2 | 1 | 3 | – | 4 |
| $M_3$ | 4 | 5 | 3 | 6 | 2 |
| $M_4$ | 13 | 6 | 11 | 7 | 9 |

***Solution*** Verification of condition :

Min. time on $M_1$ = 7, Min. time on $M_4 = 6$

Max. time on $M_2$ = 4, Max. time on $M_3 = 6$

∴ Min time on $M_1 \geq$ Max. time on $(M_2, M_3)$

Min time on $M_4 \geq$ Max time on $(M_2, M_3)$

∴ Required condition is satisfied

The problem thus reduess to 2 machine problem by adding processing times for $M_1$, $M_2$, $M_3$ and $M_2$, $M_3$, $M_4$ as follows

| Machines ↓ | Jobs | | | | |
|---|---|---|---|---|---|
| | *A* | *B* | *C* | *D* | *E* |
| *G* | 15 | 17 | 13 | 20 | 21 |
| *H* | 19 | 12 | 17 | 13 | 15 |

Required optimum sequence (by following the usual procedure).

| C | A | E | D | B |
|---|---|---|---|---|

Calculation of minimum elapsed time

| *Jobs* | *Machine $M_1$* | | | *Machine $M_2$* | | | *Machine $M_3$* | | | *Machine $M_4$* | | |
|---|---|---|---|---|---|---|---|---|---|---|---|---|
| | *In* | *PT* | *Out* | *In* | *PT* | *Out* | *In* | *PT* | *Out* | *In* | *PT* | *Out* |
| C | 0 | 7 | 7 | 7 | 3 | 10 | 10 | 3 | 13 | 13 | 11 | 24 |
| *A* | 7 | 9 | 16 | 16 | 2 | 18 | 18 | 4 | 22 | 24 | 13 | 37 |
| *E* | 16 | 15 | 31 | 31 | 4 | 35 | 35 | 2 | 37 | 37 | 9 | 46 |
| *D* | 31 | 14 | 45 | 45 | – | 45 | 45 | 6 | 51 | 51 | 7 | 58 |
| *B* | 45 | 11 | 56 | 56 | 1 | 57 | 57 | 5 | 62 | 62 | 6 | 68 |

∴ Total elapsed time = 68 hrs

Idle time for Machine $M_1$ = (68 – 56) = 12 hrs

Idle time for Machine $M_2$ = {(7 – 0) + (16 – 10) + (31 – 18)

+ (45 – 35) + (56 – 45) + (68 – 57)}

= 7 + 6 + 13 + 10 + 11 + 11

= 58 hrs

Idle time for Machine $M_3$ = {(10 – 0) + (18 – 13) + (35 – 22)

+ (45 – 37) + (57 – 51) + (68 – 62)}

= 10 + 5 + 13 + 8 + 6 + 6

= 48 hrs

Idle time for Machine $M_4$ = {(13 – 0) + (51 – 46) + (62 – 58)

= 13 + 5 + 4

= 22 hrs

**Example 8.7** *5 jobs are to be processed on 3 machines. The processing time (in hours) for each job on each machine is given as below. Determine the sequence of jobs that minimizes total elapsed time to complete the jobs. Also find idle time for each machine.*

| Machines ↓ | Jobs | | | | |
|---|---|---|---|---|---|
| | A | B | C | D | E |
| $M_1$ | 9 | 11 | 7 | 8 | 12 |
| $M_2$ | 6 | 7 | 3 | 4 | 5 |
| $M_3$ | 5 | 10 | 9 | 7 | 6 |

**Solution** Verifying the conditions :

Min. time on $M_1$ = 7, Min. time on $M_3$ = 5

Max. time on $M_2$ = 7

∴ Min. time on $M_1 \geq$ Max. time on $M_2$

∴ Required condition is satisfied.

The problem thus reduces to 2 machines problem by adding processing times for $M_1$, $M_2$ and $M_2$, $M_3$ as follows.

| Machines ↓ | Jobs | | | | |
|---|---|---|---|---|---|
| | A | B | C | D | E |
| G | 15 | 18 | 10 | 12 | 17 |
| H | 11 | 17 | 12 | 11 | 11 |

Required optimum sequence (by following the usual procedure)

| C | B | E | D | A |
|---|---|---|---|---|

Calculation of minimum elapsed time :

| Jobs | Machine $M_1$ | | | Machine $M_2$ | | | Machine $M_3$ | | |
|---|---|---|---|---|---|---|---|---|---|
| | IN | PT | Out | IN | PT | Out | IN | PT | Out |
| C | 0 | 7 | 7 | 7 | 3 | 10 | 10 | 9 | 19 |
| B | 7 | 11 | 18 | 18 | 7 | 25 | 25 | 10 | 35 |
| E | 18 | 12 | 30 | 30 | 5 | 35 | 35 | 6 | 41 |
| D | 30 | 8 | 38 | 38 | 4 | 42 | 42 | 7 | 49 |
| A | 38 | 9 | 47 | 47 | 6 | 53 | 53 | 5 | 58 |

∴ Total elapsed time = 58 hrs

Idle time for Machine $M_1$ = (58 – 47) = 11 hrs

Idle time for Machine $M_2$ = {(7 – 0) + (18 – 10) + (30 – 25) + (38 – 35) + (47 – 42) + (58 – 53)}

= 7 + 8 + 5 + 3 + 5 + 5

= 33 hrs

Idle time for Machine $M_3$ = {(10 – 0)+(25 – 19)+(42 – 41)+(53 – 49)}

$$= 10 + 6 + 1 + 4$$

$$= \mathbf{21\ hrs}$$ **Ans.**

**Example 8.8** *A firm can work for a maximum of 60 hours a week. It has received 7 orders to be processed on 3 machines A, B and C in the order ABC to be delivered in a week's time from now. The processing times (in hours) are given in the following table :*

| *Jobs* | *1* | *2* | *3* | *4* | *5* | *6* | *7* |
|---|---|---|---|---|---|---|---|
| *Machine A (hours) :* | 7 | 8 | 6 | 6 | 7 | 8 | 5 |
| *Machine B (hours) :* | 2 | 2 | 1 | 3 | 3 | 2 | 4 |
| *Machine C (hours) :* | 6 | 5 | 4 | 4 | 2 | 1 | 5 |

*The manager who in fairness insists on performing the jobs in the sequence in which they are received , is refusing to accept an 8th order , which requires 7, 2 and 5. hours respectively on A, B,C machines because according to him 8th order will require a total of 61 hours for processing, which exceeds the firm's capacity. What is your advice to him ?*

***Solution*** Let us find out the optimal sequence for 8 jobs to minimize the elapsed time.

| *Jobs* → | *1* | *2* | *3* | *4* | *5* | *6* | *7* | *8* |
|---|---|---|---|---|---|---|---|---|
| Machine A | 7 | 8 | 6 | 6 | 7 | 8 | 5 | 7 |
| Machine B | 2 | 2 | 1 | 3 | 3 | 2 | 4 | 2 |
| Machine C | 6 | 5 | 4 | 4 | 2 | 1 | 5 | 5 |

Minimum time on Machine *A* is 5 which is greater than maximum time on Machine *B*.

There by changing above into two machine problem by introducing two ficticious machines *G* and *H*.

| *Jobs* | *1* | *2* | *3* | *4* | *5* | *6* | *7* | *8* |
|---|---|---|---|---|---|---|---|---|
| *Machine G* | 9 | 10 | 7 | 9 | 10 | 10 | 9 | 9 |
| *Machine H* | 8 | 7 | 5 | 7 | 5 | 3 | 9 | 7 |

Optimal sequence using Johnson's algorithm is :

| 7 | 1 | 4 | 8 | 2 | 3 | 5 | 6 |
|---|---|---|---|---|---|---|---|

| *Jobs* ↓ | *Machine A* | | | *Machine B* | | | *Machine C* | | |
|---|---|---|---|---|---|---|---|---|---|
| | *In* | *PT* | *Out* | *In* | *PT* | *Out* | *In* | *PT* | *Out* |
| 7 | 0 | 5 | 5 | 5 | 4 | 9 | 9 | 5 | 14 |
| 1 | 5 | 7 | 5 | 9 | 2 | 11 | 14 | 6 | 20 |
| 4 | 12 | 6 | 18 | 18 | 3 | 21 | 21 | 4 | 25 |

*Contd.*

| Jobs ↓ | Machine A | | | Machine B | | | Machine C | | |
|---|---|---|---|---|---|---|---|---|---|
| | In | PT | Out | In | PT | Out | In | PT | Out |
| 8 | 18 | 7 | 25 | 25 | 2 | 27 | 27 | 5 | 32 |
| 2 | 25 | 8 | 33 | 33 | 2 | 35 | 35 | 5 | 40 |
| 3 | 33 | 6 | 39 | 39 | 1 | 40 | 40 | 4 | 44 |
| 5 | 39 | 7 | 46 | 46 | 3 | 49 | 49 | 2 | 47 |
| 6 | 46 | 8 | 54 | 54 | 2 | 56 | 56 | 1 | 57 |

∴ Total elapsed time for 8 jobs is 57 hours which is within capacity of the firm. So manage may be advised to take all the eight jobs.

### 8.2.3 Sequencing of 2 jobs on *m* machines (Graphical Method)

The procedure of sequencing of 2 jobs on *m* machines is given in the following steps.

**Step 1 :** Draw a Cartesian co-ordinate system with the processing times coressponding to the first job on the horizontal axix and that to second job on vertical axis.

**Step 2 :** On each axis, mark the operation times in the order in which the operations must be performed for that job.

**Step 3 :** Block out areas corresponding to each machine at the intersection of the intervals marked for that machine on the two axis.

**Step 4 :** Determine a path from the origin to the end of the final block that does not cross any of the blocks and that minimizes the vertical movement. Movement is allowed only in three directions horizontal, vertical and 45° diagonaly. The path with minimum vertical distance will indicate the optimal solution. Note that this will be same as the path with minimum horizontal distance.

**Example 8. 9** *Using Graphical Method, calculate the time needed to process the following jobs on the machines shown below. Also calculate idle time in the process.*

| | | | | | | |
|---|---|---|---|---|---|---|
| *Job 1* | *Sequence :* | A | B | C | D | E |
| | *Time (hrs) :* | 3 | 4 | 2 | 6 | 2 |
| *Job 2* | *Sequence :* | B | C | A | D | E |
| | *Time (hrs) :* | 5 | 4 | 3 | 2 | 6 |

**(April, 2012 IPU)**

***Solution*** By constructing the 2 D graph, where $x$-axis represeuts Job 1 and $y$-axis represents Job 2.

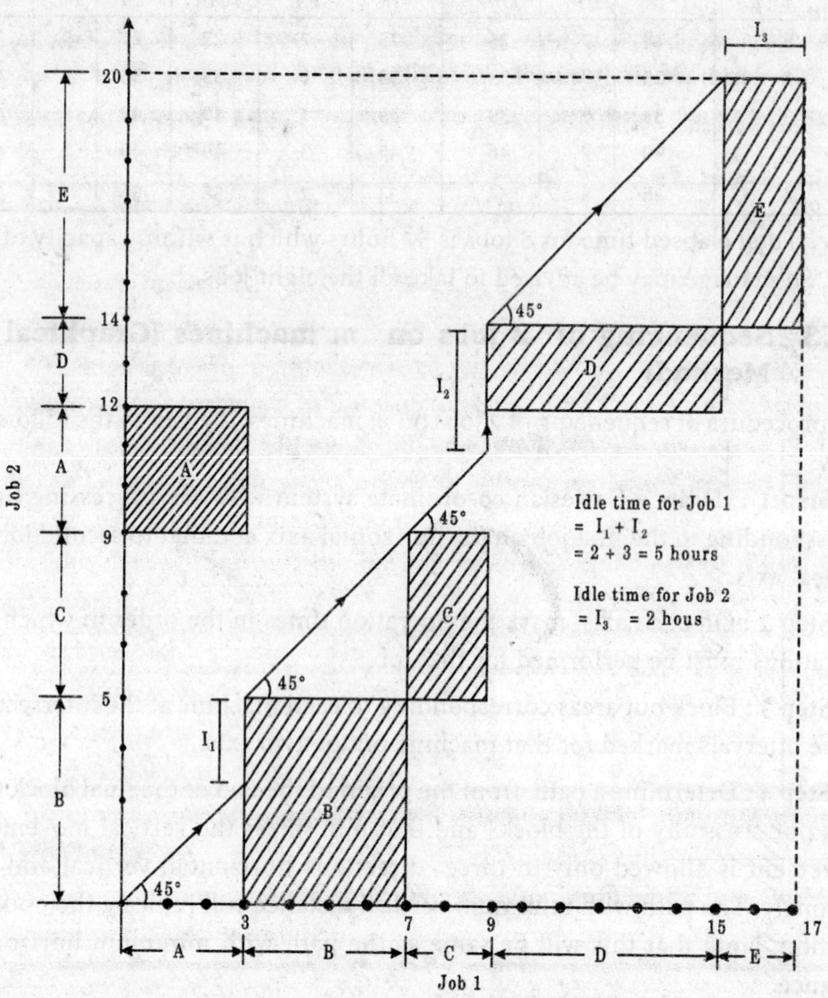


Total elapsed time

= Processing time of Job 1 + Idle time for Job 1

$$= 17 + (\overset{I_1}{2} + \overset{I_2}{3}) = 22 \text{ hrs}$$

or

= Processing time of Job 2 + Idle time for Job 2

$$= 20 + (\overset{I_3}{2}) = 22 \text{ hrs}$$

**Ans.**

**Example 8.10** *Using Graphical Method calculate the minimum time needed to process Job (1) and Job (2) on five machines A, B, C, D and E. For each machine also find which job should be done first.*

| | | | | | | |
|---|---|---|---|---|---|---|
| *Job 1* | *Sequence :* | A | B | C | D | E |
| | *Time (hrs) :* | 1 | 2 | 3 | 5 | 1 |
| *Job 2* | *Sequence :* | C | A | D | E | B |
| | *Time (hrs) :* | 3 | 4 | 2 | 1 | 5 |

***Solution*** Construct a 2 D graph, where $x$-axis represents Job 1 and $y$-axis Job 2.

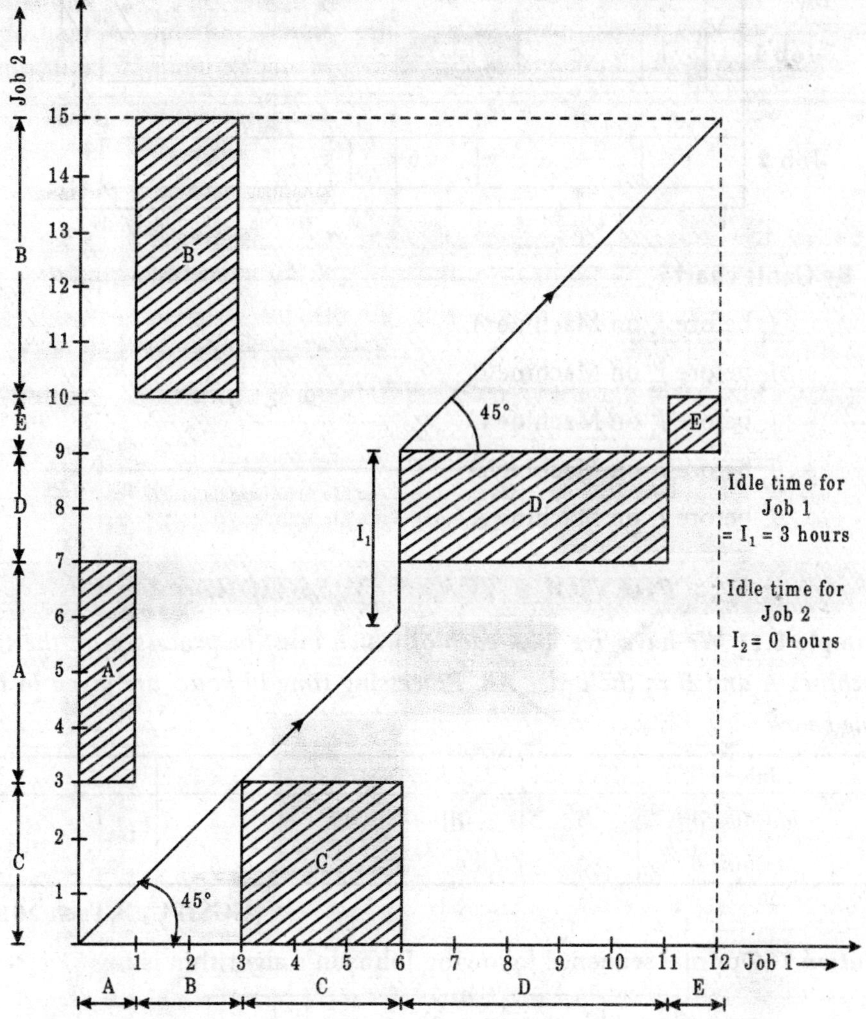


Total elapsed time

= Processing time of Job 1 + Idle time for (Job 1) i.e. $I_1$

$= \quad 12 + 3 = 15$ hrs

or

$=$ Processing time of Job 2 + Idle time for (Job 2)

$= \quad 15 + 0 = 15$ hrs

**Gantt chart**

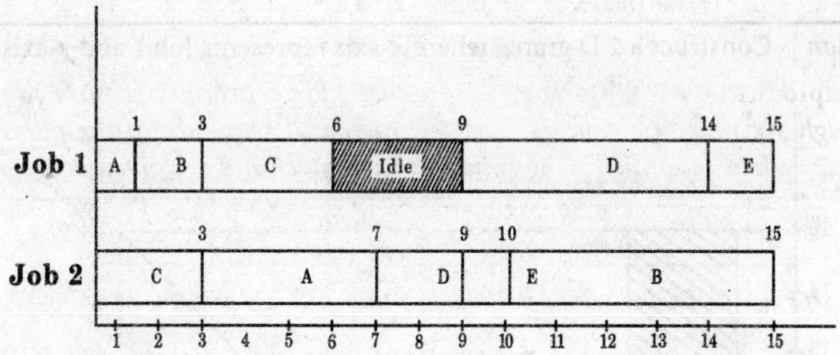


**By Gantt chart :**

$J_1$ before $J_2$ on Machine $A$

$J_1$ before $J_2$ on Machine $B$

$J_2$ before $J_1$ on Machine $C$

$J_2$ before $J_1$ on Machine $D$

$J_2$ before $J_1$ on Machine $A$

## PREVIOUS YEARS QUESTIONS

**Example 8.11** *We have five jobs, each of which must be processed on the two machines A and B in the order AB. Processing time in hours are given in the table below :*

| *Job →* | *1* | *2* | *3* | *4* | *5* |
|---|---|---|---|---|---|
| *Machine A* | *5* | *1* | *9* | *3* | *10* |
| *Machine B* | *2* | *6* | *7* | *8* | *4* |

**[GGSIPU, B.Tech 2014]**

***Solution*** Optimal sequence following Johnson's algorithm is :

| 2 | 4 | 3 | 5 | 1 |
|---|---|---|---|---|

Computation of total elapsed time.

| Sequence | Machine A | | | Machine B | | |
|---|---|---|---|---|---|---|
| | IN | PT | OUT | IN | PT | OUT |
| 2 | 0 | 1 | 1 | 1 | 6 | 7 |
| 4 | 1 | 3 | 4 | 7 | 8 | 15 |
| 3 | 4 | 9 | 13 | 15 | 7 | 22 |
| 5 | 13 | 10 | 23 | 23 | 4 | 27 |
| 1 | 23 | 5 | 28 | 28 | 2 | 30 |

Total elapsed time = 30 hrs

**Example 8.12** *Two major parts $P_1$ and $P_2$ for a product require processing through six machine centers. The technological sequence of the parts on six machines and manufacturing times on each machines are given below :*

| Part $P_1$ | Machine sequence : | C | A | E | F | D | B |
|---|---|---|---|---|---|---|---|
| | Time (hrs) : | 2 | 3 | 4 | 5 | 6 | 1 |
| Part $P_2$ | Machine sequence : | B | A | E | F | C | D |
| | Time (hrs) : | 3 | 2 | 5 | 3 | 2 | 3 |

*What would be the optimal scheduling to minimize the total processing time for these two parts ? Find also the total elapsed time. For each machine specify the job that sohould be done first.* **[GGSIPU, B.Tech. 2014]**

***Solution*** Costruct a 2 D graph where $x$-axis represents part 1 and $y$-axis part 2.

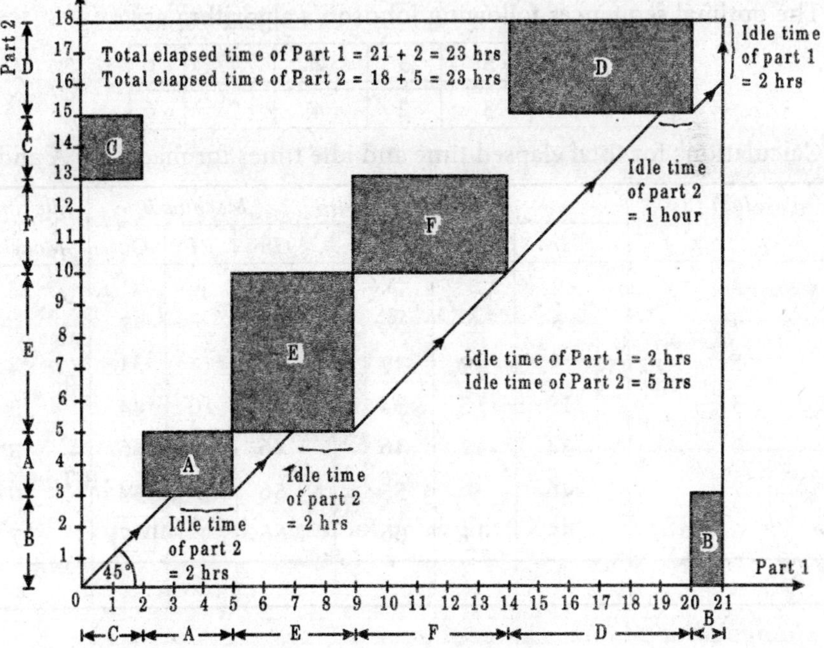

Total elapsed time

= Processing time of part 1 + Idle time (Part 1)

= 21 + 2 = 23 hrs

or

= Processing time of part 2 + Idle time (Part 2)

= 18 + (2 + 2 + 1) = **23 hrs** **Ans.**

**Example 81.3** *There are seven jobs each of which has to go through the machines A and B in the order AB. Processing times in hours are gives as :*

| *Jobs* | *1* | *2* | *3* | *4* | *5* | *6* | *7* |
|---|---|---|---|---|---|---|---|
| *Machine A* | *3* | *12* | *15* | *6* | *10* | *11* | *9* |
| *Machine B* | *8* | *10* | *10* | *6* | *12* | *1* | *3* |

*Determine a sequence of these jobs that will minimize that total elapsed time T.* **[GGSIUP, B.Tech, 2013]**

***Solution***

| *Jobs* | *1* | *2* | *3* | *4* | *5* | *6* | *7* |
|---|---|---|---|---|---|---|---|
| Machine *A* | 3 | 12 | 15 | 6 | 10 | 11 | 9 |
| Machine *B* | 8 | 10 | 10 | 6 | 12 | 1 | 3 |

The optimal sequences following Johnson's algorithm are :

| 1 | 4 | 5 | 3 | 2 | 7 | 6 |
|---|---|---|---|---|---|---|
| 1 | 5 | 3 | 2 | 4 | 7 | 6 |

Calculations for total elapsed time and idle times for machines A and B.

| *Job* | *Machine A* | | | *Machine B* | | | *Idle time for* |
|---|---|---|---|---|---|---|---|
| | *In* | *PT* | *Out* | *In* | *PT* | *Out* | *Mahcine B* |
| 1 | 0 | 3 | 3 | 3 | 8 | 11 | 3 |
| 4 | 3 | 6 | 9 | 11 | 6 | 17 | 0 |
| 5 | 9 | 10 | 19 | 19 | 12 | 31 | 2 |
| 3 | 19 | 15 | 34 | 34 | 10 | 44 | 3 |
| 2 | 34 | 12 | 46 | 46 | 10 | 56 | 2 |
| 7 | 46 | 9 | 55 | 56 | 3 | 59 | 0 |
| 6 | 55 | 11 | 66 | 66 | 1 | 67 | 7 |
| | | | | | | | 17 |

The minimum elapsed time is 67 hrs.

Idle time for Machine *A* is 1 hour

Idle time for Machine *B* is 17 hrs. **Ans.**

**Example 8.14** *A machine operator has to perform two operations, turning and threading on a number of different jobs. The time required to perform these operations for each job is known. Determine the order in which the jobs should be processed in order to minimize the total time required to turn out all the jobs.*

| *Job* | *Time for turning* | *Time for threading* |
|---|---|---|
| *1* | *3* | *8* |
| *2* | *12* | *10* |
| *3* | *5* | *9* |
| *4* | *2* | *6* |
| *5* | *9* | *3* |
| *6* | *11* | *1* |

**[GGSIPU, B. Tech, 2013]**

***Solution***

| *Job* | *1* | *2* | *3* | *4* | *5* | *6* |
|---|---|---|---|---|---|---|
| Time for turning : | 3 | 12 | 5 | 2 | 9 | 11 |
| Time for threading : | 8 | 10 | 9 | 6 | 3 | 1 |

Optimal sequence following Johnson's algorithm is :

| 4 | 1 | 3 | 2 | 5 | 6 |
|---|---|---|---|---|---|

Calculation of total elapsed time and idle time for each machine :

| *Job* | *Turning* | | | *Threading* | | |
|---|---|---|---|---|---|---|
| | *In* | *P.T.* | *Out* | *In* | *P.T.* | *Out* |
| 4 | 0 | 2 | 2 | 2 | 6 | 8 |
| 1 | 2 | 3 | 5 | 8 | 8 | 16 |
| 3 | 5 | 5 | 10 | 16 | 9 | 25 |
| 2 | 10 | 12 | 22 | 25 | 10 | 35 |
| 5 | 22 | 9 | 31 | 35 | 3 | 38 |
| 6 | 31 | 11 | 42 | 42 | 1 | 43 |

The minimum elapsed time is 43 hours

Idle time for Turning Machine is 1 hour

Idle time for Threading Machine is 2 + 4 = **6 hours** **Ans.**

**Example 8.15** *A book binder has one printing press, one binding machine and the manuscripts of a number of different books. The time required to perform the printing and binding operations for each book is shown below. Determine the order in which books should be processed in order to minimize the total time required to turn out all the books.*

| *Book* | *1* | *2* | *3* | *4* | *5* | *6* |
|---|---|---|---|---|---|---|
| *Printing time (hrs)* | *30* | *120* | *50* | *20* | *90* | *100* |
| *Binding time (hrs)* | *80* | *100* | *90* | *60* | *30* | *10* |

**[GGSIPU, B.Tech, 2012]**

***Solution*** In this problem book will first go to printing press then on to the binding machine. Optimal sequences following Johnson's algorithm is obtained as :

Printing time → | 4 | 1 | 3 | 2 | 5 | 6 | ← Binding time

**Calculation for minimum elapsed time :**

| | *Printing time* | | | *Binding time* | | | |
|---|---|---|---|---|---|---|---|
| *Jobs* | *In* | *Pt* | *Out* | *In* | *Pt* | *Out* | *Idle time* |
| 4 | 0 | 20 | 20 | 20 | 60 | 80 | 20 |
| 1 | 20 | 30 | 50 | 80 | 80 | 160 | — |
| 3 | 50 | 50 | 100 | 160 | 90 | 250 | — |
| 2 | 100 | 120 | 220 | 250 | 100 | 350 | — |
| 5 | 220 | 90 | 310 | 350 | 30 | 380 | — |
| 6 | 310 | 100 | 410 | 410 | 10 | 420 | 30 |

Total elapsed time is 420 hours

Idle time for book printing = (420 – 410) hrs

= 10 hours

Idle time for book binding = 50 hours

**Example 8.16** *A manufacturing company processes 6 different on two machines A and B. Number of units of each job and its processing times on A and B are given in the following table. Find the optimal sequence and total minimum elapsed time.* **[GGSIPU, 2011]**

| *Job No.* | *No. of units of each job* | *Processing time Machine A (minutes)* | *Processing time Machine B (minutes)* |
|---|---|---|---|
| *1* | *3* | *5* | *8* |
| *2* | *4* | *16* | *7* |
| *3* | *2* | *6* | *11* |
| *4* | *5* | *3* | *5* |
| *5* | *2* | *9* | *7.5* |
| *6* | *3* | *6* | *14* |

***Solution*** *Optimal sequence following Johnson's algorithm is*

| 4 | 1 | 3 | 6 | 5 | 2 |
|---|---|---|---|---|---|

For minimum clased time we have

| Job | No. of units | *Machine A* | | | *Machine B* | | | *Idle time* |
|---|---|---|---|---|---|---|---|---|
| | | *In* | *Pt* | *Out* | *In* | *Pt* | *Out* | *Machine B* |
| 4 | 5 | 0 | $3 \times 5 = 15$ | 15 | 15 | $5 \times 5 = 25$ | 40 | 15 |
| 1 | 3 | 15 | $5 \times 3 = 15$ | 30 | 40 | $8 \times 3 = 24$ | 64 | — |
| 3 | 2 | 30 | $6 \times 2 = 12$ | 42 | 64 | $11 \times 2 = 22$ | 86 | — |
| 6 | 3 | 42 | $6 \times 3 = 18$ | 60 | 86 | $14 \times 3 = 42$ | 128 | — |
| 5 | 2 | 60 | $9 \times 2 = 18$ | 78 | 128 | $7.5 \times 2 = 15$ | 143 | — |
| 2 | 4 | 78 | $16 \times 4 = 64$ | 142 | 143 | $7 \times 4 = 28$ | 171 | — |

Total minimum elapsed time = 171 minutes

Idle time for machine A = 171 − 142 = 29 minutes

Idle time for machine B = 15 minutes

**Example 8.17** *There are five jobs, each of which is to be processed through three machines A, B and C in the order ABC, processing times in hours are*

| *Job* | *A* | *B* | *C* |
|---|---|---|---|
| *1* | *3* | *4* | *7* |
| *2* | *8* | *5* | *9* |
| *3* | *7* | *1* | *5* |
| *4* | *5* | *2* | *6* |
| *5* | *4* | *3* | *10* |

**[GGSIPU, B.Tech, 2010]**

*Determine the optimim sequence for the five jobs and the minimum elapsed time.*

***Solution*** Min $A = 3$, max $B = 5$, min $C = 5$

Since min $C \geq$ max $B$ is satisfied

The given problem is converted in to 5 jobs, 2 machine problem

| *Job* | *1* | *2* | *3* | *4* | *5* |
|---|---|---|---|---|---|
| $G = A + B$ | $3 + 4 = 7$ | $8 + 5 = 13$ | $7 + 1 = 8$ | $5 + 2 = 7$ | $4 + 3 = 7$ |
| $H = B + C$ | $4 + 7 = 11$ | $5 + 9 = 14$ | $1 + 5 = 6$ | $2 + 6 = 8$ | $3 + 10 = 13$ |

The optimal sequences following Johnson's algorithm are obtained as :

$G \rightarrow$

| 1 | 5 | 4 | 2 | 3 |
|---|---|---|---|---|

$\leftarrow H$

Optimal or

| 1 | 5 | 2 | 4 | 3 |
|---|---|---|---|---|

For total elasped time :

| Job | Machine A | | Machine B | | Machine C | |
|---|---|---|---|---|---|---|
| | In | Out | In | Out | In | Out |
| 1 | 0 | 3 | 3 | 3 + 4 = 7 | 7 | 7 + 7 = 14 |
| 5 | 3 | 3 + 4 = 7 | 7 | 7 + 3 = 10 | 14 | 14 + 10 = 24 |
| 4 | 7 | 7 + 5 = 12 | 12 | 12 + 2 = 14 | 24 | 24 + 6 = 30 |
| 2 | 12 | 12 + 8 = 20 | 20 | 20 + 5 = 25 | 30 | 30 + 9 = 39 |
| 3 | 20 | 20 + 7 = 27 | 27 | 27 + 1 = 28 | 39 | 39 + 5 = 44 |

∴ Minimum total elapsed time is 44 hours.

## MULTIPLE CHOICE QUESTIONS

8.1. Johnson's algorithm is used for which type of sequencing problem:

(a) $n$ jobs on two machines (b) $n$ jobs on $m$ machines

(c) $n$ jobs on three machines (d) 2 jobs on $m$ machines.

8.2. In sequencing of $n$ jobs on $m$ machines the following condition must satisfied :

(a) Min $M_1 \geq$ Max $(M_2....M_{m-1})$
Min $M_m \geq$ Max $(M_2....M_{m-1})$

(b) Max $M_1 \geq$ Min $(M_2....M_{m-1})$
Max $M_m \geq$ Min $(M_2....M_{m-1})$

(c) Min $M_1 \geq$ Min $(M_2....M_{m-1})$
Min $M_m \geq$ Min $(M_2....M_{m-1})$

(d) Max $M_1 \leq$ Max $(M_2....M_{m-1})$
Max $M_m \leq$ Max $(M_2....M_{m-1})$

8.3. In sequencing of $n$ jobs through $m$ machines, we convert the $m$ machines problem into — machine problem (Fictitious machines).

(a) One (b) Two

(c) Three (d) Four

8.4. In sequencing, graphical Method used in which types of problem :

(a) $n$ jobs on two machines (b) $n$ jobs on $m$ machines

(c) $n$ jobs on three machines (d) 2 jobs on $m$ machines

8.5. In sequencing of 2 jobs on $m$ machine problem, horizontal axis ($x$-axis) indicates the processing timing of .

(a) Machines (b) job 1

(c) job 2 (d) None of these

8.6. In graphical method movement of path (line) is allowed only in — directions.

(a) 1 (b) 2

(c) 3 (d) None of these

8.7. The path (line) from origin to the end of the block (final) is started with on angle of

(a) 30° (b) 45°

(c) 60° (d) 75°

8.8. The path (line) which is at an angle of 45° will indicate the idle time for
(a) job 1 (b) job 2
(c) Both job 1 and job 2 (d) None of these

8.9. The path (line) with vertical distance on graph indicates the idle time for
(a) Machines (b) job 2
(c) job 1 (d) Both job 1 and job 2

8.10. The horizontal distance (lines) on graph indicates the idle time for
(a) job 1 (b) job 2
(c) Both job 1 and job 2 (d) Machines

## ANSWERS

1. (a) 2. (a) 3. (b) 4. (d) 5. (b) 6. (c) 7. (b)
8. (c) 9. (c) 10. (b)

## SUMMARY

**Types of sequencing problems**

(i) **'*n*' jobs on 'two' machines (Johnson's Algorithm)**

(a) First of all find the smallest machines time for each job from the columns of machine 1 and 2.

(b) If smallest processing time of job lies in machine 1 then insert the job on L.H.S. and delete it from table.

(c) If smallest processing time of job coresponds to machine 2 then insert the job in R.H.S. and delete it from the table.

(d) Repeat the steps b and c till you left with single column.

(ii) **'*n*' jobs on '*m*' machines**

Johnson's Algorithm can also be used for '*n*' jobs through '*m*' machines problems, if the problem is converted into two machine problem (fictitious). And for converting '*n*' jobs through '*m*' machines into 'two' machines, following are the conditions :

(a) The minimum time on machine $M_1 \geq$ the maximum time on machines $M_2$ ... $M_{m-1}$.

(b) The minimum time on machine $M_m \geq$ the maximum time on machines $M_2$ ....$M_{m-1}$. Otherwise method fails.

(iii) **2 jobs on *m* machines**

Graphical Method is used for solving sequencing problem of '2' jobs on '*m*' machines.

CHAPTER 9

# Network Analysis : PERT and CPM

## 9.1 INTRODUCTION

A project is a piece of planned activity to achieve a particular aim, over a period of time. A project may be big or small like construction of bridges, factories, dams, houses or even to design a curriculum for a school or college.

CPM (Critical Path Method) and PERT (Project Evaluation and Review Techniques) are two important techniques of project management and scheduling in a given time frame.

### Activity

Every project consists of number of jobs or tasks which are called activities. An activity is represented by an arrow in the network diagram. An activity is usually named in capital alphabets like *A*, *B*, *C* etc.

### Event

An event is a specific instant of time which indicates the beginning or end of any activity. An event is specified by a number and is given inside a circle.

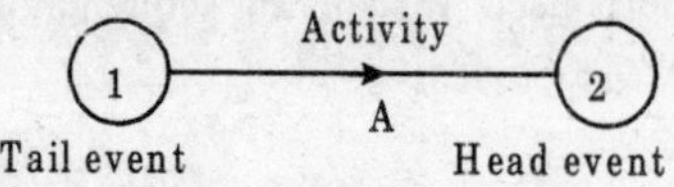


### Dummy Activities

Dummy (or zero time) activities neither consume time nor resources, shown by a dotted lines and are used in following cases :

**1. To produce unique representation of concurrent activities in direction of work flow :**

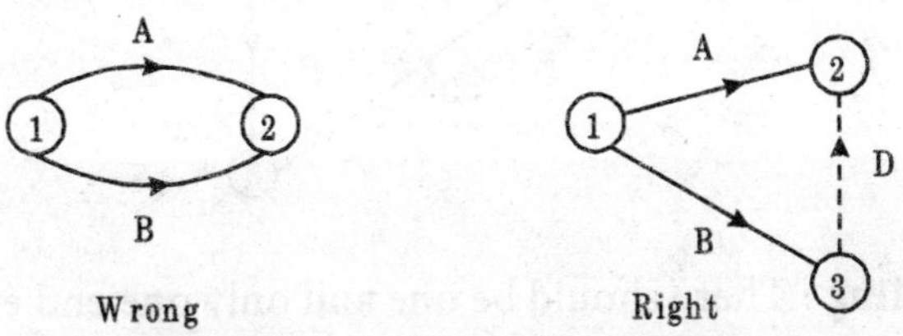


**2. To ensure correct precedence relationship**

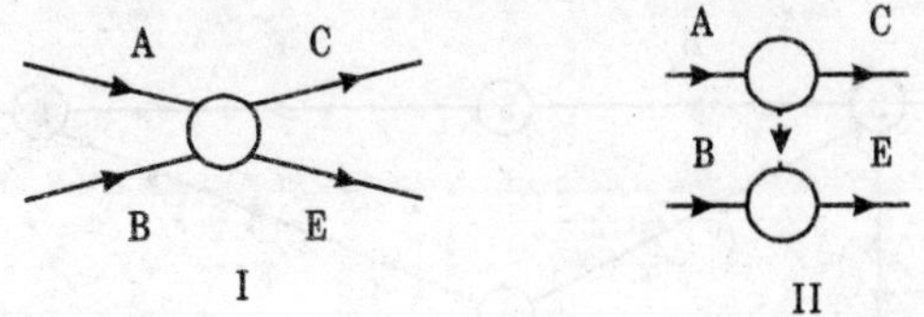


I. shows *A* and *B* both should be completed before *C* or *E* can start.

**II. Depicts following restrictions**

(i) Activity *C* can start immediately after *A* is completed, it has nothing to do with completion of *B*.

(ii) Activity *E* can be started only after *A* and *B* both are completed.

## 9.2 RULES FOR NETWORK CONSTRUCTION

1. Each activity should be represented by one and only one arrow in the network.
2. The length of the arrow does not indicate the time which the activity takes.
3. The direction of the arrow indicates the direction of work flow and is from left to right.
4. Dummy activities should be introduced if absolutely necessary.
5. Path of activities cannot form a loop that returns to any previously accomplished event.
6. Head event should be numbered greater than its tail and numbering should be carried out from left to right.

## 9.3 COMMON ERRORS IN NETWORK CONSTRUCTION

1. **Formation of a loop :** As represented by given figure a closed would produce an endless cycle in computer programmes. Thus a correctly constructed network is non-cyclic.

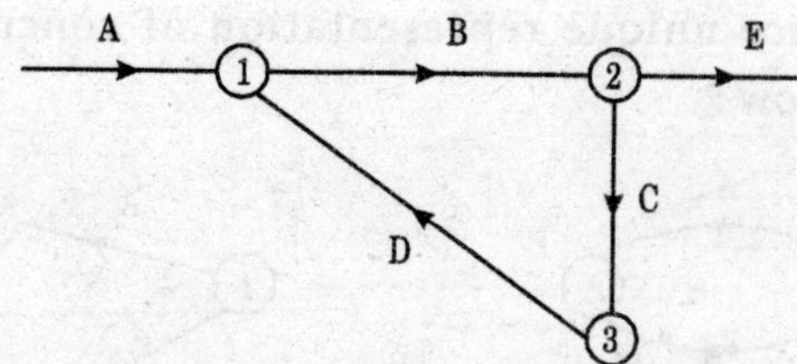


2. **Dangling :** There should be one and only one end event. No activity should end without being joined to the end event.

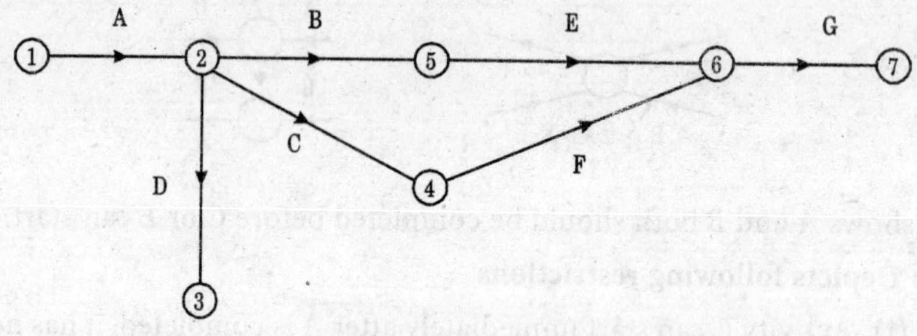


3. **Redundancy :** If a dummuy activity is the only activity emerging from an event, it can be eliminated.

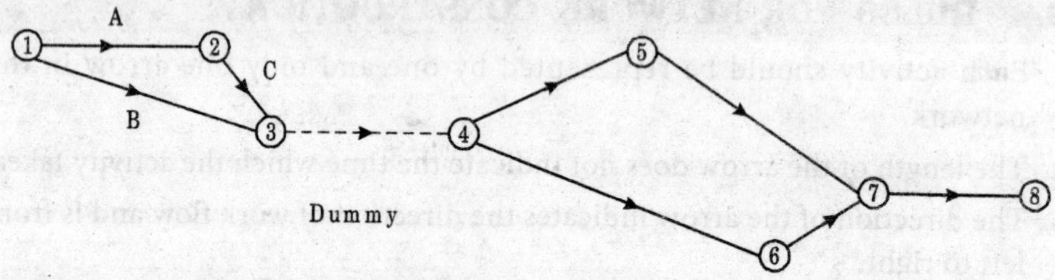


Here dummy activity is redundant, which can be eliminated and network re-drawn by merging events 3 and 4.

**Example 9.1 :** *Draw a network diagram according to given set of activities:*

| *Activity* | : *A* | *B* | *C* | *D* | *E* | *F* | *G* |
|---|---|---|---|---|---|---|---|
| *Predecessor Activity* | : — | *A* | *B* | *A* | *D* | *D* | *C,E* |

***Solution*** Indentify the start activity *A* which has no predecessor activity. Also determine end activities *F* and *G*, which are not predecessors to any activity junction point which is at activities *C* and *E* and draw the network diagram.

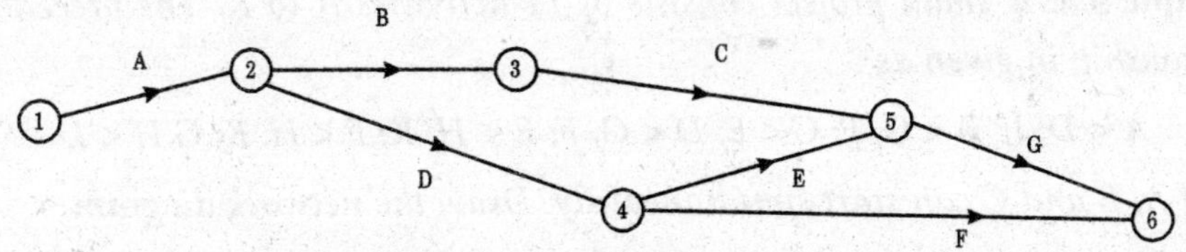


**Example 9.2** *Draw the following network diagram and label it.*

| *Activity* : | *A* | *B* | *C* | *D* | *E* | *F* | *G* | *H* | *I* | *J* | *K* |
|---|---|---|---|---|---|---|---|---|---|---|---|
| *Predecessor Activity* : | – | *A* | – | *C* | *B, D* | *B, D* | *C* | *E* | *F, H* | *G, I, E* | *J* |

**Solution** Identify start activities *A* and *C*

Identify junction at (*B, D*), (*F, H*) and (*G, I, E*)

Determine end activity which is *K*.

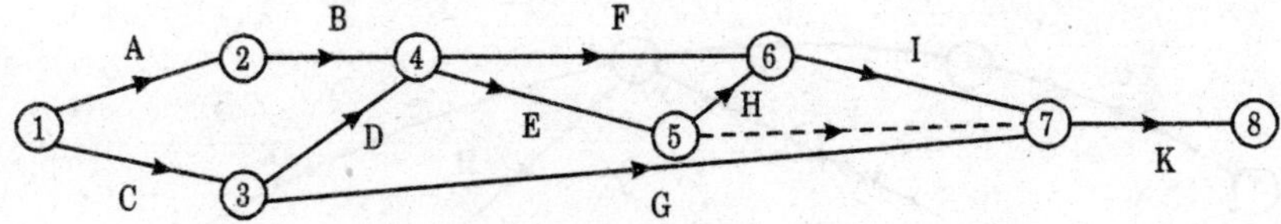


**Example 9.3** *Draw the following network diaram :*

| *Activity* | *A* | *B* | *C* | *D* | *E* | *F* | *G* | *H* | *I* | *J* | *K* | *L* |
|---|---|---|---|---|---|---|---|---|---|---|---|---|
| *Predecessor Activity* | – | – | – | *C* | *D, G* | *B* | *A, F* | *A, F* | *B* | *H, I* | *D, G* | *K,J* |

**Solution** Start activities are *A*, *B* and *C*

Junction are at (*D, G*), (*A, F*), (*H, I*) and (*K, J*)

End activities are *E* and *L*.

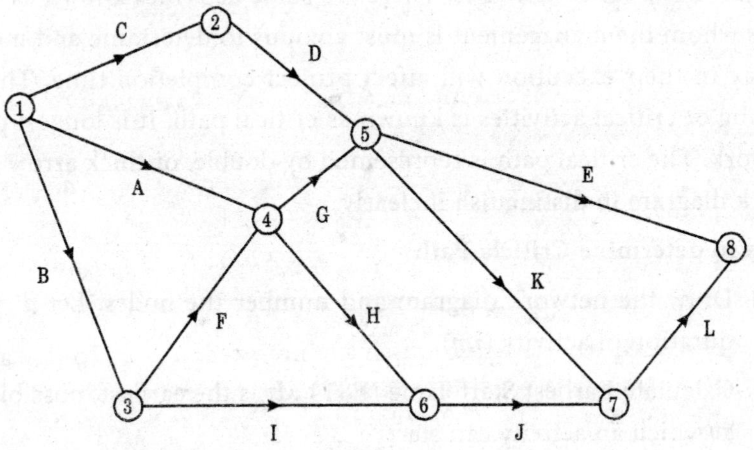

**Example 9.4** *A small project consists of 11 activities A to K. The precedence relationship in given as :*

*A < D, I; B < G, F; C < E; D < G, F; E < H, K; F < H, K; G,H < J.*

*If A, B and C can start simultaneously. Draw the network diagram.*

***Solution*** Rewriting activities in following way :

| Predecessor Activity | A | B | C | D | E | F | G,H |
|---|---|---|---|---|---|---|---|
| Acitivity | D, I | G,F | E | G, F | H, K | H, K | J |

Starting activities are A, B, C

End acitivities are I, J, K

Junction are at (D, I), (G, F), (H, K)

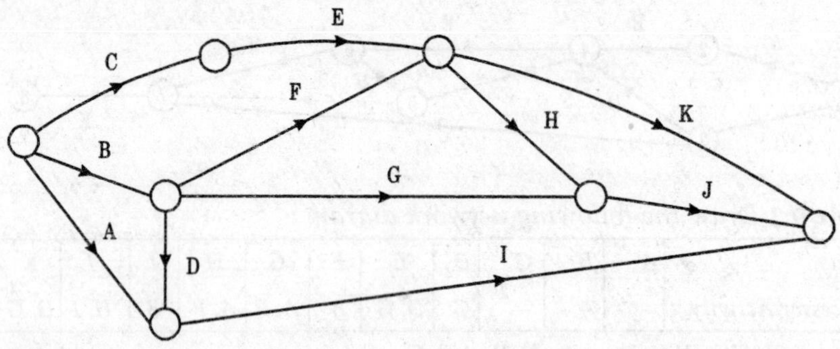


## 9.4 CRITICAL PATH METHOD (C.P.M.)

Any project has a set of activities. There are some activities known as critical activities whom the management is most anxious to determine and monitor. Any delay in their execution will affect project completion time. The path comprising of critical activities is known as critical path. It is longest path in the network. The critical path is represented by double or thick arrow line in a network diagram to distinguish it clearly.

**Steps to determine Criticla Path**

(i) Draw the network diagram and number the nodes. Let $d_{ij}$ be the duration of activity $(i, j)$.

(ii) Calculate Earliest Start Time (EST) : It is the earliest possible time at which an activity can start.

Let $\quad \square_i$ = Earliest start time of event $i$

$$\square_j = \text{Maximum } [\square_i + d_{ij}]$$

Calculation of EST for each event is done in forward pass i.e. from left to right.

(iii) Calculate Latest Finesh Time (LFT) : It is the latest time by which an activity must be completed so that the scheduled date for the completion of the project may not be delayed.

Let $\quad \Delta_j$ = Latest finish time of event $j$

$$\Delta_i = \text{Minimum } [\Delta_j - d_{ij}]$$

Calculation of LFT for each event is done in backward pass, i.e. from end event to starting event. LFT for end event is same as its EST and LFT for start event is zero.

(iv) Calculate Earliest Finish Time (EFT) :

$$\text{EFT} = \text{EST} + d_{ij}$$

(v) Calculate Latest Start Time (LST)

$$\text{LST} = \text{LFT} - d_{ij}$$

(vi) Calculate Total Float (TF) :

$$\text{TF} = \text{LST} - \text{EST} = \text{LFT} - \text{EFT}$$

Total float is the time by which an activity may be delayed or extended without affecting the total project duration.

**Example 9.5** *A project shedule has following characteristics :*

| *Activity* : | *1–2* | 1–4 | 1–7 | 2–3 | 3–6 | 4–5 | 4–8 | 5–6 | 6–9 | 7–8 | 8–9 |
|---|---|---|---|---|---|---|---|---|---|---|---|
| *Duration (Days)* : | 4 | 3 | 2 | 4 | 3 | 4 | 6 | 7 | 5 | 5 | 4 |

*Evaluate EST, EFT, LST, LFT and total float for each activity and also the project duration and critical path.*

***Solution*** Drawing network diagram and calculating EST and LFT as

$$\square_j = [\text{EST}]_{\text{for event } j}$$

$$= \text{Maximum } [\text{EST at event } i + d_{ij}]$$

$$\Delta_i = [\text{LFT}]_{\text{for event } i}$$

$$= \text{Minimum } [\text{LFT}_{\text{at event } j} - d_{ij}]$$

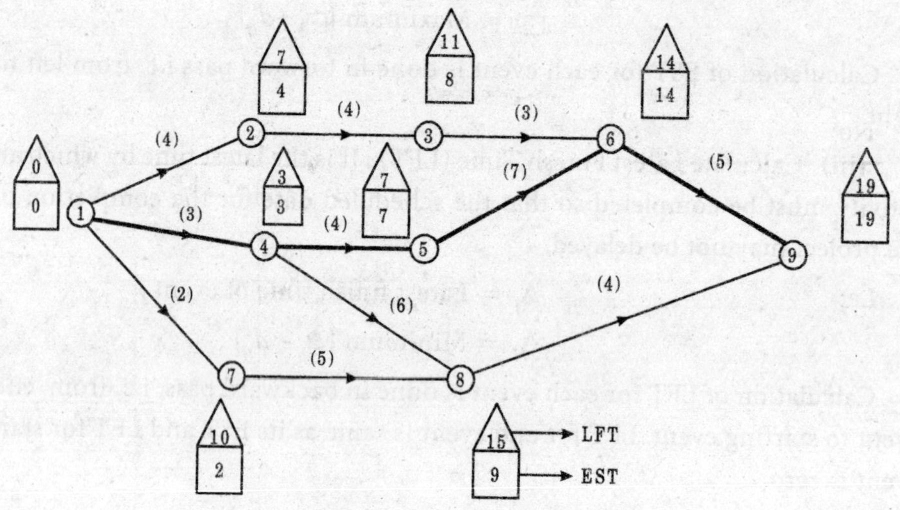


EST calculation are done in forward pass

$$\square_1 = EST]_{\text{at event1}} = 0$$

$$\square_2 = EST]_{\text{at event2}} = 0+4 = 4$$

$$\square_3 = EST]_{\text{at event3}} = 4+4 = 8$$

$$\square_4 = EST]_{\text{at event4}} = 0+3 = 3$$

$$\square_5 = EST]_{\text{at event5}} = 3+4 = 7$$

$$\square_6 = EST]_{\text{at event6}} = \max\{11, 14\} = 14$$

$$\square_7 = EST]_{\text{atevent7}} = 0+2 = 2$$

$$\square_8 = EST]_{\text{at event8}} = \max\{7, 9\} = 9$$

$$\square_9 = EST]_{\text{at event9}} = \max\{19, 13\} = 19$$

LFT calculation are done in backward pass

$$\Delta_9 = [LFT]_{\text{at event9}} = 19$$

$$\Delta_8 = [LFT]_{\text{at event8}} = 19-4 = 15$$

$$\Delta_7 = [LFT]_{\text{at event7}} = 15-5 = 10$$

$$\Delta_6 = [LFT]_{\text{at event6}} = 19-5 = 14$$

$$\Delta_5 = [LFT]_{\text{at event5}} = 14-7 = 7$$

$$\Delta_4 = [LFT]_{\text{at event4}} = \min(3, 9) = 3$$

$$\Delta_3 = [LFT]_{at\ event3} = 14-3 = 11$$
$$\Delta_2 = [LFT]_{at\ event2} = 11-4 = 7$$
$$\Delta_1 = [LFT]_{at\ event1} = \min\{3, 0, 8\} = 0$$

Now to find EFT, LST and total float

| *Activity* (*i–j*) | *Duration* (*dij*) | *EST* | *EFT* $EST+d_{ij}$ | *LFT* | *LST* $LFT-d_{ij}$ | *Total float* *LST–EST or LFT – EFT* |
|---|---|---|---|---|---|---|
| 1-2 | 4 | 0 | 4 | 7 | 3 | 3 |
| 1-4 | 3 | 0 | 3 | 3 | 0 | 0 |
| 1-7 | 2 | 0 | 2 | 10 | 8 | 8 |
| 2-3 | 4 | 4 | 8 | 11 | 7 | 3 |
| 3-6 | 3 | 8 | 11 | 14 | 11 | 3 |
| 4-5 | 4 | 3 | 7 | 7 | 3 | 0 |
| 4-8 | 6 | 3 | 9 | 15 | -9 | 6 |
| 5-6 | 7 | 7 | 14 | 14 | 7 | 0 |
| 6-9 | 5 | 14 | 19 | 19 | 14 | 0 |
| 7-8 | 5 | 2 | 7 | 15 | 10 | 8 |
| 8-9 | 4 | 9 | 13 | 19 | 15 | 6 |

Critical path is the path where total float is zero, i.e. 1–4–5–6–9

Total project duration is 19 days.

**Example 9.6** *A project consists of activities A to K with precedence relationship and duration as given below :*

*A < C, D, I; B < G, F; C < G, F; G < H,K; F, H < J; I, J, K < E.*

| *Activity* | : A | B | C | D | E | F | G | H | I | J | K |
|---|---|---|---|---|---|---|---|---|---|---|---|
| *Duration (Days)* | : 6 | 4 | 9 | 23 | 7 | 5 | 10 | 8 | 9 | 7 | 7 |

*(i) Draw the network diagram*

*(ii) What is minimum project duration?*

*(iii) Find the critical path and critical activities.*

***Solution*** Drawing network diagram and finding EST and LFT for each event :

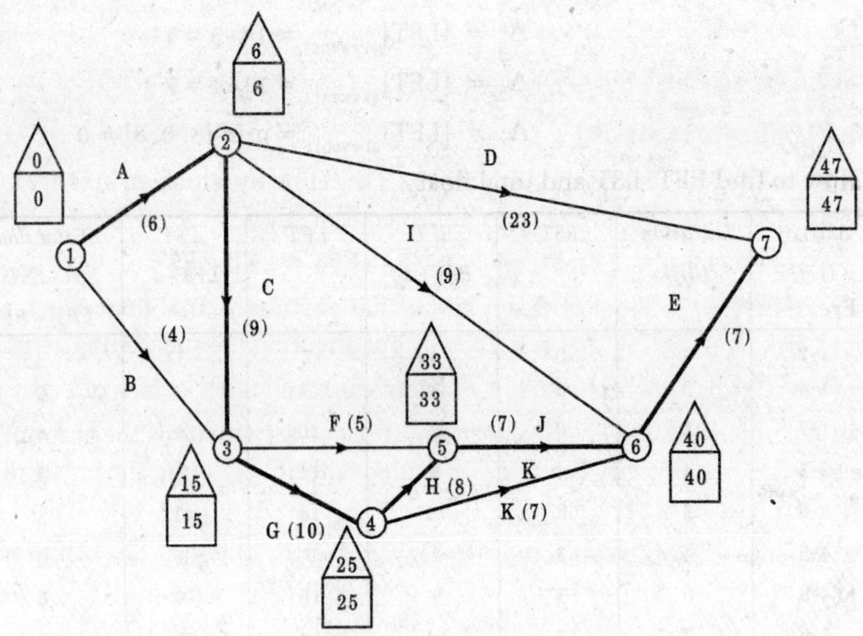


Calculating EST for each event in forward pass and LFT in backward pass.

| | |
|---|---|
| $\square_1 = 0$ | $\Delta_7 = 47$ |
| $\square_2 = 0+6 = 6$ | $\Delta_6 = 47 - 7 = 40$ |
| $\square_3 = \max\{15, 4\} = 15$ | $\Delta_5 = 40 - 7 = 33$ |
| $\square_4 = 15+10 = 25$ | $\Delta_4 = \min\{25, 33\} = 25$ |
| $\square_5 = \max\{20, 33\} = 33$ | $\Delta_3 = \min\{28, 15\} = 15$ |
| $\square_6 = \max\{40, 32\} = 40$ | $\Delta_2 = \min\{24, 31, 6\} = 6$ |
| $\square_7 = \max\{29, 47\} = 47$ | $\Delta_1 = \min\{0, 11\} = 0$ |

| *Activity* | *Duration* ($d_{ij}$) | *EST* | *EFT* | *LFT* | *LST* | *Total float* |
|---|---|---|---|---|---|---|
| A (1–2) | 6 | 0 | 6 | 6 | 0 | 0 |
| B (1–3) | 4 | 0 | 4 | 15 | 11 | 11 |
| C (2 – 3) | 9 | 6 | 15 | 15 | 6 | 0 |
| D (2–7) | 23 | 6 | 29 | 47 | 24 | 18 |
| E (6–7) | 7 | 40 | 47 | 47 | 40 | 0 |
| F (3–5) | 5 | 15 | 20 | 33 | 28 | 13 |
| G (3–4) | 10 | 15 | 25 | 25 | 15 | 0 |
| H (4–5) | 8 | 25 | 33 | 33 | 25 | 0 |
| I (2–6) | 9 | 6 | 15 | 40 | 31 | 25 |
| J (5–6) | 7 | 33 | 40 | 40 | 33 | 0 |
| K (4–6) | 7 | 25 | 32 | 40 | 33 | 8 |

Critical path is 1–2–3–4–5–6–7

Minimum project duration is 47 days.

### 9.4.1 Floats (Slack)

**Total Float (TF) :** Total float of an activity is the time by which an activity may be delayed without affecting the total project duration.

$$\text{Total float} = \text{LFT} - \text{EFT} = \text{LST} - \text{EST}$$

**Free Float (FF) :** Float of an activity is that portion of total float which can be used for rescheduling that activity without affecting the succeeding activity.

$$\text{Free float of activity } [i\text{–}j] = [\text{Total float of } (i\text{–}j)] - \text{Slack of event } j$$

**Independent Float (IF) :** Independent float of an activity is the amount of time by which the activity can be rescheduled without affecting the preceeding or succeeding activity.

Independent Float of an activity $(i\text{–}j)$ = (Free float of $[i\text{-}j]$) - Slack of event $i$

**Note :** (i) The negative value of independent float is considered to be zero.

(ii) $\text{IF} \leq \text{FF} \leq \text{TF}$

**Example 9.7** *Calculate the total float, free float and independent float for the project whose activities are given below :*

| *Activity* | *1–2* | *1–3* | *2–3* | *2–4* | *3–4* | *3–5* | *4–5* | *4–6* | *5–6* |
|---|---|---|---|---|---|---|---|---|---|
| *Duration (Days)* | *5* | *6* | *4* | *9* | *10* | *7* | *5* | *4* | *8* |

***Solution*** Drawing network diagram and finding EST and LFT for each event.

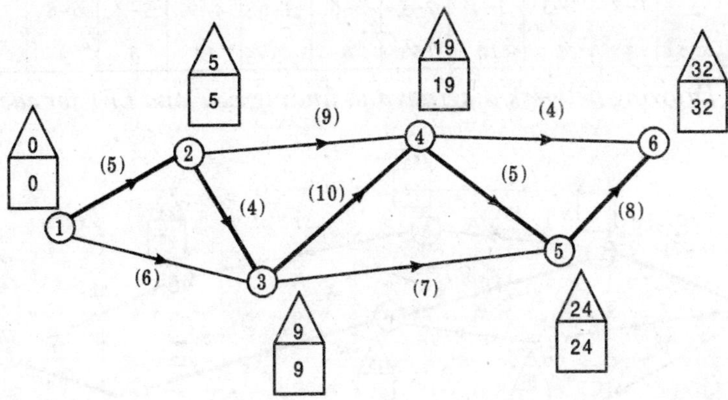


Calculating EST for each event in forward pass and LFT in backwark pass.

$\square_1 = 0$ $\Delta_6 = 32$

$\square_2 = 0+5 = 5$ $\Delta_5 = 32\text{-}8 = 24$

$\square_3$ = Max {6, 9} = 9 $\Delta_4$ = Min {28, 19} = 19

$\square_4$ = Max {14, 19} = 19 $\Delta_3$ = Min {9, 17} = 9

$\square_5$ = Max {24, 16} = 24 $\Delta_2$ = Min {10, 5} = 5

$\square_6$ = Max {23, 32} = 32 $\Delta_1$ = Min {0, 3} = 0

| *Activity (i–j)* | *Duration dij* | *EST* | *EFT* | *LFT* | *LST* | *TF* | *FF* | *IF* |
|---|---|---|---|---|---|---|---|---|
| 1–2 | 5 | 0 | 5 | 5 | 0 | 0 | 0 | 0 |
| 1–3 | 6 | 0 | 6 | 9 | 3 | 3 | 3 | 3 |
| 2–3 | 4 | 5 | 9 | 9 | 5 | 0 | 0 | 0 |
| 2–4 | 9 | 5 | 14 | 19 | 10 | 5 | 5 | 5 |
| 3–4 | 10 | 9 | 19 | 19 | 9 | 0 | 0 | 0 |
| 3–5 | 7 | 9 | 16 | 24 | 17 | 8 | 8 | 8 |
| 4–5 | 5 | 19 | 24 | 24 | 19 | 0 | 0 | 0 |
| 4–6 | 4 | 19 | 23 | 32 | 28 | 9 | 9 | 9 |
| 5–6 | 8 | 24 | 32 | 32 | 24 | 0 | 0 | 0 |

Critical path is 1–2–3–4–5–6

All the floats are equal because all events are on critical path and hence their slack is zero.

**Example 9.8** *Calculate total float, free float and independent float for the activities given below for completion of a project. Also find minimum project duration and critical path.*

| *Activity* | *1–2* | *1–3* | *1–4* | *2–5* | *2–6* | *3–6* | *4–9* | *5–7* | *6–8* | *7–9* | *8–9* |
|---|---|---|---|---|---|---|---|---|---|---|---|
| *Duration (Days)* | *7* | *5* | *10* | *5* | *8* | *5* | *12* | *8* | *9* | *5* | *4* |

**Solution** *Drawing network diagram and finding EST and LFT for each event.*

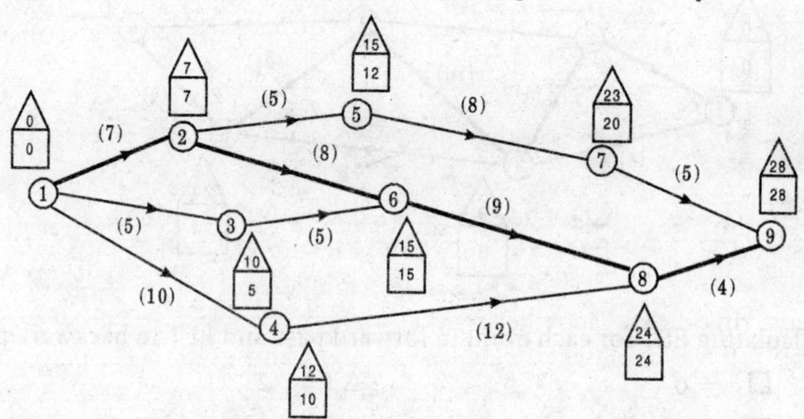

Calculation of EST in forward pass :

$\square_1 = 0$ $\square_6 = \text{Max}\{15, 10\} = 15$

$\square_2 = 0+7 = 7$ $\square_7 = 12+8 = 20$

$\square_3 = 0+5 = 5$ $\square_8 = \text{Max}\{24, 22\} = 24$

$\square_4 = 0+10 = 10$ $\square_9 = \text{Max}\{25, 28\} = 28$

$\square_5 = 7+5 = 12$

Calculation of LFT in backward pass:

$\Delta_9 = 28$ $\Delta_4 = 24-12 = 12$

$\Delta_8 = 28-4 = 24$ $\Delta_3 = 15-5 = 10$

$\Delta_7 = 28-5 = 23$ $\Delta_2 = \text{Min}\{10, 7\} = 7$

$\Delta_6 = 24-9 = 15$ $\Delta_1 = \text{Min}\{0, 5, 2\} = 0$

$\Delta_5 = 23-8 = 15$

| *Activity* | *Duration* | *EST* | *EFT* | *LFT* | *LST* | *TF* | *FF* | *IF* |
|---|---|---|---|---|---|---|---|---|
| 1–2 | 7 | 0 | 7 | 7 | 0 | 0 | 0 | 0 |
| 1–3 | 5 | 0 | 5 | 10 | 5 | 5 | 0 | 0 |
| 1–4 | 10 | 0 | 10 | 12 | 2 | 2 | 0 | 0 |
| 2–5 | 5 | 7 | 12 | 15 | 10 | 3 | 0 | 0 |
| 2–6 | 8 | 7 | 15 | 15 | 7 | 0 | 0 | 0 |
| 3–6 | 5 | 5 | 10 | 15 | 10 | 5 | 5 | 0 |
| 4–9 | 12 | 10 | 22 | 28 | 16 | 6 | 6 | 4 |
| 5–7 | 8 | 12 | 20 | 23 | 15 | 3 | 0 | 0 |
| 6–8 | 9 | 15 | 24 | 24 | 15 | 0 | 0 | 0 |
| 7–9 | 5 | 20 | 25 | 28 | 23 | 3 | 3 | 0 |
| 8–9 | 4 | 24 | 28 | 28 | 24 | 0 | 0 | 0 |

Minimum project duration is 28 days.

Critical path is 1–2–6–8–9.

## 9.5 PERT (PROJECT EVALUATION AND REVIEW TECHNIQUE)

PERT is a technique used for planning, scheduling and controlling, the projects whose activities are subject to uncertainity in the performance time. It is a probablistic model whereas CPM is a deterministic model. PERT gives a way of indentifying critical activities right at planning stage, so that management is aware of these potentially troublesome areas and exercise proper control over them, so as to complete the project within the scheduled time. It can be employed in projects where activities cannot be easily defined in terms of time or resources required due to lack of past data and uncertanities involved in working. To account for there uncertainities, PERT uses 3 time estimates for each activity.

(i) **Optimistic Time ($t_o$) :** It is the shortest time in which an activity can be completed assuming that everything goes exceptionally well. It has low probability of occurrence.

(ii) **Most likely time ($t_m$) :** It is the most likely time required to complete an activity taking into consideration all favourable and unfavourable circumstances.

(iii) **Pessimistic time ($t_p$) :** The completion time of an activity under adverse circumstances is called pessimistic time.

**PERT procedure :**

(i) Draw the network diagram.

(ii) Compute expected duration of each activity

$$t_e = \left(\frac{t_0 + 4t_m + t_p}{6}\right)$$

(iii) Determine the critical path and identify critical activities.

(iv) Compute the expected variance $\sigma^2 = \left(\frac{t_p - t_o}{6}\right)^2$

(v) Compute variance of the critical path $\sigma_c^2$ which is the sum of variances of all critical activities.

(vi) Calculate expected standard deviation of project length $\sigma_c$ and calculate the standard normal.

Variate $$Z = \frac{t_s - t_e}{\sigma_c}$$

where $t_s$ = Scheduled time

$t_e$ = Expected time

$\sigma_c$ = S.D. of the project

(vii) Using Z, one can estimate the probability of completing the project within a specified time by using the normal curve (area) tables.

**Example 9.9** *A project has following activities and other characteristics :*

| *Activity* | *1–2* | *1–3* | *2–4* | *2–5* | *2–7* | *3–7* | *4–6* | *5–6* | *6–8* | *7–8* |
|---|---|---|---|---|---|---|---|---|---|---|
| *$T_o$ (Days)* | *3* | *2* | *1* | *3* | *5* | *3* | *3* | *4* | *2* | *4* |
| *$T_m$ (Days)* | *4* | *5* | *4* | *6* | *11* | *6* | *5* | *8* | *4* | *7* |
| *$T_p$ (Days)* | *5* | *8* | *7* | *9* | *17* | *15* | *13* | *12* | *6* | *16* |

*(a) Draw the network diagram.*

*(b) Identify the critical path.*

*(c) Determine expected project duration.*

*(d) Find the probability that project is completed in 30 days.*

*(e) Probability that project is completed in 18 days.*

*(f) What project duration will have 98% confidence of completion?*

**Solution**

| *Activity* | $T_o$ | $T_m$ | $T_p$ | $T_e = \frac{T_o + 4T_m + T_p}{6}$ | $\sigma^2 = \left(\frac{T_p - T_o}{6}\right)^2$ |
|---|---|---|---|---|---|
| 1–2 | 3 | 4 | 5 | $\frac{3+16+5}{6} = 4$ | $\left(\frac{5-3}{6}\right)^2 = 0.11$ |
| 1–3 | 2 | 5 | 8 | $\frac{2+20+8}{6} = 5$ | $\left(\frac{8-2}{6}\right)^2 = 1$ |
| 2-4 | 1 | 4 | 7 | $\frac{1+16+7}{6} = 4$ | $\left(\frac{7-1}{6}\right)^2 = 1$ |
| 2–5 | 3 | 6 | 9 | $\frac{3+24+9}{6} = 6$ | $\left(\frac{9-3}{6}\right)^2 = 1$ |
| 2–7 | 5 | 11 | 17 | $\frac{5+44+17}{6} = 11$ | $\left(\frac{17-5}{6}\right)^2 = 4$ |
| 3–7 | 3 | 6 | 15 | $\frac{3+24+15}{6} = 7$ | $\left(\frac{15-3}{6}\right)^2 = 4$ |
| 4–6 | 3 | 5 | 13 | $\frac{3+20+13}{6} = 6$ | $\left(\frac{13-3}{6}\right)^2 = 2.78$ |
| 5–6 | 4 | 8 | 12 | $\frac{4+32+12}{6} = 8$ | $\left(\frac{12-4}{6}\right)^2 = 1.78$ |
| 6–8 | 2 | 4 | 6 | $\frac{2+16+6}{6} = 4$ | $\left(\frac{6-2}{6}\right)^2 = 0.44$ |
| 7–8 | 4 | 7 | 16 | $\frac{4+28+16}{6} = 8$ | $\left(\frac{16-4}{6}\right)^2 = 4$ |

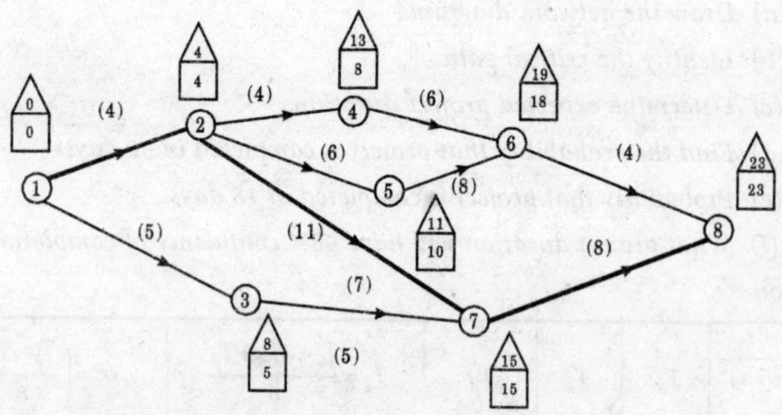


Calculation of EST in forward pass;

| | |
|---|---|
| $\square_1 = 0$ | $\square_5 = 4+6 = 10$ |
| $\square_2 = 0+4 = 4$ | $\square_6 = \text{Max}\ \{14, 18\} = 18$ |
| $\square_3 = 0+5 = 5$ | $\square_7 = \text{Max}\ \{15, 12\} = 15$ |
| $\square_4 = 4+4 = 8$ | $\square_8 = \text{Max}\ \{22, 23\} = 23$ |

Calculation LFT in backward pass ;

| | |
|---|---|
| $\Delta_8 = 23$ | $\Delta_4 = 19-6 = 13$ |
| $\Delta_7 = 23-8 = 15$ | $\Delta_3 = 15-7 = 8$ |
| $\Delta_6 = 23-4 = 19$ | $\Delta_2 = \text{Min}\ \{9, 5, 4\} = 4$ |
| $\Delta_5 = 19-8 = 11$ | $\Delta_1 = \text{Min}\ \{0, 3\} = 0$ |

(b) Critical path : 1–2–7–8

$$\sigma^2_c = 0.11 + 4 + 4 = 8.11$$

$$\sigma_c = 2.85 \text{ approx.}$$

(c) Expected project durection is 23 days.

(d) To find the probability that project is completed in 30 days :

$$t_s = 30\ ,\ t_e = 23,\ \sigma_c = 2.85$$

$$Z = \frac{t_s - t_e}{\sigma_c}$$

$$= \frac{30-23}{2.85} = 2.46 \text{ approx.}$$

$$P(Z \leq 2.46) = 0.5 + 0.4930$$

$$= 0.9930$$

$$= 99.3\%$$

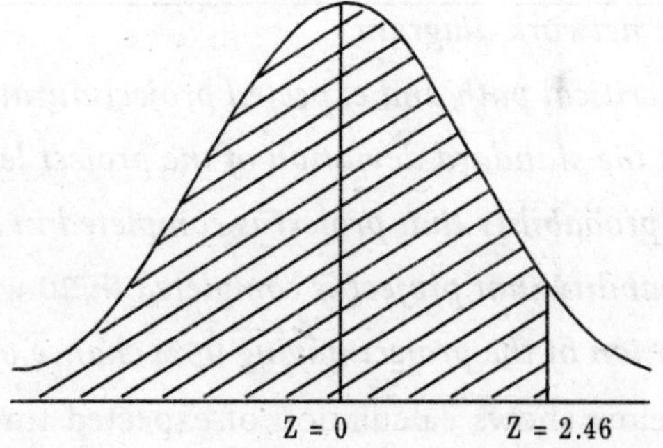


(e) To find the probability that project is completed in 18 days :

$$t_s = 18, t_e = 23, \sigma_c = 2.85$$

$$Z = \frac{t_s - t_e}{\sigma c}$$

$$= \frac{18-23}{2.85} = -1.75 \text{ apprrox}$$

$$P(Z \leq -1.75) = P(Z \geq 1.75)$$

$$= 0.5 - 0.4599$$

$$= 0.0401$$

$$= 4.01\%$$

Z = −1.75 Z = 0 Z = 1.75

Therefore there is 4.01% chance that project is completed in 18 days.

**Example 9.10** *Following table shows the activities of a network along with their time estimates (in weeks) :*

| *Activity* | *1–2* | *1–6* | *2–3* | *3–4* | *3–5* | *4–5* | *5–7* | *6–7* |
|---|---|---|---|---|---|---|---|---|
| $T_o$ | 2 | 3 | 2 | 1 | 2 | 3 | 2 | 4 |
| $T_m$ | 5 | 9 | 5 | 4 | 5 | 5 | 4 | 5 |
| $T_p$ | 14 | 27 | 8 | 7 | 14 | 7 | 6 | 12 |

*(a) Draw the network diagram*

*(b) Find the critical path and expected project duration.*

*(c) Compute the standard deviation of the project length.*

*(d) Find the probability that project is completed in 25 weeks.*

*(e) The probability that project is completed in 20 weeks.*

*(f) The duration of the project having 95% chance of being completed.*

***Solution*** Table below shows calculation of expected time and variance for each activities :

| Activity | $T_o$ | $T_m$ | $T_p$ | $T_e = \dfrac{T_o + 4T_m + T_p}{6}$ | $\sigma^2 = \left(\dfrac{T_p - T_o}{6}\right)^2$ |
|---|---|---|---|---|---|
| 1–2 | 2 | 5 | 14 | 6 | $\left(\dfrac{14-2}{6}\right)^2 = 4$ |
| 1–6 | 3 | 9 | 27 | 11 | $\left(\dfrac{27-3}{6}\right)^2 = 16$ |
| 2–3 | 2 | 5 | 8 | 5 | $\left(\dfrac{8-2}{6}\right)^2 = 1$ |
| 3–4 | 1 | 4 | 7 | 4 | $\left(\dfrac{7-1}{6}\right)^2 = 1$ |
| 3–5 | 2 | 5 | 14 | 6 | $\left(\dfrac{14-2}{6}\right)^2 = 4$ |
| 4–5 | 3 | 5 | 7 | 5 | $\left(\dfrac{7-3}{6}\right)^2 = 0.44$ |
| 5–7 | 2 | 4 | 6 | 4 | $\left(\dfrac{6-2}{6}\right)^2 = 0.44$ |
| 6–7 | 4 | 5 | 12 | 6 | $\left(\dfrac{12-4}{6}\right)^2 = 1.78$ |

(a) Figure shows calculation of EST and LFT for each event.

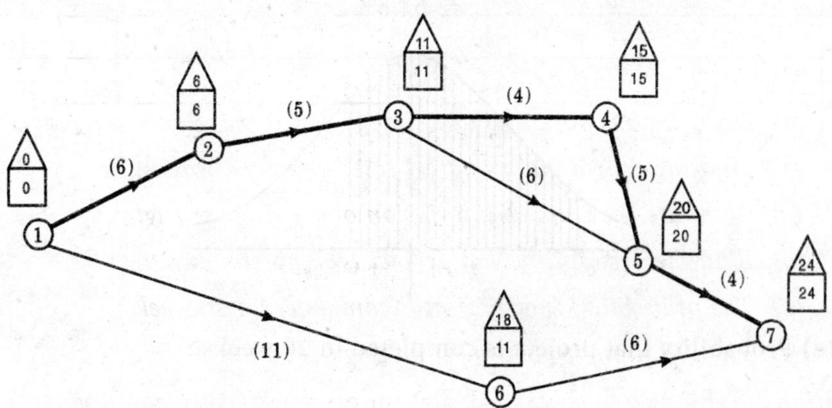


Calculation of EST in forward pass and LFT in backward pass,

| | |
|---|---|
| $\square_1 = 0$ | $\Delta_7 = 24$ |
| $\square_2 = 0 + 6 = 6$ | $\Delta_6 = 24 - 6 = 18$ |
| $\square_3 = 6 + 5 = 11$ | $\Delta_5 = 24 - 4 = 20$ |
| $\square_4 = 11 + 4 = 15$ | $\Delta_4 = 20 - 5 = 15$ |
| $\square_5 = \text{Max}\{20, 17\} = 20$ | $\Delta_3 = \text{Min}\{11, 14\} = 11$ |
| $\square_6 = 0 + 11 = 11$ | $\Delta_2 = 11-5 = 6$ |
| $\square_7 = \text{Max}\{27, 17\} = 27$ | $\Delta_1 = \text{Min}\{0, 7\} = 0$ |

(b) Critical path is 1–2–3–4–5–7

Expected project duration is 24 weeks.

(c) Variance of the project length is :

$$\sigma^2_c = 4+1+1+0.44+0.44 = 6.88$$

Standard deviation of project length

$$\sigma_c = \sqrt{6.88} = 2.62$$

(d) To find the probability that project is completed in 25 weeks

$$t_s = 25,\ t_e = 24,\ \sigma_c = 2.62$$

$$Z = \frac{t_S - t_e}{\sigma_c} = \frac{25-24}{2.62} = \frac{1}{2.62} = 0.38$$

$$P(Z \leq 0.38) = 0.5+0.1480$$

$$= 0.6480$$
$$= 64.8\%$$

$Z = 0$ $\quad Z = 0.38$

(e) Probability that project is completed in 20 weeks :

$$t_s = 20,\ t_e = 24,\ s_c = 2.62$$

$$Z = \frac{t_S - t_e}{\sigma_c} = \frac{20-24}{2.62} = -1.53$$

$$P(Z \leq -1.53) = P(Z \geq 1.53)$$
$$= 0.5 - P(Z \leq 1.53)$$
$$= 0.5 - 00.4370$$
$$= 0.063 = 6.3\%$$

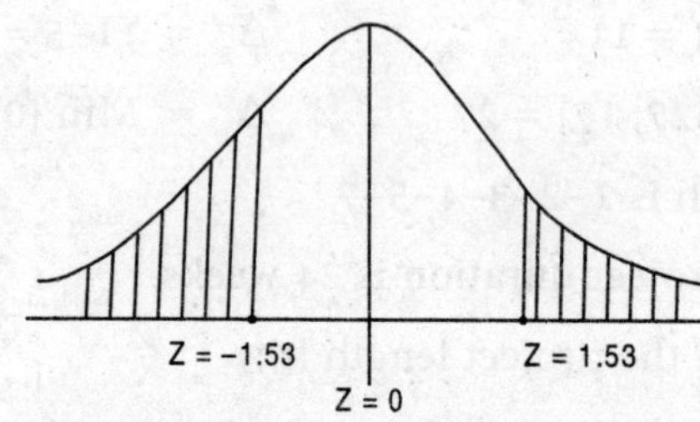


(f) Given that :

$$P\left(Z \leq \frac{T_s - T_e}{\sigma_c}\right) = 0.95$$

Seeing $Z$ value at 0.95 probability

$$\Rightarrow \qquad 1.64 = \frac{T_s - 24}{2.62}$$

$$\Rightarrow \qquad T_s = (1.64)(2.62) + 24 = 28.3 \text{ weeks approx.}$$

## 9.6 CRASHING OF NETOWRK (TIME-COST RELATIONSHIP)

Project crashing is a process fo shortening a project duration by reducing time of one or more activities on the critical path by adding extra resources to concerned activities.

$$\text{Project cost} = \text{Direct cost} + \text{Indirect cost}$$

**Direct cost :** The expenditure which is directly allocated to different activities in a project like labour, material etc. is called direct cost.

**Indirect cost :** It includes overheads, depreciation, insurance, supervisory cost etc. Indirect costs are directly related to project duration, longer the project takes to complete, the higher are the indirect costs.

**Optimum project duration :** It is the length of the project which corresponds to minimum project cost.

**Steps Involved in Project Crashing**

**Step 1 :** Calculate normal project duration and crashed project duration by drawing network diagrams.

**Step 2 :** compute cost slope (Cost of crashing per day) for each activity.

$$\text{Cost slope} = \frac{\text{Crash cost - Normal cost}}{\text{Normal time - Crash time}}$$

**Step 3 :** Crash the activities on the critical path which have minimum cost slope by maximum time at which project duration is actually reduced.

**Step 4 :** Compute new project cost

$$\text{Total cost} = \text{New direct cost} + \text{Indirect cost}$$

**Step 5 :** Crash the activities on critical path (New) untill minimum (Crashed) project duration is reached.

**Step 6 :** Identify the optimum project duration against the minimum project cost.

**Example 9.11** *A small marketing project consists of jobs as shown below, with each job, normal and crash time estimates are listed :*

| *Job* | *Normal time (Days)* | *Normal cost cost (Rs.)* | *Crash Time (Days)* | *Crash cost (Rs.)* |
|---|---|---|---|---|
| 1–2 | 8 | 200 | 6 | 300 |
| 1–3 | 4 | 150 | 2 | 350 |
| 2–4 | 2 | 50 | 1 | 100 |

*Contd.*

*Contd.*

| *Job* | *Normal time (Days)* | *Normal cost cost (Rs.)* | *Crash Time (Days)* | *Crash cost (Rs.)* |
|---|---|---|---|---|
| 2–5 | 10 | 200 | 5 | 500 |
| 3–4 | 5 | 200 | 1 | 300 |
| 4–5 | 3 | 60 | 1 | 100 |

(a) *If the indirect cost in Rs. 100 per day, find the last cost shedule.*

(b) *What is the minimum duration of the project?*

***Solution*** Figure below A and B show the network diagrams for normal duration and minimum (Crash duration) respectively.

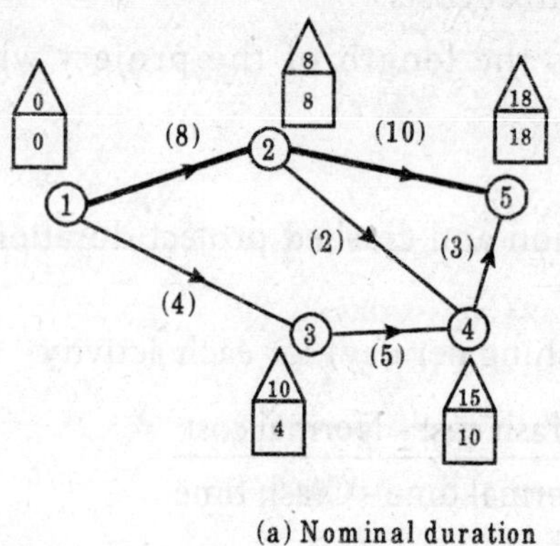


(a) Nominal duration

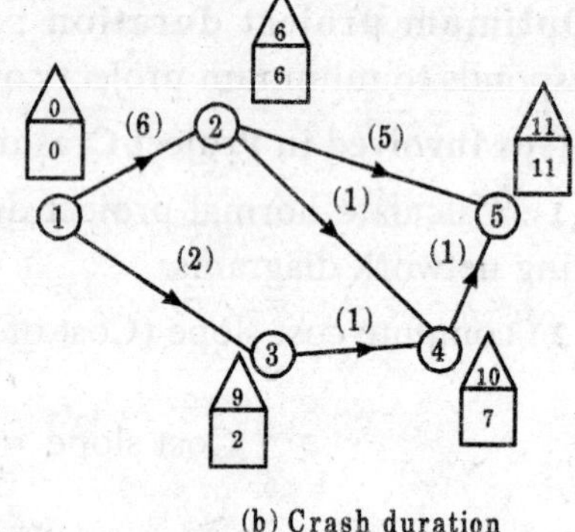


(b) Crash duration

Normal project duration = 18 days | Minimum project duration = 11 days

Critical path = 1–2–5

Following Table shows calculation of cost slope (Cost of crashing each activity per day) :

| *Activity* | *N.T. (Days)* | *N.C. (₹)* | *C.T. (Days)* | *C.C.* | *N.T.=C.T. (Δt)* | *Cost slope = (C.C.–N.C.)/(N.T.–C.T.)* |
|---|---|---|---|---|---|---|
| 1–2 | 8 | 200 | 6 | 300 | 2 | (300–200)/2 = 50 |
| 1–3 | 4 | 150 | 2 | 350 | 2 | (350–150)/2 = 100 |
| 2–4 | 2 | 50 | 1 | 100 | 1 | (100–50)/1 = 50 |
| 2–5 | 10 | 200 | 5 | 500 | 5 | (500–200)/5 = 60 |
| 3–4 | 5 | 200 | 1 | 300 | 4 | (300–200)/4 = 25 |
| 4–5 | 3 | 60 | 1 | 100 | 2 | (100–60)/2 = 20 |
| | | ₹ 860 | | | | |

Crash the activities on the critical path whose cost slope in less than indirect cost per day.

Taking minimum cost slope of activities, on critical path :

$$\text{Min }\{50, 60\} = 50$$

So crashing 1–2 by 2 days (which is the maximum time by which it can be crashed) as shown in the following figure.

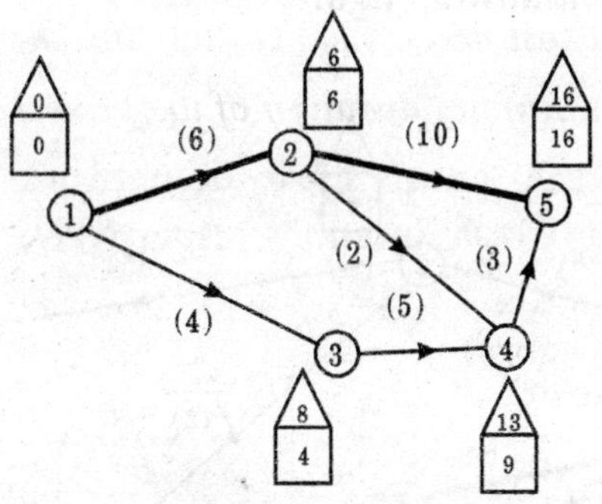


Direct cost = Normal cost + (Cost of crashing per day)×(No of days)
= ₹ 860 + ₹ 100 = ₹ 960

Indirect cost = (Indirect cost per day) × (No. of days)
= 100 × 16 = Rs. 1600

Total cost = ₹ 2560

Again the critical path is 1–2–5

So crashing activity 2–5 by 4 days as crashing by 5 days won't reduce project duration by 5 days, i.e. even if we crash the activity 2–5 by 5 days total project duration would still be reduce by 4 days.

Figure given below shows project duration after crashing activity 2–5 by 4 days.

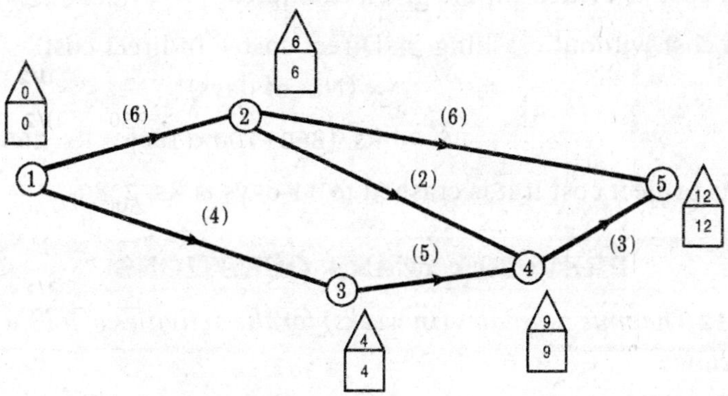

$$\text{Direct cost} = ₹\,[960 + 60\,(4)] = ₹\,1200$$

$$\text{Indirect cost} = (100 \times 12) = ₹\,1200$$

$$\text{Total cost} = ₹\,2400$$

Now all the activities are on critical path.

Crashing parallel activities 2–5 and 4–5 by 1 day each, the new network diagram is given in the following figure.

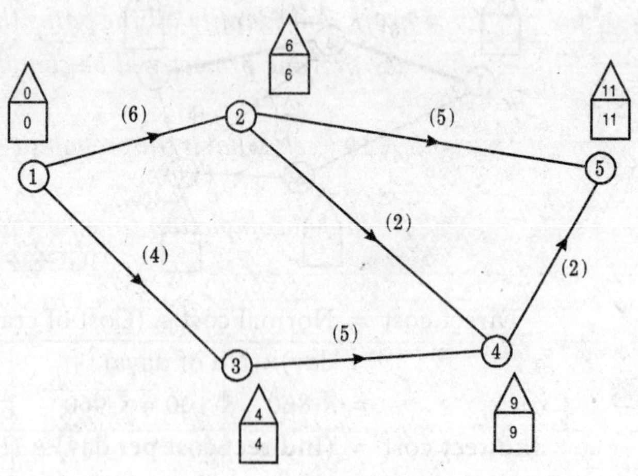


$$\text{Direct cost} = \text{Rs. } [1200 + 60(1) + 20\,(1)] = \text{Rs. } 1280$$

$$\text{Indirect cost} = \text{Rs } (100 \times 11) = 1100$$

$$\text{Total cost} = \text{Rs. } 2380$$

∴ Optimal cost is Rs. 2380 having an optimal project duration of 11 days.

**Note :** Clearly project crashing not only reduces time duration but also cuts project cost as evident in the given example.

$$\text{Project cost without crashing} = \text{Direct cost} + \text{Indirect cost} \times (\text{No. of days})$$

$$= \text{Rs. } [860 + 100 \times (18)] = \text{Rs. } 2660$$

where as project cost if it is crashed to 11 days is Rs. 2380.

## PREVIOUS YEARS QUESTIONS

**Example 9.12** *The time estimates (in weeks) for the activities of a PERT network are given below :*

| *Activity* | $t_o$ | $t_m$ | $t_p$ |
|---|---|---|---|
| *1–2* | *1* | *1* | *7* |
| *1–3* | *1* | *4* | *7* |
| *1–4* | *2* | *2* | *8* |
| *2–5* | *1* | *1* | *1* |
| *3–5* | *2* | *5* | *14* |
| *4–6* | *2* | *5* | *8* |
| *5–6* | *3* | *6* | *15* |

*(a) Draw the project network and identify all the paths through it.*

*(b) What is the probability that the project will be completed at least 4 weeks earlier than expected time.*

*(c) If the project due date is 19 week, what is the probability of not meeting the due date?*

*(d) What should be the scheduled completion time for the probability of completion to be 90%?* **[GGSIPU 2010, 2011]**

***Solution***

| *Activity* | $t_o$ | $t_m$ | $t_p$ | $t_e = \frac{t_o+4t_m+t_p}{6}$ | $\sigma^2 = \left(\frac{t_p-t_o}{6}\right)^2$ |
|---|---|---|---|---|---|
| 1–2 | 1 | 1 | 7 | 2 | 1 |
| 1–3 | 1 | 4 | 7 | 4 | 1 |
| 1–4 | 2 | 2 | 8 | 3 | 1 |
| 2–5 | 1 | 1 | 1 | 1 | 0 |
| 3–5 | 2 | 5 | 14 | 6 | 4 |
| 4–6 | 2 | 5 | 8 | 5 | 1 |
| 5–6 | 3 | 6 | 15 | 7 | 4 |

(a)

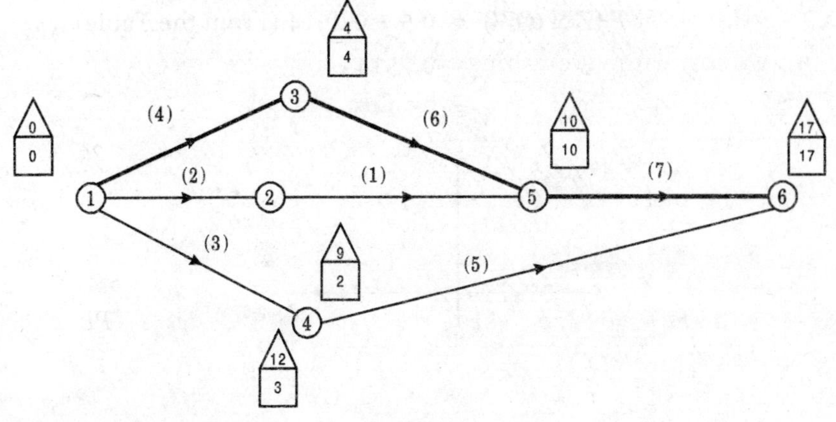

Critical path is 1→3→5→6

(Variance) $\sigma_c^2 = 1+4+4$

$\sigma_c = 3$

Expected project duration is 17 weeks

(b) Probability that the project will be completed at least 4 weeks earlier than expected time is given by

$$Z = \frac{t_S - t_e}{\sigma_C}$$

$$= \left(\frac{\text{Due date} - \text{Expected date of completion}}{\sqrt{\text{Variance}}}\right)$$

$$Z = \frac{13-17}{3} = -\frac{4}{3} = -1.33$$

$\therefore$ $P(Z \leq -1.33) = 0.5 - 0.4082$ (from the Table)

$= 0.0918$

$= 9.18\%$

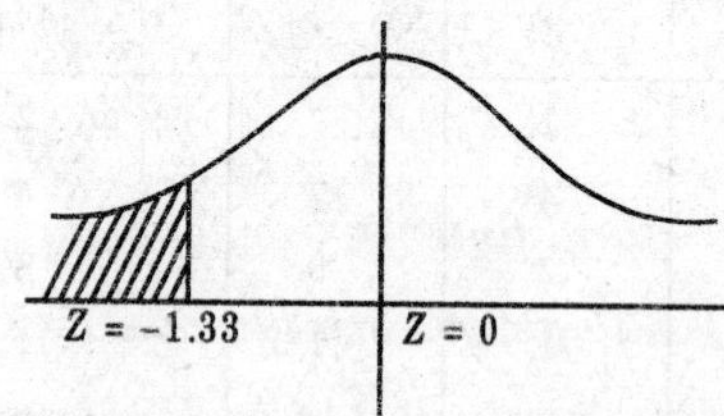


(c) Probability when due date is 19 weeks

$$Z = \frac{t_S - t_e}{\sigma_C} = \frac{19-17}{3} = \frac{2}{3} = 0.666$$

$\therefore$ $P(Z \leq 0.67) = 0.5 + 0.2514$ (From the Table)

$= 0.7514$

$= 75.14\%$

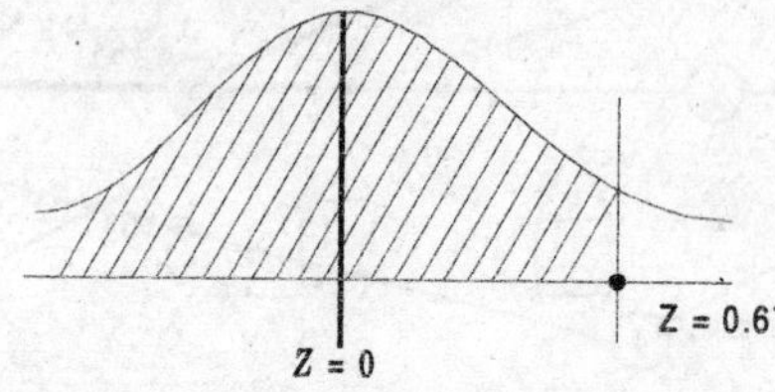

(d) Given that :

$$P\left(Z \leq \frac{t_S - t_e}{\sigma_C}\right) = 0.90$$

Now from the Table $Z$ value at 0.90 probability

$$1.29 = \frac{t_S - 17}{3}$$

$$T_s = 21 \text{ weeks}$$

(i.e. four weeks later than expected time)

**Example 9.13** *A small project consists of seven activities for which the relevant data are give below :*

| *Activity* | *Preceding activities* | *Activity duration (Days)* |
|---|---|---|
| A | – | 4 |
| B | – | 7 |
| C | – | 6 |
| D | A, B | 5 |
| E | A, B | 7 |
| F | C, D, E | 6 |
| G | C, D, E | 5 |

(a) *Draw the network and find the project completion time.*

(b) *Calculate the total float for each of the activities.* **[GGSIPU - 2012]**

***Solution*** (a) Figure shows the calculation of EST and LFT for each event

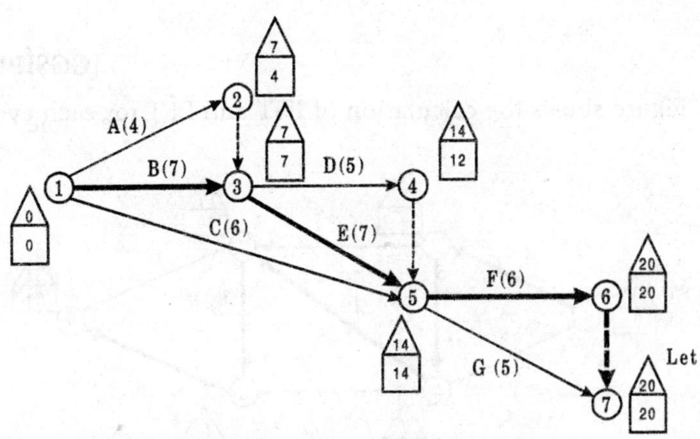

Critical path is 1–3–5–6–7

Project completion time is 20 days.

(b)

| *Activity (i-j)* | *Duration ($d_{ij}$)* | *EST* | *EFT (EST + $d_{ij}$)* | *LFT* | *LST (LFT–$d_{ij}$)* | *Total float LST–EST or LFT–EFT* |
|---|---|---|---|---|---|---|
| 1–2 | 4 | 0 | 4 | 7 | 3 | 3 |
| 1–3 | 7 | 0 | 7 | 7 | 0 | 0 |
| 1–5 | 6 | 0 | 6 | 14 | 8 | 8 |
| 2–3 | 0 | 4 | 4 | 7 | 7 | 3 |
| 3–4 | 5 | 7 | 12 | 14 | 9 | 2 |
| 3–5 | 7 | 7 | 14 | 14 | 7 | 0 |
| 4–5 | 0 | 12 | 12 | 14 | 14 | 2 |
| 5–6 | 6 | 14 | 20 | 20 | 14 | 0 |
| 5–7 | 5 | 14 | 19 | 20 | 15 | 1 |
| 6–7 | 0 | 20 | 20 | 20 | 20 | 0 |

**Example 9.14** *For the network shown below, the scheduled completion time is 32 days. Determine the slack times for the events and identify the critical path.*

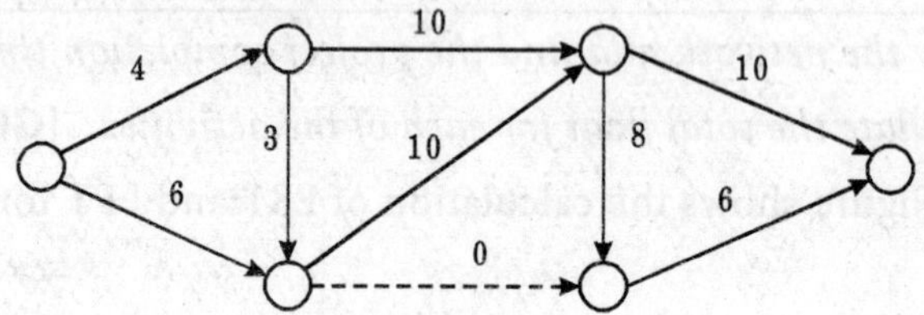


**[GGSIPU - 2013]**

***Solution*** Figure shows the calculation of EST and LFT for each event

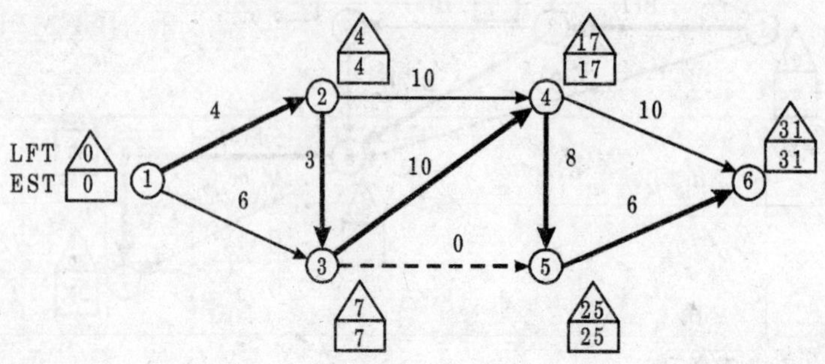

Critical path is 1–2–3–4–5–6

| *Activity (i-j)* | *Duration ($d_{ij}$)* | *EST* | *EFT* | *LFT* | *LST* | *T. F.* LST–EST LFT–EFT | *F.F.* T.F. (i–j) –Slack of j | *I.F.* F.F. (i–j) –Slack of i |
|---|---|---|---|---|---|---|---|---|
| 1–2 | 4 | 0 | 4 | 4 | 0 | 0 | 0 | 0 |
| 1–3 | 6 | 0 | 6 | 7 | 1 | 1 | 1 | 1 |
| 2–3 | 3 | 4 | 7 | 7 | 4 | 0 | 0 | 0 |
| 2–4 | 10 | 4 | 14 | 17 | 7 | 3 | 3 | 3 |
| 3–4 | 10 | 7 | 17 | 17 | 7 | 0 | 0 | 0 |
| 3–5 | 0 | 7 | 7 | 25 | 25 | 18 | 18 | 18 |
| 4–5 | 8 | 17 | 25 | 25 | 17 | 0 | 0 | 0 |
| 4–6 | 10 | 17 | 27 | 31 | 21 | 4 | 4 | 4 |
| 5–6 | 6 | 25 | 31 | 31 | 25 | 0 | 0 | 0 |

Slack of an event = (Upper – Lower)

i.e. = (LFT – EST)

∴ Slack for each node is zero. **Ans.**

**Example 9.15** *The utility data for a network are given below. Determine the total, free, independent and interfering floats and identify the critical path.*

| *Activity* | *Duration* |
|---|---|
| *0–1* | *2* |
| *1–2* | *8* |
| *1–3* | *10* |
| *2–4* | *6* |
| *2–5* | *3* |
| *3–4* | *3* |
| *3–6* | *7* |
| *4–7* | *5* |
| *5–6* | *2* |
| *6–7* | *8* |

**[GGSIPU - 2014]**

***Solution***

| *Activity (i-j)* | *Duration ($d_{ij}$)* | *EST* | *EFT* | *LFT* | *LST* | *T. F.* | *F.F.* | *I.F.* | *Int. F (T.F. – F.F.)* |
|---|---|---|---|---|---|---|---|---|---|
| 0–1 | 2 | 0 | 2 | 0 | 2 | 0 | 0 | 0 | 0 |
| 1–2 | 8 | 2 | 10 | 4 | 12 | 2 | 0 | 0 | 2 |
| 1–3 | 10 | 2 | 12 | 2 | 12 | 0 | 0 | 0 | 0 |

*Contd.*

*Contd.*

| *Activity (i-j)* | *Duration ($d_{ij}$)* | *EST* | *EFT* | *LFT* | *LST* | *T. F.* | *F.F.* | *I.F.* | *Int. F (T.F. – F.F.)* |
|---|---|---|---|---|---|---|---|---|---|
| 2–4 | 6 | 10 | 16 | 16 | 22 | 6 | 0 | –2=0 | 6 |
| 2–5 | 3 | 10 | 13 | 14 | 17 | 4 | 0 | –2=0 | 4 |
| 3–4 | 3 | 12 | 15 | 19 | 22 | 7 | 1 | 1 | 6 |
| 3–6 | 7 | 12 | 19 | 12 | 19 | 0 | 0 | 0 | 0 |
| 4–7 | 5 | 16 | 21 | 22 | 27 | 6 | 6 | 0 | 0 |
| 5–6 | 2 | 13 | 15 | 17 | 19 | 4 | 4 | 0 | 0 |
| 6–7 | 8 | 19 | 27 | 19 | 27 | 0 | 0 | 0 | 0 |

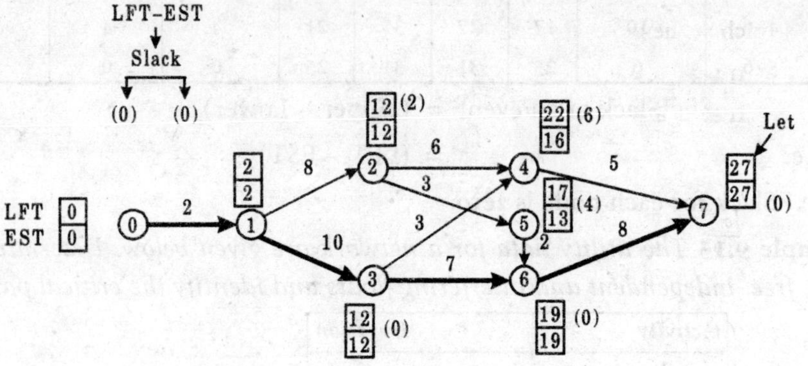


**Critical path is 0→1→3→6→7** **Ans.**

$$\text{T.F.} = \text{LST} - \text{EST}$$

$$= \text{LFT} - \text{EFT}$$

$$\text{F.F.} = \text{T.F.} - \text{Slack of event } j$$

$$\text{I.F.} = \text{F.F.} - \text{Slack of event } i$$

(Interfering float) Int. F.= T.F. – F.F.

$$\text{LST} = \text{LFT} - d_{ij}$$

$$\text{EFT} = \text{EST} + d_{ij}$$

$$\text{I.F.} \leq \text{F.F.} \leq \text{T.F.}$$

Interfering float for are activity is equal to the slack for the head event of that activity

$$\boxed{f_{I.f.} = T.f. - f.f.}$$

## MULTIPLE CHOICE QUESTIONS

9.1. Earliest finish time (EFT) :

(a) EST+$d_{ij}$ (b) LFT−$d_{ij}$

(c) EST−$d_{ij}$ (d) LST−EST

9.2. Latest start time (LST) :

(a) LFT+$d_{ij}$ (b) LFT−EFT

(c) LFT−$d_{ij}$ (d) All of the above

9.3. Total float (TF) :

(a) LST−EST (b) LFT−EFT

(c) EST−$d_{ij}$ (d) LFT−$d_{ij}$

9.4. Which sequence is correct :

(a) If ≤ FF ≥ TF (b) If ≥ FF ≤ TF

(c) If ≥ FF ≥ TF (d) If ≤ FF ≤ TF

9.5. Expected duration of each activity ($T_e$) :

(a) $\frac{t_o + 4t_m + t_p}{6}$ (b) $\left(\frac{t_p - t_o}{6}\right)$

(c) $\left(\frac{t_p - t_o}{6}\right)^2$ (d) $\frac{t_o + 6t_m + t_p}{6}$

9.6. Cost slope (Cost of crashing) for each activity :

(a) $\frac{\text{Crash cost} - \text{Normal cost}}{\text{Normal time} - \text{Crash time}}$ (b) $\frac{\text{Normal time} - \text{Crash time}}{\text{Crash cost} - \text{Normal cost}}$

(c) $\frac{\text{Normal time} - \text{Normal cost}}{\text{Crash cost} - \text{Crash time}}$ (d) $\frac{\text{Crash time} - \text{Normal cost}}{\text{Crash cost} - \text{Normal time}}$

9.7. CPM is :

(a) Critical Project Method (b) Critical Path Method

(c) Critical Project Management (d) Crash Project Method

9.8. Mark the wrong statement :

(a) CPM is even oriented (b) PERT is Probabilistic in nature

(c) CPM is deterministic (d) CPM and PERT both are similar

9.9. A dummy activity :

(a) Does not consume time and resources (b) Represented by dotted line

(c) Introduced Artifically (d) All of the above

9.10. PERT and CPM technique first developed by :

(a) UAE (b) USA

(c) New Zealand (d) UK.

## ANSWER

1. (a) 2. (c) 3. (a), (b) 4. (d) 5. (a) 6. (a) 7. (b)
8. (d) 9. (d) 10. (b)

## SUMMARY

**Activity :** Every project consists of number of jobs or tasks which are called activities. An activity is represented by an arrow in the network diagram. An activity is usually named in capital alphabets like A, B, C etc.

**Event :** An event is a specific instant of time which indicates the begining or end of any activity. An event is specified by a number and is given inside a circle.

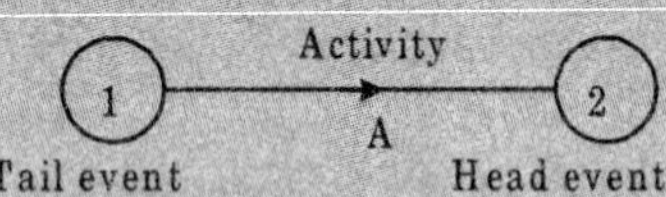


**Dummy Activities :** Dummy (or zero time) activities neither consume time nor resources and shown by a dotted lines.

**Critical Path Method (C.P.M.) :** Steps to determine critical path –

1. Draw the network diagram and number the nodes. Lit $d_{ij}$ be the duration of activity $(i, j)$.
2. Calculate Earliest Start Time (EST) : It is the earliest possible time at which an activity can start.

Let $\square_i$ = Earliest Start Time of event $i$

$\square_j$ = Maximum ($\square_i + dij$)

Calculation of EST for each event is done in forward pass *i.e.* from left to right.

3. Calculate Latest Finish Time (LFT) : It is the latest time by which on activity must be completed so that the scheduled date for the completion of the project may not be delayed.

Let $\Delta_j$ = Latest finish time of event $j$

$\Delta_i$ = Minimum ($\Delta_j - dij$)

Calculation of LFT for each event is done in backward pass, i.e. from end event to starting event. LFT for end event is same as its EST and LFT for start event is zero.

4. Calculate Earliest Finish Time (EFT) :

$$EFT = EST + dij$$

5. Calculate Latest Start Time (LST) :

$$LST = LFT - dij$$

6. Calculate Total Float (T.F.) :

$$T.F. = LST - EST = LFT - EFT$$

Total float is the time by which an activity may be delayed or extended without affecting the total project duration.

**Floats (Slack) :**

**Total Float (T.F.) :** Total float of an activity is the time by which an activity may be delayed without affecting the total project duration.

$$\text{Total float} = \text{LFT} - \text{EFT} = \text{LST} - \text{EST}$$

**Free Float (F.F.) :** Free float of an activity is that portion of total float which can be used for resheduling that activity without affecting the suceeding activity.

Free float of activity $(i–j)$ =[Total float of $(i–j)$] – slack of event $j$

**Independent Float (I.F.) :** Independent float of an activity is the amount of time by which the acitivity can be resheduled without affecting the preceding or succeeding activity.

Independent float of an activity $(i–j)$

= [Free float of$(i–j)$] – slack of event $i$

**Programme Evaluation and Review Techniques (PERT) :**

(i) **Optimistic time ($t_o$)** : It is the shortest time in which an activity can be completed.

(ii) **Most likely time ($t_m$)** : It is the time required to complete an activity taking into consideration all favourable and unfarourable circumstances.

(iii) **Pessimistic time ($t_p$)** : It is the longest time in which an activity can be completed (Every thing goes wrong).

**PERT procedure :**

(i) Draw the network diagram.

(ii) Compiete expected duration of each activity

$$t_e = \left(\frac{t_o + 4t_m + t_p}{6}\right)$$

(iii) Determine the critical path and idendtify critical activities.

(iv) Compute the expected varance $\sigma^2 = \left(\frac{t_p - t_o}{6}\right)^2$

(v) Compiete variance of the critical path $\sigma^2 c$ which is the sum of variance of all critical activities.

(vi) Calculate expected standard deviation of project length $\sigma_c$ and calculate the standard normal variate;

$$Z = \frac{t_s - t_e}{\sigma_c}$$

where $t_s$ = Scheduled time

$t_e$ = Expected time

$\sigma_c$ = S.D. of the project

(vii) Using Z, one can estimate the probability of completing the project within a specified time by using the normal curve (area) table.

CHAPTER 10

# Game Theory

## 10.1 INTRODUCTION

In every business, there are conflicts and competitive situations. A game represents a competitive situation in which two or more players participate with an aim to win, over a period of time.

## 10.2 TWO PERSON ZERO SUM GAME

A game with only two players is called a two persons zero sum game. Here algebraic sum of gains and losses of two players is zero i.e. amount won by a player is exactly equal to amount lost by the other. We shall discuss two persons zero sum games in this Chapter.

### 10.2.1 Strategy

A plan of action adopted by a player, designed to achieve victory is called strategy. In terms of game theory, strategies can be of two types :

(i) **Pure strategy** : In a game if any player always selects the same strategy irrespective of other person's strategies, then it is known as 'pure strategy'. Pure strategies are deterministic.

(ii) **Mixed strategy** : The game in which different strategies are used in different situations, it is called mixed strategy. A mixed strategy is probabilistic. It is more advantageous since the opponent is always kept guessing.

**Pay off matrix** : The outcomes resulting from a game, represented in the form of table is called 'payoff matrix'.

Player $Y$

| | | | |
|---|---|---|---|
| I | $s_{11}$ | $-s_{12}$ | $s_{13}$ |
| Player $X$ II | $s_{21}$ | $s_{22}$ | $-s_{23}$ |
| III | $-s_{31}$ | $s_{32}$ | $s_{33}$ |

A +ve entry indicates gain to $X$ (*i.e.* payment from $Y$ to $X$), while a –ve entry indicates loss to $X$ (i.e. payment form $X$ to $Y$).

### 10.2.2 Minimax and Maximin Principle

This means minimization of maximum losses and maximization of minimum profit.

(i) **Saddle point :** A saddle point of a pay-off matirx is that element in the matrix, where the maximum of row minima coincides with the minimum of column maxima, i.e. minimax and maximin intersect. A game may have no saddle point or more, than one saddle points at different positions. A game is strictly determinable if saddle point exists. We determine saddle point of a game as follows :

1. Identify the minimum element (maximin value) in each row of the payoff matrix and enclose it within a square.
2. Identify the maximum element (minimax value) in each column of the payoff matrix and enclose it within a circle.
3. Saddle point exists at the intersection of row minima and column maxima. If there is no such value, saddle point does not exist.

(ii) **Value of the game :** The payoff at the saddle point is the value of the game. If there is no saddle point, value of game is determined using other methods. A game is called fair if the value of game is zero. A +ve value represents gain to first player (loss to second player) and a – ve value indicates loss to first player (gain to second player).

**Example 10.1** *Consider a game with the following pay off matrix :*

*Player B*

$$\textit{Palyer A} \quad \begin{array}{c} \\ A_1 \\ A_2 \end{array} \begin{array}{c} \begin{array}{cc} B_1 & B_2 \end{array} \\ \begin{bmatrix} 2 & 5 \\ -3 & \lambda \end{bmatrix} \end{array}$$

(i) *Show that game is strictly determinable, whatever 'λ' may be.*

(ii) *Determine the value of the game.*

**Solution** Marking minimum value in first row within a square and maximum value within first column within a circle

*Player B*

| *Player A* | | $B_1$ | $B_2$ |
|---|---|---|---|
| | $A_1$ | ② | 5 |
| | $A_2$ | −3 | λ |

(i) Saddle point is 2 which is unique. Hence game is strictly determinable.

(ii) Value of game is 2.

**Example 10.2** *Find the range of values of 'p' and 'q' which will render the element* $a_{22}$ *as saddle point for the game whose pay off matrix is given below :*

*Player B*

| *Player A* | | $B_1$ | $B_2$ | $B_3$ |
|---|---|---|---|---|
| | $A_1$ | −1 | 2 | 3 |
| | $A_2$ | 8 | 5 | $q$ |
| | $A_3$ | 3 | $p$ | 7 |

***Solution*** Considering $a_{22}$ as sadddle point and ignoring values of $p$ and $q$.

*Player B*

| *Player A* | | $B_1$ | $B_2$ | $B_3$ |
|---|---|---|---|---|
| | $A_1$ | −1 | 2 | 3 |
| | $A_2$ | 8 | ⑤ | $q$ |
| | $A_3$ | 3 | $p$ | 7 |

For 5 to be saddle point , $p \leq 5$ and $q \geq 5$.

**Example 10.3** *For the game with pay-off matrix*

$$\begin{array}{cc} & \textit{Player B} \\ & \begin{array}{ccc} B_1 & B_2 & B_3 \end{array} \\ \textit{Player A} \begin{array}{c} A_1 \\ A_2 \end{array} & \begin{bmatrix} 1 & 2 & -1 \\ 4 & 6 & -4 \end{bmatrix} \end{array}$$

Determine the best strategies for players A and B. Also determine the value of the game. Is this game

(i) Fair?

(ii) Strictly determinable?

***Solution*** Exclosing minimum element of each row in a square and maximum element of each column in a circle :

$$\begin{array}{cc} & \textit{Player B} \\ & \begin{array}{ccc} B_1 & B_2 & B_3 \end{array} \\ \textit{Player A} \begin{array}{c} A_1 \\ A_2 \end{array} & \begin{bmatrix} 1 & 2 & \boxed{-1} \\ 4 & 6 & -4 \end{bmatrix} \end{array}$$

(i) Game is not fair as value of game is – 1.

A game is fair if value of game is zero.

(ii) Saddle point exists at (1, 3) position and value of game is –1. Hence the game is strictly determinable.

### 10.2.3 Summary of Systematic Methods for Solving Rectangular Games

**Step 1 :** Search for a saddle point. If it is found, the problem is readily solved. Strategies are pure strategies.

**Step 2 :** If no saddle point is found, in this case use the concept of dominance to reduce the size of the matrix game.

| | | |
|---|---|---|
| If size of reduced matrix becomes 2×2 with no saddle point, solve by arithmetic or algebraic methods. | If size is (2×$n$) or ($m$×2), use graphical method or subgame method to reduce it to (2×2) matrix and then solve it by arithmetic or algebraic method | If size of matrix (3×3) or higher, then algebraic method, method of matrices or method of linear programming may be used |

## 10.3 GAMES WITHOUT SADDLE POINT

When a game does not possess a saddle point, both the players must determine their mixed strateties to maximize their gains or to minimize their losses.

### 10.3.1 Solution of 2 × 2 Game Without Saddle Point

When saddle point does not exist for a game, best strategies are mixed strateties.

(i) ***Algebraic Method :*** For any 2 person zero sum game, when the optimal strategies are not pure strategies (i.e. there is no saddle point) and for which $A$'s pay off matrix is

| | | B | |
|---|---|---|---|
| | | I | II |
| A | I | $a_{11}$ | $a_{12}$ |
| | II | $a_{21}$ | $a_{22}$ |

Let $A$'s probability of selection of strategies I and II be $p_1$ and $p_2$, $p_1+p_2=1$. Similarly if $B$'s probabilities of selection of strategies I and II be $q_1$ and $q_2$, $q_1+q_2=1$.

Then
$$p_1 = \frac{a_{22}-a_{21}}{(a_{11}+a_{22})-(a_{12}+a_{21})}, \; p_2 = 1 - p_1$$

$$q_1 = \frac{a_{22}-a_{12}}{(a_{11}+a_{22})-(a_{12}+a_{21})}, \; q_2 = 1 - q_1$$

and value of game to player $A$ is given by

$$V = \frac{(a_{11}\cdot a_{22})-(a_{12}\cdot a_{21})}{(a_{11}+a_{22})-(a_{12}+a_{21})}$$

(ii) ***Arithmetic method for (2×2) game***

**Step 1 :** Find the difference (Oddments) of two numbers in column I and put it under column II.

**Step 2 :** Find the difference (Oddments) of two numbers in column II and put it under column I.

**Step 3 :** Repeat the above two steps for rows also.

**Step 4 :** Divide these differences by row or column total.

**Step 5 :** Value of game is obtained by multiplying row/column oddments by any one row/column of given 2×2 matrix and then dividing by sum.

**Note :** Game is not solvable if row sum ≠ column sum.

**Example 10.4** *Two players A and B without showing each other, put on a table a coin, with head or tail up. A wins ₹ 8 when both coins show head and ₹ 2 when both are tails. B wins ₹ 3 when coins do not match. Solve the game.*

***Solution***

$$A\begin{array}{c} \\ H \\ T \end{array}\overset{\displaystyle B}{\begin{array}{cc} H & T \\ \end{array}}\begin{bmatrix} 8 & -3 \\ -3 & 2 \end{bmatrix}$$

Saddle points does not exist for given game. Let $p_1, p_2$ be probabilities by which $A$ throws head and tail respectively and $q_1, q_2$ be probabilities by which B throws head and tail.

Using Algebraic Method

$$p_1 = \frac{2+3}{(8+2)-(-3-3)}$$

$$= \frac{5}{16}, \; p_2 = 1 - p_1 = \frac{11}{16}$$

$$q_1 = \frac{2+3}{(8+2)-(-3-3)}$$

$$= \frac{5}{16}, \; q_2 = 1 - q_1 = \frac{11}{16}$$

$$V = \frac{(8\cdot 2)-[(-3)\cdot(-3)]}{(8+2)-(-3-3)} = \frac{16-9}{16} = \frac{7}{16}$$

∴ Value of game to player $A$ is $\frac{7}{16}$.

Solving same problem by Arithmetic method :

| | | B: H | B: T | |
|---|---|---|---|---|
| A | H | 8 | −3 | 5 |
| | T | −3 | 2 | 11 |
| | | 5 | 11 | 16 |

*[Writting difference of elements of calumn I under calumn II and difference of elements of column II under calumn I. And repeating the procedure for rows.]*

Sum of differences is 16 both ways

$$p_1 = \frac{5}{16},\ p_2 = \frac{11}{16},\ q_1 = \frac{5}{16},\ q_2 = \frac{11}{16}$$

$$V = \frac{5.8+11\cdot(-3)}{16} = \frac{40-33}{16} = \frac{7}{16}$$

$\therefore$ (i) Optimal strategy for $A$ is $\left(\frac{5}{16}, \frac{11}{16}\right)$.

(ii) Otimal strategy for $B$ is $\left(\frac{5}{16}, \frac{11}{16}\right)$.

(iii) Value of game to $A$ is $\frac{7}{16}$.

### 10.3.2 Dominance Rules

Dominance rules are used to reduce the size of a pay-off matrix.

**Rule 1 : Row dominance :** If all the elements of $i^{th}$ row in a pay off matrix are greater than or equal to corresponding elements of $j^{th}$ row, then we say $i^{th}$ row dominates the $j^{th}$ row and $j^{th}$ row is deleted as player '$A$' will never use $j^{th}$ strategy.

**Rule 2 : Column dominance :** If all the elements in $r^{th}$ column are greater than or equal to corresponding elements in $s^{th}$ column, we say $s^{th}$ column dominates over $r^{th}$ column and $r^{th}$ column is deleted from the pay-off matrix as $B$ will never use $r^{th}$ column.

**Rule 3 :** A pure strategy may be dominated, if it is interior to an average of two or more other pure strategies.

**Example 10.5** *Solve the following 3×3 game*

| | | $P_2$ | | |
|---|---|---|---|---|
| | | *I* | *II* | *III* |
| | *I* | 1 | 7 | 2 |
| $P_1$ | *II* | 6 | 2 | 7 |
| | *III* | 5 | 1 | 6 |

*Saddle point does not exist.*

***Solution*** All the elements of row II are greater than corresponding elements of row III,

∴ Row III can be deleted.

| | | $P_2$ | | |
|---|---|---|---|---|
| | | I | II | III |
| $P_1$ | I | 1 | 7 | 2 |
| | II | 6 | 2 | 7 |

Again all the elements o column III are greater than corresponding elements of column I.

∴ Column I dominates over column III

Hence column III may be deleted.

| | | $P_2$ | |
|---|---|---|---|
| | | I | II |
| $P_1$ | I | 1 | 7 |
| | II | 6 | 2 |

Reduced 2×2 matrix can be solved by any of algebraic or airthmetic method.

**Using algebraic method**

Let $p_1$ and $p_2$ be the probabilities by which $p_1$ chooses strategies I and II and $q_1$, $q_2$ be the probabilities by which $P_2$ chooses his strategies I and II.

$$p_1 = \frac{2-6}{(1+2)-(7+6)}$$

$$= \frac{-4}{-10} = \frac{2}{5} \Rightarrow p_2 = \frac{3}{5}$$

$$q_1 = \frac{2-7}{-10} = \frac{1}{2} = \Rightarrow q_2 = \frac{1}{2}$$

$$V = \frac{(1.2)-(7.6)}{-10} = 4$$

∴ Optimal strategies for $P_1$ are $\left(\frac{2}{5}, \frac{3}{5}, 0\right)$.

Optimal strategies for $P_2$ are $\left(\frac{1}{2}, \frac{1}{2}, 0\right)$.

Valuse of games to $A$ is 4.

**Example 10.6** *A and B play a game in which each has 3 coins, a 1 rupee coin, 2 rupee coin and 5 rupee coin. Each player selects a coin without the knowledge of other's choice. If the sum of coins is an odd amount, A wins B's coin and if the sum is even, B wins A's coin. Find the best strategies for each player and also value of the game.*

***Solution***

B

| A \ B | I (1) | II (2) | III (5) |
|---|---|---|---|
| I (1) | −1 | 2 | −1 |
| II (2) | 1 | −2 | 5 |
| III (5) | −5 | 2 | −5 |

*Saddle point does not exist.*

All the elements of row I ≥ corresponding element of row III.

∴ Row III may be deleted.

B

| A \ B | I(1) | II(2) | III(5) |
|---|---|---|---|
| I (1) | −1 | 2 | −1 |
| II (2) | 1 | −2 | 5 |

All the elements of column III ≥ corresponding elements of column I

∴ Column III may be deleted.

B

| A \ B | I(1) | II(2) |
|---|---|---|
| I (1) | −1 | 2 |
| II (2) | 1 | −2 |

Solving the game by arithmetic method. Let $p_1, p_2$ be probabilities by which A selects strategies I and II, $q_1, q_2$ be probabilities by which $B$ selects I and II.

| | | B | | |
|---|---|---|---|---|
| | | *I* (1) | *II*(2) | |
| A | *I* (1) | −1 | 2 | 3 |
| | *II* (2) | 1 | −2 | 3 |
| | | 4 | 2 | 6 |

$$p_1 = \frac{3}{6} = \frac{1}{2}, p_2 = \frac{1}{2}$$

$$q_1 = \frac{4}{6} = \frac{2}{3}, q_2 = \frac{1}{3}$$

$$\therefore \qquad V = \frac{3(-1)+3(1)}{6} = 0$$

∴ Value of game is zero, i.e. its a fair game.

Optimal strategies for $A$ are $\left(\frac{1}{2}, \frac{1}{2}, 0\right)$.

Optical strategies for $B$ are $\left(\frac{2}{3}, \frac{1}{3}, 0\right)$.

### 10.3.3 Graphical Methods for 2 × n or m × 2 Games

(i) For $2 \times n$ matrix game which is not reducible by dominance rules.

**Working Rule**

**Step 1 :** Draw two vertical lines one unit apart and make scale on each. These two lines represent the two strategies available to first player.

**Step 2 :** Mark points of 1st row in pay-off matrix on 1st scale and that of 2nd row on 2nd scale. Join the corresponding points.

**Step 3 :** Determine the highest point on the lowest envelope which is maximin point.

**Step 4 :** The two or more lines passing through maximin point determine the required $2 \times 2$ pay-off matrices.

**Step 5 :** Solve this $2 \times 2$ game by usual methods.

**Example 10.7** *Solve following 2×4 game by graphical method.*

$$A \begin{matrix} & B \\ & \begin{matrix} I & II & III & IV \end{matrix} \\ \begin{matrix} I \\ II \end{matrix} & \begin{bmatrix} 2 & 2 & 5 & -1 \\ 4 & 3 & 2 & 6 \end{bmatrix} \end{matrix}$$

***Solution*** *This 2 × 4 game has no saddle point*

Also this matrix can not be further reduced using dominance rules.

∴ Using graphical mehod to reduce this matrix game to 2×2.

Let $p_1$, $p_2$ be the probabilities by which player $A$ uses his pure strategies I and II and $q_1$, $q_2$, $q_3$, $q_4$ be probabilities by which player $B$ uses his pure strategies I, II, III and IV.

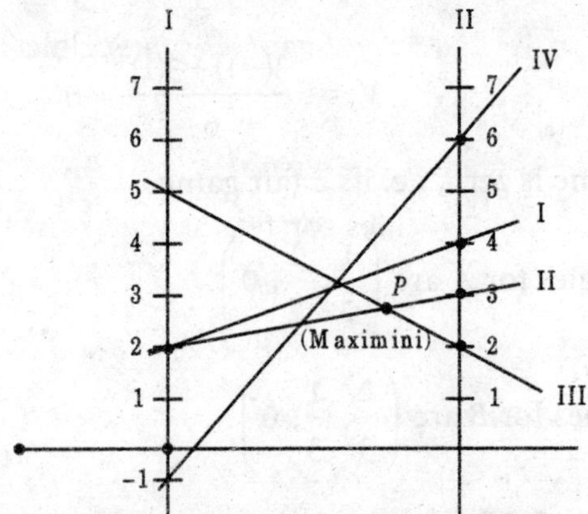


Maximin occuss at $P$ which is highest point on lowest envelope. This is point of intersection of two pay off lines II and III.

Thus, the game reduces to following 2 × 2 game.

$$A \begin{matrix} & B \\ & \begin{matrix} II & III \end{matrix} \\ \begin{matrix} I \\ II \end{matrix} & \boxed{\begin{matrix} 2 & 5 \\ 3 & 2 \end{matrix}} \end{matrix}$$

Solving by algebraic method :

$$p_1 = \frac{2-3}{(2+2)-(3+5)} = \frac{1}{4}, \; p_2 = \frac{3}{4}$$

$$q_2 = \frac{2-5}{(2+2)-(3+5)} = \frac{3}{4}, \quad q_3 = \frac{1}{4}$$

$$V = \frac{(2\cdot 2)-(3\cdot 5)}{(2+2)-(3+5)} = \frac{11}{4}$$

$\therefore$ Optimal strategy for $A$ is $\left(\frac{1}{4}, \frac{3}{4}\right)$.

Optimal strategy for $B$ is $\left(0, \frac{3}{4}, \frac{1}{4}, 0\right)$.

Value of game to A is $\frac{11}{4}$.

II Graphical solution for $m \times 2$ game not reducible by dominance rules.

**Working Rules**

Procedure is same as that of $2 \times n$ game.

Here we take two vertical lines for two strategies available to 2nd player and take minimax.

**Example 10.8** *Solve the given game graphically :*

| | | B: *I* | *II* |
|---|---|---|---|
| | *I* | 2 | 7 |
| A | *II* | 3 | 5 |
| | *III* | 11 | 2 |

***Solution*** This 3×2 game has no saddle point. Let $p_1$, $p_2$, $p_3$ be the probabilities by which player $A$ uses his strategies I, II and III and $q_1$, $q_2$ be the probabilities by which player $B$ uses his strategies I, II.

$$\therefore \qquad p_1+p_2+p_3 = 1 \text{ and } q_1+q_2=1$$

Drawing two vertical lines for $B$'s pure st tegies I and II and plotting points (2, 7), (3, 5) and (11, 2).

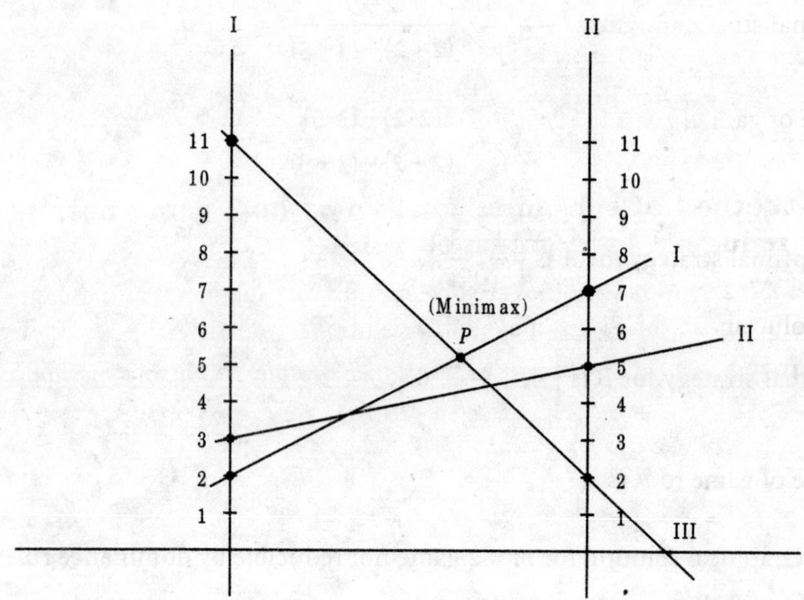


Clearly minimax occuss at $P$ which lies on intersection of strategies I and III.

∴ Given game reduces to following 2×2 game.

Solving by airthmatic method :

| | | B | | |
|---|---|---|---|---|
| | | *I* | *II* | |
| A | *I* | 2 | 7 | 9 |
| | *II* | 11 | 2 | 5 |
| | | 5 | 9 | 14 |

$$p_1 = \frac{9}{14},\ p_3 = \frac{5}{14},\ q_1 = \frac{5}{14},\ q_2 = \frac{9}{14}$$

$$V = \frac{(2)(9)+(11)(5)}{14} = \frac{73}{14}$$

∴ Optimal strategy for $A$ are $\left(\frac{9}{14}, 0, \frac{5}{14}\right)$.

Optimal strategy for $B$ is $\left(\frac{5}{14}, \frac{9}{14}\right)$.

Value of game to $A$ is $\frac{73}{14}$.

### 10.3.4 Method of subgames for 2×*n* or *m*×2 game not reducible by dominance rules

In this method 2×*n* or *m*×2 games are broken into 2×2 subgames and the most suitable solution is found, out of available solutions.

**Example 10.9** *Solve the 2 × 4 game given in Example 10.7 using method of subgames.*

| A \ B | I | II | III | IV |
|---|---|---|---|---|
| I | 2 | 2 | 5 | −1 |
| II | 4 | 3 | 2 | 6 |

***Solution*** Breaking in to $4_{C_2}$ *i.e.* 6 subgames and solving each by arithmetic method :

Subgame I

| A \ B | I | II | |
|---|---|---|---|
| I | 2 | 2 | 1 |
| II | 4 | 3 | 0 |
| | 1 | 2 | × |

Subgame II

| A \ B | I | II | |
|---|---|---|---|
| I | 2 | 5 | 2 |
| II | 4 | 2 | 3 |
| | 3 | 2 | 5 |

Subgame III

| A \ B | I | II | |
|---|---|---|---|
| I | 2 | −1 | 2 |
| II | 4 | 6 | 3 |
| | 7 | 2 | × |

Subgame I

| A \ B | I | II | |
|---|---|---|---|
| I | 2 | 5 | 1 |
| II | 3 | 2 | 3 |
| | 3 | 1 | 4 |

Subgame II

| A \ B | I | II | |
|---|---|---|---|
| I | 2 | −1 | 3 |
| II | 3 | 6 | 3 |
| | 7 | 1 | × |

Subgame III

| A \ B | I | II | |
|---|---|---|---|
| I | 5 | −1 | 4 |
| II | 2 | 6 | 6 |
| | 7 | 3 | 10 |

Clearly subgames I, III and V are not solvable as row sum of oddments is not equal to column oddment sum. Practically subgames whose lines for strategies posses same signs for slopes in graphical solution must not be taken together. Solving subgames II, IV and VI.

**For subgames II :**

$$p_1 = \frac{2}{5},\ p_2 = \frac{3}{5}$$

$$q_1 = \frac{3}{5},\ q_3 = \frac{2}{5}$$

$$V = \frac{(2)(2)+(4)(3)}{5} = \frac{16}{5}$$

Optimal strategy for $A\left(\frac{2}{5}, \frac{3}{5}\right)$.

Optimal strategy for $A\left(\frac{3}{5}, 0, \frac{2}{5}, 0\right)$.

Value of game to $A$ is $\frac{16}{5}$.

**For subgames IV :**

$$p_1 = \frac{1}{4},\ p_2 = \frac{3}{4}$$

$$q_1 = \frac{3}{4},\ q_3 = \frac{1}{4}$$

$$V = \frac{(2)(1)+(3)(3)}{4} = \frac{11}{4}$$

Optimal strategy for $A\left(\frac{1}{4}, \frac{3}{4}\right)$.

Optimal strategy for $B\left(0, \frac{3}{4}, \frac{1}{4}, 0\right)$.

Value of game to $A$ is $\frac{11}{4}$.

**For subgames VI :**

$$p_1 = \frac{2}{5},\ p_2 = \frac{3}{5}$$

$$q_3 = \frac{7}{10},\ q_3 = \frac{3}{10}$$

$$V = \frac{(5)(4)+(2)(6)}{10} = \frac{16}{5}$$

Optimal strategy for $A\left(\frac{2}{5}, \frac{3}{5}\right)$.

Optimal strategy for $B\left(0, 0, \frac{7}{10}, \frac{3}{10}\right)$.

Value of game to $A$ is $\frac{16}{5}$.

Now as two strategies for $A$ are fixed and $B$ can choose any of his strategies I, II, III and IV, subgame IV is most suitable for $B$ as he would like to minimize his losses.

∴ Optimal solution is solution of subgame IV

i.e. Optimal strategies for $A$ are $\left(\frac{1}{4}, \frac{3}{4}\right)$.

Optimal strategies for $B$ are $\left(0, \frac{3}{4}, \frac{1}{4}, 0\right)$.

Value of game to is $\frac{11}{4}$.

**Example 10.10** *Solve the game given in Example 10.9 using method of subgames.*

| | | B | |
|---|---|---|---|
| | | *I* | *II* |
| A | *I* | 2 | 7 |
| | *II* | 3 | 5 |
| | *III* | 11 | 2 |

***Solution*** Forming $^3C_2$ *i.e.* 3 subgames and soling by arithmatic method :

**Subgame I**

| | | B: I | B: II | |
|---|---|---|---|---|
| A | I | 2 | 7 | 2 |
| | II | 3 | 5 | 5 |
| | | 2 | 1 | × |

Subgame I

**Subgame II**

| | | B: I | B: II | |
|---|---|---|---|---|
| A | I | 2 | 7 | 9 |
| | II | 11 | 2 | 5 |
| | | 5 | 9 | 14 |

Subgame II

**Subgame III**

| | | B: I | B: II | |
|---|---|---|---|---|
| A | I | 3 | 5 | 9 |
| | II | 11 | 2 | 2 |
| | | 3 | 8 | 11 |

Subgame III

Subgame I is not feasible as row sum of oddments is not equal to column sum of oddments.

**For subgame II**

$$p_1 = \frac{9}{14},\ p_2 = \frac{5}{14}$$

$$q_1 = \frac{5}{14},\ q_2 = \frac{9}{14}$$

$$V = \frac{(2)(9)+(11)(5)}{14} = \frac{73}{14}$$

**For subgame III**

$$p_2 = \frac{9}{11},\ p_3 = \frac{2}{11}$$

$$q_1 = \frac{3}{11},\ q_2 = \frac{8}{11}$$

$$V = \frac{(3)(9)+(11)(2)}{11} = \frac{49}{11}$$

Now player $A$ would like to maximize his gains as much as possible, therefore subgame II is better for him due to more value of game.

Hence optimal strategy for $A$ is $\left(\frac{9}{14}, 0, \frac{5}{14}\right)$.

Optimal strategy for $B$ is $\left(\frac{5}{14}, \frac{9}{14}\right)$.

Value of game to $A$ is $\frac{73}{14}$.

### 10.3.5 Algebraic Method for (3×3) or higher order matrix game.

**Working Rule**

**Step 1 :** Let $(p_1, p_2, p_3)$ and $(q_1, q_2, q_3)$ be the probabilities by which the two players select their respective strategies for a $3 \times 3$ matrix game and let $V$ be the value of game to first player.

**Step 2 :** Form 3 equations in $p_1, p_2, p_3$ so as to maximize value of the game as per first player's notion.

**Step 3 :** Form 3 equation in $q_1, q_2, q_3$ so as to minimize the value of game as per second player's criteria.

**Step 4 :** Solve for $p_1, p_2, p_3$ and $q_1, q_2, q_3$ using the above 6 equations and

$$p_1 + p_2 + p_3 = 1, q_1 + q_2 + q_3 = 1.$$

**Example 10.11** *Solve the following 3 × 3 game using algebraic method.*

| | | B | | |
|---|---|---|---|---|
| | | *I* | *II* | *III* |
| | *I* | 1 | −1 | −1 |
| A | *II* | −1 | −1 | 3 |
| | *III* | −1 | 2 | −1 |

***Solution*** Let $(p_1, p_2, p_3)$ and $(q_1, q_2, q_3)$ be the probabilities by which player $A$ and $B$ select their mixed strategies I, II, III respectively.

For player $A$

$$p_1 - p_2 - p_3 \geq V$$
$$-p_1 - p_2 + 2p_3 \geq V$$
$$-p_1 + 3p_2 - p_3 \geq V$$

Similarly for player $B$

$$q_1 - q_2 - q_3 \leq V$$
$$-q_1 - q_2 + 3q_3 \leq V$$
$$-q_1 + 2q_2 - 3q_3 \leq V$$

Also

$$p_1 + p_2 + p_3 = 1$$
$$q_1 + q_2 + q_3 = 1$$

To solve for $p_1, p_2, p_3$ and $q_1, q_2, q_3$, let the above inequalities hold as equalities

$$p_1 - p_2 - p_3 = V \quad ...(1)$$
$$-p_1 - p_2 + 2\,p_3 = V \quad ...(2)$$
$$-p_1 + 3\,p_2 - p_3 = V \quad ...(3)$$
$$q_1 - q_2 - q_3 = V \quad ...(4)$$
$$-q_1 - q_2 + 3\,q_3 = V \quad ...(5)$$
$$-q_1 + 2\,q_2 - q_3 = V \quad ...(6)$$

$(1) + (2) \Rightarrow \quad -2\,p_2 + p_3 = 2V \quad ...(7)$

$(2) - (3) \Rightarrow \quad -4\,p_2 + 3\,p_3 = 0 \quad ...(8)$

$(7) \quad \Rightarrow \quad 4\,p_2 - 2\,p_3 = -4V \quad ...(9)$

Solving (8) and (9) :

$$p_3 = -4V \quad ...(10)$$

Using (10) in (9) :

$$p_2 = -3V$$

Using $p_2$, $p_3$ in (1), $p_1 = -6V$

Also $\quad p_1 + p_2 + p_3 = 1$

$$\Rightarrow \quad -6V - 3V - 4V = 1 \Rightarrow V = -\frac{1}{13}$$

Similarly solving (4), (5) , (6)

$$q_1 = -6V, q_2 = -4V, q_3 = -3V$$

$\therefore$ Solution of given (3 × 3) game is

Optimal mixed strategies for $A$ is $\left(\frac{6}{13}, \frac{3}{13}, \frac{4}{13}\right)$.

Optimal mixed strategies for $B$ is $\left(\frac{6}{13}, \frac{4}{13}, \frac{3}{13}\right)$.

Value of game to player $A$ is $-\frac{1}{13}$.

### 10.3.6 Matrix Method (Arithmetic method) or Method of Oddment for (3×3) Game

This method is an extenstion of Arithmetic method for (2 × 2) game. It can be extended to ($n \times n$) game.

**Step 1 :** Subtract entries in column 2 from column 1 and column 3 from column 2 and wirte the result on right of matrix.

**Step 2 :** Repeat procedure for rows and write the result below the matrix.

**Step 3 :** Compute oddments for $A_i$'s (Strategies of $A$) and $B_i$'s (Strategies of $B$) by hiding $i^{th}$ row/column difference and finding the remaining 2×2 determinant.

**Step 4 :** Write the absolute values of oddments against their respective rows/ columns.

**Step 5 :** Add the row and column oddments. If row sum = column sum, optimal strategy can be obtained by dividing each oddment by row/column sum.

**Step 6 :** Value of game is obtained by multiplying row/column oddments by any one row/column of 3×3 matrix and then dividing by row/column sum.

**Note.** If row sum of oddments ≠ column sum of oddments, given game cannot be solved by this method.

**Example 10.12** *Solve the game given in Example 10.11 by arithmetic method.*

B

| | | $B_1$ | $B_2$ | $B_3$ |
|---|---|---|---|---|
| A | $A_1$ | 1 | −1 | −1 |
| | $A_2$ | −1 | −1 | 3 |
| | $A_3$ | −1 | 2 | −1 |

***Solution*** Given game does not have any saddle point. Also it can not be reduced using dominance rules. For solving it using Arithmetic method, writing difference of column 2 from column 1 and column 3 from column 2 on right of the matrix. Also writing row differences below matrix.

B

| | | $B_1$ | $B_2$ | $B_3$ | | |
|---|---|---|---|---|---|---|
| A | $A_1$ | 1 | −1 | −1 | 2 | 0 |
| | $A_2$ | −1 | −1 | 3 | 0 | −4 |
| | $A_3$ | −1 | 2 | −1 | −3 | 3 |
| | | 2 | 0 | −4 | | |
| | | 0 | −3 | 4 | | |

To find oddments :

$$\text{Oddment for } A_1 = \begin{vmatrix} 0 & -4 \\ -3 & 3 \end{vmatrix} = -12$$

$$\text{Oddment for } A_2 = \begin{vmatrix} 2 & 0 \\ -3 & 3 \end{vmatrix} = 6$$

$$\text{Oddment for } A_3 = \begin{vmatrix} 2 & 0 \\ 0 & -4 \end{vmatrix} = -8$$

$$\text{Oddment for } B_1 = \begin{vmatrix} 0 & -4 \\ -3 & 4 \end{vmatrix} = -12$$

$$\text{Oddment for } B_2 = \begin{vmatrix} 2 & -4 \\ 0 & 4 \end{vmatrix} = 8$$

$$\text{Oddment for } B_3 = \begin{vmatrix} 2 & 0 \\ 0 & -3 \end{vmatrix} = -6$$

Now taking absolute values of oddments

| | | B | | | |
|---|---|---|---|---|---|
| | | $B_1$ | $B_2$ | $B_3$ | |
| | $A_1$ | 1 | −1 | −1 | 12 |
| A | $A_2$ | −1 | −1 | 3 | 6 |
| | $A_3$ | −1 | 2 | −1 | 8 |
| | | 12 | 8 | 6 | 26 |

If $(p_1, p_2, p_3)$ be probabilties by which player $A$ selects his strategies $(A_1, A_2, A_3)$ and $(q_1, q_2, q_3)$ be probabilities by which player B selects his strategies $(B_1, B_2, B_3)$ :

$$p_1 = \frac{12}{26} = \frac{6}{13},\ p_2 = \frac{6}{26} = \frac{3}{13},\ p_3 = \frac{8}{26} = \frac{4}{13}$$

$$q_1 = \frac{12}{26} = \frac{6}{13},\ q_2 = \frac{8}{26} = \frac{4}{13},\ q_3 = \frac{6}{26} = \frac{3}{13}$$

$$V = \frac{(12)(1)+(6)(-1)+8(-1)}{26} = -\frac{1}{13}$$

$\therefore$ Optimal strategy for $A$ is $\left(\frac{6}{13}, \frac{3}{13}, \frac{4}{13}\right)$

Optimal strategy for $B$ is $\left(\frac{6}{13}, \frac{4}{13}, \frac{3}{13}\right)$

Value of the game to $A$ is $-\frac{1}{13}$

### 10.3.7 L.P. method for solving $m \times n$ matrix game

If the pay-off matrix is of order $m \times n$, and it has no saddle point, method of linear programming may be used to solve such a game.

**Working Rule**

**Step 1 :** Find minimax ($\overline{V}$) and maximin ($\underline{V}$) values,

$$\underline{V} < V < \overline{V}$$

**Step 2 :** Add a constant $C > -\underline{V}$ so that all the elements in the matrix become +ve.

**Step 3.** Formulate as L.P.P. in maximization for second player where $y_i = \frac{q_i}{V}$.

**Step 4 :** Solution for first player is obtained from the simplex table as dual L.P.P.

**Note :** By adding C to all the elements of pay-off matrix, value of game is increased by C. i.e. $V' = V + C$.

**Example 10.13** *Solve the game given in example 10.11 using L.P. method*

| | | Player B | | | |
|---|---|---|---|---|---|
| | | I | II | III | Row minima |
| | I | 1 | −1 | −1 | -1 |
| Player A | II | −1 | −1 | 3 | -1 Maximum ($\underline{V}$) = -1 |
| | III | −1 | 2 | −1 | - 1 |
| Column maxima | | 1 | 2 | 3 | Minimax($\overline{V}$) = 1 |

Adding 2 (≥1) to all the elements of the matrix so that all the elements become +ve, new matrix is

*Player B*

| | I | II | III |
|---|---|---|---|
| **Player A** I | 3 | 1 | 1 |
| II | 1 | 1 | 5 |
| III | 1 | 4 | 1 |

Formulating as L.P.P. for player *B*, $y_i = \frac{q_i}{V'}$

$$\text{Maximize } Z_q = \left(\frac{1}{V'}\right) = y_1 + y_2 + y_3$$

Subject to

$$3y_1 + y_2 + y_3 \leq 1$$
$$y_1 + y_2 + 5y_3 \leq 1$$
$$y_1 + 4y_2 + y_3 \leq 1$$
$$y_1, y_2, y_3 \geq 0$$

Hence $\quad y_1 + y_2 + y_3 = \frac{q_1}{V'} + \frac{q_2}{V'} + \frac{q_3}{V'} = \frac{1}{V'}$

Also as by algebraic method

$$3q_1 + q_2 + q_3 \leq V' \Rightarrow 3y_1 + y_2 + y_3 \leq 1$$

Adding slack variables $s_1, s_2, s_3$, standard form of L.P.P. is

$$\text{Maximize } Z_q = \frac{1}{V'} = y_1 + y_2 + y_3 + 0s_1 + 0s_2 + 0s_3$$

Subject to

$$3y_1 + y_2 + y_3 + s_1 = 1$$
$$y_1 + y_2 + 5y_3 + s_2 = 1$$
$$y_1 + 4y_2 + y_3 + s_3 = 1$$
$$y_1, y_2, y_3, s_1, s_2, s_3 \geq 0$$

Forming the simplex table.

| B.V. | $C_B$ | $Y_B$ | $y_1$ | $y_2$ | $y_3$ | $s_1$ | $s_2$ | $s_3$ | Min $\frac{Y_B}{y_K}$, $y_K > 0$ |
|---|---|---|---|---|---|---|---|---|---|
| | $C_j \rightarrow$ | | 1 | 1 | 1 | 0 | 0 | 0 | |
| $s_1$ | 0 | 1 | $\boxed{3}$ | 1 | 1 | 1 | 0 | 0 | $\frac{1}{3} \rightarrow$ |
| $s_2$ | 0 | 1 | 1 | 1 | 5 | 0 | 1 | 0 | 1 |
| $s_3$ | 0 | 1 | 1 | 4 | 1 | 0 | 0 | 1 | 1 |
| | $Z_q = 0$ | | $-1\uparrow$ | $-1$ | $-1$ | 0 | 0 | 0 | $\leftarrow \Delta_j$ |
| | | | | $y_1$ enters and $s_1$ leaves | | | | | |
| $y_1$ | 1 | $\frac{1}{3}$ | 1 | $\frac{1}{3}$ | $\frac{1}{3}$ | $\frac{1}{3}$ | 0 | 0 | $\frac{1}{3} \times 3 = 1$ |
| $s_2$ | 0 | $\frac{2}{3}$ | 0 | $\frac{2}{3}$ | $\frac{14}{3}$ | $-\frac{1}{3}$ | 1 | 0 | $\frac{2}{3} \times \frac{3}{2} = 1$ |
| $s_3$ | 0 | $\frac{2}{3}$ | 0 | $\boxed{\frac{11}{3}}$ | $\frac{2}{3}$ | $-\frac{1}{3}$ | 0 | 1 | $\frac{2}{3} \times \frac{3}{11} = \frac{2}{11} \rightarrow$ |
| | $Z_q = \frac{1}{3}$ | | 0 | $-\frac{2}{3}\uparrow$ | $-\frac{2}{3}$ | $\frac{1}{3}$ | 0 | 0 | |
| | | | | $y_2$ enters and $s_3$ leaves | | | | | |
| $y_1$ | 1 | $\frac{3}{11}$ | 1 | 0 | $\frac{3}{11}$ | $\frac{4}{11}$ | 0 | $-\frac{1}{11}$ | $\frac{3}{11} \times \frac{11}{3} = 1$ |
| $s_2$ | 0 | $\frac{6}{11}$ | 0 | 0 | $\boxed{\frac{50}{11}}$ | $-\frac{3}{11}$ | 1 | $-\frac{2}{11}$ | $\frac{6}{11} \times \frac{11}{50} = \frac{6}{50} \rightarrow$ |
| $y_2$ | 1 | $\frac{2}{11}$ | 0 | 1 | $\frac{2}{11}$ | $-\frac{1}{11}$ | 0 | $\frac{3}{11}$ | $\frac{2}{11} \times \frac{11}{2} = 1$ |
| | $Z_q = \frac{5}{11}$ | | 0 | 0 | $-\frac{6}{11}\uparrow$ | $\frac{3}{11}$ | 0 | $\frac{2}{11}$ | $\Delta_j$ |
| | | | | $y_3$ enters and $s_2$ leaves | | | | | |
| $y_1$ | 1 | $\frac{6}{25}$ | 1 | 0 | 0 | $\frac{19}{50}$ | $-\frac{3}{50}$ | $-\frac{2}{50}$ | – |
| $y_3$ | 1 | $\frac{3}{25}$ | 0 | 0 | 1 | $-\frac{3}{50}$ | $\frac{11}{50}$ | $-\frac{1}{25}$ | – |
| $y_2$ | 1 | $\frac{4}{25}$ | 0 | 1 | 0 | $-\frac{2}{25}$ | $-\frac{1}{25}$ | $\frac{7}{25}$ | – |
| | $Z_q = \frac{13}{25}$ | | 0 | 0 | 0 | $\frac{6}{25}$ | $\frac{3}{25}$ | $\frac{4}{25}$ | $\Delta_j \geq 0$ |

$$\therefore \quad \text{Maximum } Z_q = \frac{13}{25} = \frac{1}{V'} \Rightarrow V' = \frac{25}{13}$$

$$\therefore\ V \text{ (Original value of game)} = V' - C = \frac{25}{13} - 2 = -\frac{1}{13}$$

Optimal mixed strategies for player $B$ are given by

$$q_1 = y_1 V' = \left(\frac{25}{13}\right)\cdot\left(\frac{6}{25}\right) = \frac{6}{13}$$

$$q_2 = y_2 V' = \left(\frac{25}{13}\right)\cdot\left(\frac{4}{25}\right) = \frac{4}{13}$$

$$q_3 = y_3 V' = \left(\frac{25}{13}\right)\cdot\left(\frac{3}{25}\right) = \frac{3}{13}$$

Optimal strategies for $A$ are obtained from solution of dual of the above problem, i.e. $\Delta_j$ values for $s_1$, $s_2$, and $s_3$

$$x_1 = \frac{6}{25}, x_2 = \frac{3}{25}, x_3 = \frac{4}{25}$$

$$\therefore \quad p_1 = x_1 V' = \frac{6}{13}$$

$$p_1 = x_2 V' = \frac{3}{13}$$

$$p_1 = x_3 V' = \frac{4}{13}$$

$\therefore$ Optimal strategy for $A$ is $\left(\frac{6}{13}, \frac{3}{13}, \frac{4}{13}\right)$.

Optimal strategy for $B$ is $\left(\frac{6}{13}, \frac{4}{13}, \frac{3}{13}\right)$.

Value of game to player $A$ is $-\frac{1}{13}$.

## PREVIOUS YEARS QUESTIONS

**Example 10.14** *A and B play a game in which each has three coins a 5p, a 10p and a 20p. Each player selects a coin with the knowledge of the other;s choice. If the sum of the coins is an odd amount. A wins B's coin, if the sum is even, B wins A's coin, find the best strategy for each player and the value of the game.*

**[GGSIPU, 2010]**

***Solution*** Pay off matrix is

Player *B*

| | | | |
|---|---|---|---|
| | −5 | 10 | 20 |
| Player *A* | 5 | −10 | −10 |
| | 5 | −20 | −20 |

Game has no saddle point

Third strategy of *B* is dominated by his second strategy

$$\therefore \text{ Reduced pay off matrix } A \overset{B}{\begin{bmatrix} -5 & 10 \\ 5 & -10 \\ 5 & -20 \end{bmatrix}}$$

Second strategy of player *A* is dominated by third strategy

$$\therefore \text{ Reduced pay off matrix } A \overset{\text{B}}{\begin{bmatrix} -5 & 10 \\ 5 & -10 \end{bmatrix}}$$

**By Algebraic Method**

$$p_1 = \frac{a_{22} - a_{21}}{(a_{11} + a_{22}) - (a_{12} + a_{21})}$$

$$= \frac{-10-5}{(-15-10)-(10+5)} = \frac{-15}{-15-15} = \frac{-15}{-30} = \frac{1}{2}$$

$$p_2 = 1 - p_1 = \frac{1}{2}$$

$$q_1 = \frac{a_{22} - a_{12}}{(a_{11} + a_{22}) - (a_{12} + a_{21})}$$

$$= \frac{-10-10}{-15-15} = \frac{-20}{-30} = \frac{2}{3}$$

$$q_2 = 1 - q_1 = \frac{1}{3}$$

$$\text{Game value } (V) = \frac{(a_{11} \cdot a_{22}) - (a_{12} \cdot a_{21})}{(a_{11} + a_{22}) - (a_{12} + a_{21})}$$

$$= \frac{(-5 \times -10) - (10 \times 5)}{-15-15}$$

$$= \frac{50-50}{-30} = 0$$

Hence, the optimum strategies are that, none of the players should select the 20p coin. Player $A$, should seletct the 5p coin and 10p coin equally while player $B$, should select the 5p coin for $\frac{2}{3}$ of the time and 10p coin for $\frac{1}{3}$ of the time. The value of game $V = 0$.

**Example 10.15** *Reduce the following game by dominance and find the game value.*

| | | Player B | | | |
|---|---|---|---|---|---|
| | | I | II | III | IV |
| | I | 3 | 2 | 4 | 0 |
| Player A | II | 3 | 4 | 2 | 4 |
| | III | 4 | 2 | 4 | 0 |
| | IV | 0 | 4 | 0 | 8 |

[GGSIPU, 2011]

*Solution* Game has no saddle point

**Step 1 :** First row $R_1$ is dominated by third row because all the element of first row are less than or equal to third row $R_3$. ($R_1$ is deleted) and column one ($C_1$) is dominated to column third ($C_3$) ($\therefore$ $C_1$ is deleted), then reduced pay off matrix is

| | II | III | IV |
|---|---|---|---|
| II | 4 | 2 | 4 |
| III | 2 | 4 | 0 |
| IV | 4 | 0 | 8 |

**Step 2 :** Average of column III-IV be

$$\left(\frac{2+4}{2}, \frac{4+0}{2}, \frac{0+8}{2}\right) \text{ i.e. } (3, 2, 4)$$

All the elements of II column are greater than or equal to average strategy of III-IV column. ($\therefore$ II column ($C_2$) is deleted.

Then reduced pay off matrix is

| | III | IV |
|---|---|---|
| II | 2 | 4 |
| III | 4 | 0 |
| IV | 0 | 8 |

**Step 3 :** Average of rows III-IV is $\left(\frac{4+0}{2}, \frac{0+8}{2}\right)$ i.e. (2, 4) equal to II row ($R_2$). Hence row II is dominated by average of rows III-IV ($\therefore$ Row II is deleted).

Now, Reduced pay off matrix (2 × 2) is

Player *B*

$$\text{Player } A \quad \begin{array}{c} \\ \text{III} \\ \text{IV} \end{array} \begin{array}{c} \begin{array}{cc} \text{III} & \text{IV} \end{array} \\ \begin{bmatrix} 4 & 0 \\ 0 & 8 \end{bmatrix} \end{array}$$

Solving by Arithmetic Method

| | III | IV | |
|---|---|---|---|
| III | 4 | 0 | 8 |
| IV | 0 | 8 | 4 |
| | 8 | 4 | 12 |

Sum of the differences is 12 both ways

$$\therefore \quad p_1 = \frac{8}{12} = \frac{2}{3}, p_2 = \frac{4}{12} = \frac{1}{3}$$

$$q_1 = \frac{8}{12} = \frac{2}{3}, q_2 = \frac{4}{12} = \frac{1}{3}$$

and $$V = \frac{4 \times 8 + 0 \times 4}{12} = \frac{8}{3}$$

$\therefore$ (i) Optimal strategy for player $A$ is $\left(0, 0, \frac{2}{3}, \frac{1}{3}\right)$.

(ii) Optimal strategy for player $B$ is $\left(0, 0, \frac{2}{3}, \frac{1}{3}\right)$.

(iii) Value of game $(V) = \frac{8}{3}$.

**Example 10.16** *Obtain the optimal strategies for both persons and the value of the game for zero-sum two person game whose pay off matrix is as follows :*

| | | B | |
|---|---|---|---|
| | | *I* | *II* |
| | *I* | 1 | −3 |
| | *II* | 3 | 5 |
| | *III* | −1 | 6 |
| A | *IV* | 4 | 1 |
| | *V* | 2 | 2 |
| | *VI* | −5 | 0 |

[GGSIPU, 2012]

***Solution*** This 6 × 2 game has no saddle point, let $p_1, p_2, p_3, p_4, p_5$ and $p_6$ be the probabilities by which player *A* uses his strategies I, II, III, IV, V and VI and $V_1$, $V_2$ be the probabilities by which player *B* uses his strategies I, II

$$\therefore \quad p_1 + p_2 + p_3 + p_4 + p_5 + p_6 = 1$$

and

$$V_1 + V_2 = 1$$

Drawing two vertical lines for B's pure strategies I and II, and plotting point (1, −3), (3, 5), (−1, 6), (4, 1), (2, 2) and (−5, 0).

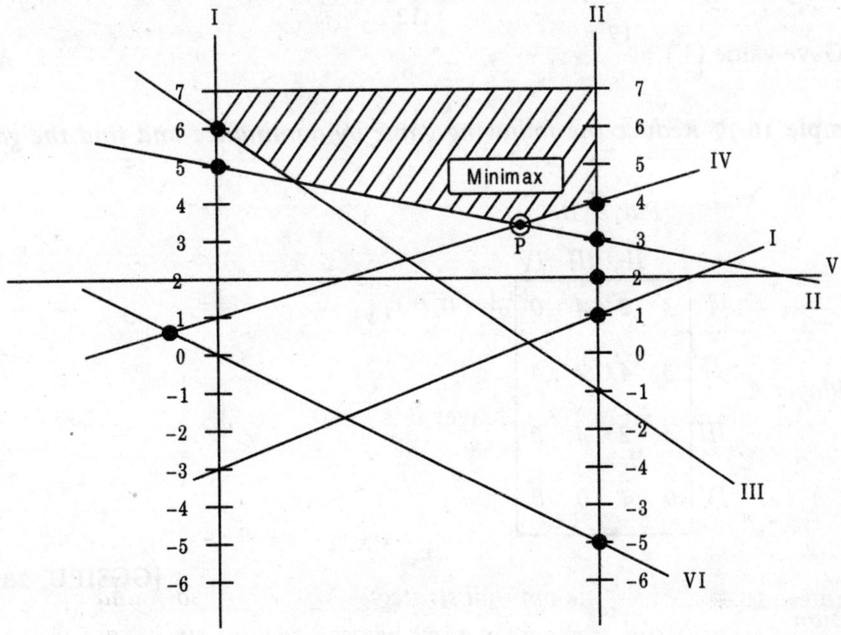


Clearly minimax occurs at *P* which lies on intersection of strategies II and III.

∴ Given game reduces to following 2×2 game.

Solving by arithmetic method :

| | | *B* | | |
|---|---|---|---|---|
| | | I | II | |
| *A* | II | 3 | 5 | 3 |
| | IV | 4 | 1 | 2 |
| | | 4 | 1 | 5 |

$$\therefore \quad P_1 = \frac{4}{5}, P_2 = \frac{1}{5} \text{ and } q_1 = \frac{3}{5}, q_2 = \frac{2}{5}$$

$$V = \frac{(4\times4)+(1\times1)}{5} = \frac{17}{5}$$

Optimal strategies for $A$ are $\left(0, \frac{3}{5}, 0, \frac{2}{5}, 0, 0\right)$.

Optimal strategies for $B$ are $\left(\frac{4}{5}, \frac{1}{5}\right)$.

Gave value $(V) = \frac{17}{5}$. **Ans.**

**Example 10.17** *Reduce the following game by dominance and find the game value.*

| | | *Player B* | | | |
|---|---|---|---|---|---|
| | | *I* | *II* | *III* | *IV* |
| *Player A* | *I* | 3 | 2 | 4 | 0 |
| | *II* | 3 | 4 | 2 | 4 |
| | *III* | 4 | 2 | 4 | 0 |
| | *IV* | 0 | 4 | 0 | 8 |

**[GGSIPU, 2013]**

***Solution***

| | | Player B | | | |
|---|---|---|---|---|---|
| | | I | II | III | IV |
| Player A | I | 3 | 2 | 4 | 0 |
| | II | 3 | 4 | 2 | 4 |
| | III | 4 | 2 | 4 | 0 |
| | IV | 0 | 4 | 0 | 8 |

It is clear that this game has no saddle point.

$\therefore$ $R_4$ dominated to $R_1$, Hence $R_1$ deleted and $C_1$ dominated to $C_3$, Hence $C_1$ deleted.

Player $B$

| | | $B_2$ | $B_3$ | $B_4$ |
|---|---|---|---|---|
| Player $A$ | $A_2$ | 4 | 2 | 4 |
| | $A_3$ | 2 | 4 | 0 |
| | $A_4$ | 4 | 0 | 8 |

Average of $B_3$ and $B_4$ $\left(\frac{(2+4)}{2}, \frac{(4+0)}{2}, \frac{(0+8)}{2}\right) = (3, 2, 4)$

$\therefore$ $B_2$ is superior to average of $B_3$ and $B_4$. Hence $B_2$ deleted.

| | $B_3$ | $B_4$ |
|---|---|---|
| $A_2$ | 2 | 4 |
| $A_3$ | 4 | 0 |
| $A_4$ | 0 | 8 |

Now Average of $A_3$ and $A_4$ $\left(\frac{(4+0)}{2}, \frac{(0+0)}{2}\right) = (2, 4)$

$\therefore$ Player $A$ will gain the same amount even if the strategy $A_2$ is never used. Hence $A_2$ deleted.

Reduced pay-off Matrix is

Player $B$

| | | $B_3$ | $B_4$ |
|---|---|---|---|
| Player $A$ | $A_2$ | 4 | 0 |
| | $A_3$ | 0 | 8 |

This game has no saddle point. Let player $A$ choose his strategies $A_3$ and $A_4$ with probability $p_1$ and $p_1$. Similarly for player $B$ with probability $q_1$ and $q_2$.

$$\Rightarrow p_1 + p_2 = 1 ;$$

$$q_1 + q_2 = 1$$

| | |
|---|---|
| $\Rightarrow \quad 4p_1 + 0\times p_2 = 0\times p_1 + 8p_2$ | $4q_1 + 0\times q_2 = 0\times q_1 + 8\times q_2$ |
| $4p_1 = 8(1-p_1)$ | $4q_1 = 8(1-q_1)$ |
| $p_1 = \frac{2}{3}$ | $q_1 = \frac{2}{3}$ |

Optimal strategics for player $A =$ (0, 0, 2/3,1/3)

Optimal strategics for player $B =$ (0, 0, 2/3, 1/3)

| *Expected gain to A* | *Expected gain to B* |
|---|---|
| $\Rightarrow 4p_1 + 0 \times p_2$ | $\Rightarrow 4q_1 + 0\times q_2$ |
| $\Rightarrow 4\left(\frac{2}{3}\right)$ | $\Rightarrow 4\left(\frac{2}{3}\right) + 0$ |
| $\Rightarrow \frac{8}{3}$ | $\Rightarrow \frac{8}{3}$ |

## MULTIPLE CHOICE QUESTIONS

10.1. A game has saddle point if

(a) Maximin value = Minimax value

(b) Minimim = Minimax

(c) Maximin = Minimin

(d) None of these

10.2. Dominance rule

(a) Reduce the value of game

(b) Increases the size of the payoff matrix

(c) Reduce the size of a payoff matrix

(d) All of the above

10.3. The game is stable when

(a) there is more than two strategies at a time

(b) there is a saddle point

(c) value of game equal to zero

(d) values of game is less than zero

10.4. Every games theory model have

(a) Strategies (b) Saddle point

(c) Probabilities (d) Payoff

10.5. Which one is a zero sum game ?

(a) When players profit must be zero

(b) Repeated game

(c) Chess

(d) None

10.6. The outcome of strategies in game theory is

(a) Probabilities (b) Payoff

(c) Pay back (d) None

10.7. Method of oddments is also known as

(a) Graphical method (b) Algebric method

(c) Sybgames method (d) Arithmetic method

10.8. In graphical method this highest point on the lowest envelope is known as

(a) Saddle point (b) Minimax

(c) Maximini (d) All of the above

10.9. Game is not solvable in arithmatic method if

(a) Row sum ≠ Column sum (b) Row sum = Column sum

(c) Row sum > Column sum (d) Row sum < Column sum

10.10 In algebric method the game valie is

(a) $\dfrac{(a_{11} \cdot a_{21})-(a_{12} \cdot a_{22})}{(a_{11}+a_{12})-(a_{12}+a_{22})}$ (b) $\dfrac{(a_{11} \cdot a_{22})-(a_{12} \cdot a_{21})}{(a_{11}+a_{22})-(a_{12}+a_{21})}$

(c) $\dfrac{(a_{11}+a_{22})-(a_{12}-a_{21})}{(a_{11} \cdot a_{22})-(a_{12} \cdot a_{21})}$ (d) $\dfrac{(a_{11} \cdot a_{22})-(a_{12} \cdot a_{21})}{(a_{11} \cdot a_{21})-(a_{12} \cdot a_{22})}$

## ANSWER

1. (a) 2. (c) 3. (b) 4. (a) 5. (c) 6. (b) 7. (d)
8. (c) 9. (a) 10. (b)

## SUMMARY

**Two Person Zero Sum Game**

In this algebric sum of gains and losses of two players is zero i.e. amount won by a player is exactly equal to amount lost by the other.

(i) **Pure strategy :** A person always selects the same strategy irrespective of other person's strategies. Pure strategies are deterministic.

(ii) **Mixed strategy :** The game in which different strategies are used in different situations. Mixed strategies are probabilistic.

**Pay off matrix :** The outcomes resulting from a game, represented in the form of table is called 'payoff matrix'.

| | | Player Y | | |
|---|---|---|---|---|
| | I | $s_{11}$ | $-s_{12}$ | $s_{13}$ |
| Player X | II | $s_{21}$ | $s_{22}$ | $-s_{23}$ |
| | III | $-s_{31}$ | $s_{32}$ | $s_{33}$ |

**Minimax and Maximin Principle :** This means minimization of maximum losses and maximization of minimum profit.

**Saddle point :** A saddle point of a pay-off matirx is that element in the matrix, where the maximum of row minima coincides with the minimum of column maxima, i.e. minimax and maximin intersect.

**Dominance Rules :** Dominance rules are used to reduce the size of a pay-off matrix.

**Rule 1 : Row dominance :** If all the elements of $i^{th}$ row in a pay off matrix are greater than or equal to corresponding elements of $j^{th}$ row, then we say $i^{th}$ row dominates the $j^{th}$ row and $j^{th}$ row is deleted as player '*A*' will never use $j^{th}$ strategy.

**Rule 2 : Column dominance :** If all the elements in $r^{th}$ column are greater than or equal to corresponding elements in $s^{th}$ column, we say $s^{th}$ column dominates over $r^{th}$ column and $r^{th}$ column is deleted from the pay-off matrix as *B* will never use $r^{th}$ column.

**Rule 3 :** A pure strategy may be dominated, if it is interior to an average of two or more other pure strategies.

**Solution of 2 × 2 Game Without Saddle Point :** When saddle point does not exist for a game, best strategies are mixed strateties.

(i) ***Algebraic Method :*** For any 2 person zero sum game, when the optimal strategies are not pure strategies (i.e. there is no saddle point) and for which *A*'s pay off matrix is

| | | B | |
|---|---|---|---|
| | | I | II |
| A | I | $a_{11}$ | $a_{12}$ |
| | II | $a_{21}$ | $a_{22}$ |

Let *A*'s probability of selection of strategies I and II be $p_1$ and $p_2$, $p_1+p_2=1$. Similarly if *B*'s probabilities of selection of strategies I and II be $q_1$ and $q_2$, $q_1+q_2=1$.

Then
$$p_1 = \frac{a_{22}-a_{21}}{(a_{11}+a_{22})-(a_{12}+a_{21})}, p_2 = 1 - p_1$$

$$q_1 = \frac{a_{22}-a_{12}}{(a_{11}+a_{22})-(a_{12}+a_{21})}, q_2 = 1 - q_1$$

and value of game to player *A* is given by

$$V = \frac{(a_{11} \cdot a_{22}) - (a_{12} \cdot a_{21})}{(a_{11} + a_{22}) - (a_{12} + a_{21})}$$

(ii) ***Arithmetic method for (2×2) game***

**Step 1 :** Find the difference (Oddments) of two numbers in column I and put it under column II.

**Step 2 :** Find the difference (Oddments) of two numbers in column II and put it under column I.

**Step 3 :** Repeat the above two steps for rows also.

**Step 4 :** Divide these differences by row or column total.

**Step 5 :** Value of game is obtained by multiplying row/column oddments by any one row/column of given 2×2 matrix and then dividing by sum.

**Note :** Game is not solvable if row sum ≠ column sum.

# APPENDIX A

## C++ Programming (Operation Research)

### FORMULATION OF LPP/LINEAR PROGRAMMING

```
//include<iostream.h>
//include<conio.h>
void main()
{    clrscr();
     float a[10]|,b[10][10],c[10][10];
     int i,j,m,n,v[ 10],k,l, f=0;
     char ch;
     cout<<"ENTER THE NUMBER OF VARIABLES IN Z\n";
     cin>>m;
     cout<<''\nPress 'M' for maximization and 'm'
for minimization\n";cout<<"ENTER THE COEFFICIENTS OF
VARIABLES\n";
     cin>>ch;
     cout<<"ENTER THE COEFFICIENTS OF VARIABLES\n";
     for(i=0;i<m;i ++ )
     cout<<"ENTER THE NUMBER OF CONSTRAINTS\n";
     cin>>n;
     cout<<"\nENERR THE DATA IN THE FOLLOMING
  ORDER\n 1 .COEFFICIENTS OF VARIABLES\n2 INEQUALITY
SIGN(PRESS 1 FOR <= , 2 FOR >= AND 3 FOR = )\n3.EQUATING
VALUES\n";
```

```
for(i=0;i<n;i++)
{
    for(i=0;j<m+2;j++)
    cin>>b[i][j];
}
for(k=0;k<n;k++)
{
    for(l=0;l<n;l++)
    {
        if(k==l)
        c[k][l]=1
        else
        e[k][l]=0
    }
}
cout<<"\n\nTHE LINEAR PROBLEM ENTERED IS\n";
switch(ch)
{
    case 'M':
    cout<<"Maximize "; break;
    case 'm':
    cout<<" Minimize ";
    break;
    default:
    cout<<"Invalid entry";
}

cout <<"Z=";
for(i=0;i<m;i++)
{
    cout<<a[i]<<"x"<<i+1;
    if(a[i]>0&&i!=0&&i!=m-1)
    cout<<"+";
}
```

```
cout<<"\n\nTHE CONSTRAINTS ARE\n";
for(i=0;i<n;i++)
{
     for(j=0;j<m+2;j++)
     {
            if(j <n-1)
            {
                  cout<<b[i][j]<<"x"<<j+1;
                  if(j!=n-2)
                  cout<<"+";
            }
            if(j==n-1)
            {
                  if(b[i][j]==1
                  cout<<"<="
                  if(b[i][j]==2)
                  cout<<">=";
                  if(b[i][j]==3)
                  cout<<"=";
                  }
                  if(j==n)
                  cout<<b[i][j];

            }
            cout<<endl;
     }
       getch();
```

## SIMPLEX TABLE

```
#include<conio.h>
#include<iostream.h>
void main()
{
int  var,con,obj[10],i,j,a[10][10],temp=0,pos;
```

```
float b[10];
clrscr();
cout<<"Enter the no. of variables and constraints\n";
cin>>var>>con;
cout<<"Enter the constants of objective function\n";
for(i=0;i<var;i++)  "
cin»obj[i];
cout<<"Enter the constants of constraints according
to equation(axl+bx2=d)\n";
for(i=0;kcon;i++)
fo2j=0;j<=var;j++)
cin>>a[i][j];
cout<<"\nObjective function :\nZ=";
for(i=0;i<var;i++)
if(i==var-1)
cout<<obj[i]<<"x"<<i+1;
else
cout<<"\nConstraints :\n"; for(i=0;i<con;i++)
{
for(j=0;j<=var;j++)
if(j==var-1)
cout<<a[i][j]<<"x"<<j+1<<"<=";
else
     if(j==var)
     cout<<a[i][j];
     else
     cout<<a[i][j]<<"x"<<j+1<<"+";
cout<<"\n";}
cout<<"\n\n";
for(i=0;i<var;i++)
if(obj[i]>temp)
{temp=obj[i];
pos=i;}
else continue;
```

```
for(i=0;i<con;i++)
b[i]=a[i][var]/a[i][pos];
cout<<"\tcj";
for(i=0;i<var+con;i++)
if(i<var)
cout<<"\t"«obj[i];
else
cout<<"\t0";
cout<<"\nCb\tXj\t";
for(i=0;i<var;i++)
cout<<"x"<<i+1<<"\t";
for(i=0;i<con;i++)
cout<<"s"<<i+1<<"\t";
cout<<"b\tMean Ratio\n";
for(i=0;i<con;i++)
{cout<<"0\ts"<<i+1<<"\t";
for(j=0;j<var+con+2;j++)
if(j<var)
cout<<a[i]0]<<"\t";
else
    if(j==var+con+1)
    cout<<b[i]<<"\t";
    else
        if(j==con+var)
        cout<<a[i][var]<<"\t";
        else
            if(j==i+var);
            cout<<"1\t";
            else
            cout<<"0\t";
cout<<"\n";}
cout<<" \tZj\t";
for(i=0;i<var+con;i++)
cout<<"0\t";
cout<<"\n \tCj-Zj\t";
```

```
for(i=0;kvar+con;i++)
if(i<var)
cout<<obj[i]<<"\t";
else
cout<<"0\t";
getch();}
```

## SIMPLEX METHOD

```
#include<iostream.h>
#include<stdio.h>
#include<conio.h>
#include<math.h>
#include<stdlib.h>
simp(float a[2][2],float n[2][2],float b[2], float
s[2],float c[2].float z,float k[2])
{
    int  i,j;
    for(i=0;i<2;i++)
    {
        z+=(b[i]*s[i]);
    }
    for(j=0;j<2;j++)
    {
        k[j]=z-c[j];
    }
    cout<<"\n\t\t\t  "<<c[0]<<  "\t"<<c[1]<<
"\t"<<s[0]<<  "\t"<<s[1];
    cout<<"\n  cb\tbv\txb\tx1\tx2\ts1\ts2   ";
  cout<<"\n"<<s[0]<<"\ts1"<<"\t"<<b[0]<<"\t"<<a[0]
[0]<<"\t"<<a[0][1]<<"\t"<<n[0][0]<<"\t"<<n[0][1]<<
"\t";
  cout<<"\n"  <<s[1]<<"\ts2"  <<"\t"<<b[1]<<"\t"
<<a[1][0]<<"\t"<<a[1][1]<<"\t"<<n[1][0]<<"\t"<>[1][1]<<"\t";
    cout<<"\n\t\t  "<<z<<"\t"<<k[0]<<  "\t"<<k[1]<<
"\t"<<s[0]<<"\t"<<s[1];
```

```
      return 0;
   }
   void  display (char A[4],char  B[3],float  C[3][5])
   {
      cout<<"bv"<<"\t";
      for(int  i=0;i<4;i++)
            cout<<A[i]<<"\t";
      cout<<"sol."<<"\n";
      for(i=0;i<3;i++)
      {
            if(i!=2)
                  cout<<B[i];
            for(int  j=0;j<5;j++)
                  cout<<C[i][j]<<"\t";
            cout<<endl;
      }
      cout<<endl;
   }
   void main()
   {
      int i,j,k;
      float a[2][2],c[2],b[2],u[2],s[2],n[2][2];
      float z=0,k1[2];
      float m[3][5];
      char a[4],B[3];
      float C[3][5];
      clrscr();
      cout<<"\nplease enter the maximization condition
for x1 and x2:";
      for(i=0;i<2;i++)
      {
            cin>>c[i];
            s[i]=0;
   }
```

```
cout<<\nenter the cofficients of constraints
equations for x1 and x2:";
for(i=0;i<2;i++)
{
    for (j=0;j<2;j++)
        cin>>a[i][ j] ;
}
cout<<"\n enter the value of limitation for
constraint equation:";
for(i=0;i<2;i++)
    cin>>b[i];
cout<<"\n enter the value of coordinates in first
quadrant:";
for(i=0;i<2;i++)
cin>> u [i]; cout <<\n enter the coefficients of
identity matrix:";
for (i=0; i<2;i++)
{
    for(j=0;j<2;j++)
        cin>>n[i][i];
}
/* DIAPLAY OF LPP*/
i=0;
cout<<"\n maximize the obiective:";
cout<<"xl "<<c[i]<<"+ x2" <<c[i+1];
cout<<"\n subject to constraints";
j=0;
for(i=0;i<2;i++)
{
    cout<<"\n xl "<<a[i][j]<<"+ x2" <<a[i] [j+1]<<"
<="<<b[i];
}
i=0;
cout<<"\n x1>="<<u[i];
cout<<"\n x2>="<<u[i+1];
```

```
/* CONVERTIN NEGATIVE CONSTANTS"/
j=0;
for(i=0; i<2 ;i++)
{
     if(b[i]<0)
     }
                    a[i][j]=-1;
                    a[i][j+1]*=-1;
                    b[i]*=-1;
     }
}
i=0;
ccut<<"\n maximize the objective:";
cout<<"x1"<<c[i]<<"+x2"<<c[i+1]<<"+0s1+0s2";
cout<<"\n subject to constraints";
j=0;
i=0;
{
     cout<<"\n x1 "<<a[i][j]<<"+ x2  "<<a[i] [j+1]<<"+
s1="<<b[i];
     cout<<"\n  x1  "<<a[i+1][j]<<"+  x2   "<<a[i+1]
[j+1]<<"+ s1="<<b[i+1];
}
i=0;
/*DISPLAY OF FIRST SIMPLEX TABLE*/
simp(a,n,b,s,c,z,k1);
/* second table*/
cout<<"\n  Enter teh non-basic variables:";
for(i=0;i<4;i++)
     cin>>A[i] ;
cout<<"\n  Enter teh basic variables : ";
for(i=0;i<2;i++)
     cin<<B[i] ;
cout<<"\n Enter the values of first table in a
matrix:   ";
```

```
for(i=0;i<3;i++)
     for(j=0;j<5;j++)
          cin>>m[i][j]
/* DISPLAY SIMPLEX TABLE*/
displdy(A, B,C) ,
/* TO CALCULATE MINIMUM OF ZJ-CJ*/
x :
float d[4];
for(i=0;i<4;i++)
     d[i]=C[2][i];
float min=d[0];
int pos=0;
for(j=1;j<4j++)
{
     if (d[j]<min)
     {
          min=d[j];
          pos=j;
     }
}
cout<<"\n The min. value of Zj-Cj is: "<<min;
cout<<"\n The  location corresponds to c[2]
["<<pos<<"] ";
cout<<"\n The entering variable is : "<<A[pos] ;

/*TO CALCULATE THE LEAVING VARIABLE
AN ARRAY 'E' TO STORE THE VALUE OF THE SOLUTION
AN ARRAY 'F' TO STORE THE VALUE OF THE ALPHA j's
AN ARRAY 'G' TO STORE THE MINIMUM RATIO
*/
k=0;
float e[4];
for(i=0;i<2.i++)
{
```

```
        if(C[i][pos]>0)
        {
            e[k]=C[i][4]/C[i][pos];
            k++;
        }
    }
    float low=e[0];
    float flag=0;
    for(i=0;i<k;i++)
    {
        if (e[i]<low)
        {
            low=e[i];
            flag=i;
        }
    }
    cout<<"\n The minimum ratio is:"<<low;
    cout<<"\n the row is : "<<flag;
    cout<<"\n location of pivot element is : "<<C[flag]
[pos];
    cout<<"\n The leaving variable is : "<<B[flag]<<endl;
    /*  PROCEDURE  TO  SOLVE */
    float pivot=C[flag][pos];
    for(j=0;j<5;j++)
    {
        C[flag][j]=C[flag][j]/pivot;
        cout<<C[flag][j]<<"\t";
    }
    B[flag]=A[pos];
    for(i=0;i<3;i++)
    {
        float y=C[i] [pos] ;
        cout<<"\n  ROw:  "<<i+1;
        for(j=0;j<5;j++)
```

```
    {
          if(i!=flag)
          {
                c[i][j]=(C[i][j]-(C[flag][j]*y));
                cout<<"\t"<<C[i][j];
          }
    }
    cout<<"\n";
}
cout<<"\n \n ";
/* DISPLAY OF NEXT SIMPLEX TABLE */
display (A, B, C);
getch ();
int leave=0;
for (j=0;j<5; j++)
{
    if(c[2][j]<0)
    {
          Leave=1;
          goto x;
    }
    else
          leave=0;
}
if (leave==0)
{
    cout<<"\n OPTIMAL ";
    getch ();
    exit(0);
}
cout<<end]<<end];
display(A,B,C);
getch();
}
```

## PRIMAL INTO DUAL

```
#include<iostream.h>
#include<conio.h>
void main()
{
int p,q,a[10],b[10][10],c[10],d[10],i,j;
clrscr();
cout<<"Enter the number of variable of objective
function, number of constrict equation    :";
cin>>p>>q;
cout<<"\nEnter cofficient of objective : ";
for(i=1;i<=p;i++)
cin>>a[i];
int m;
cout<<"\n Enter \n 1   to minimize \n 2   to
maximize\n";
cin>>m;
cout<<"\nPlease enter constrict equations
::::::";
for(i=1;i<=q;i++)
{   cout<<"\nEnter "<<i<<" equation    ====>";
for(j=1;j<=p;j++)
  { cout<<"\nEnter coefficient "<<j;
    cin>>b[i][j];
    if(j==p)
     { cout<<"\n wheather you want to maximize or
minimize this equation:\nEnter \n1.   for <=\n2.
for  >=";
      cin>>c[i];
      cout<<"Enter constant";
      cin>>d[i];
    }
  }
}
```

```
cout<<"\n\nequation you entered are    ::::\n";
if(m==1)
cout<<"MIN   ";
else
cout<<"MAX   ";
for(i=1;i<=p;i++)
cout<<a[i]<<"x"<<i<<" + ";
cout<<"\b\nconstinct equation are\n";
for(i=1;i<=q;i++)
{
  for(j=1;j<=p;j++)
  {
    cout<<b[i][j]<<"x"<<j<<" + ";
    if(j==p)
    { cout<<"\b\b\b";
      if(c[i]==1)
      cout<<" <= "<<d[i];
      else if(c[i])
      cout<<" >= "<<d[i];
    }
  }
  cout<<"\n";
}
//now solving standard form=======================
===========================
for(i=1;i<=q;i++)
{if(c[i]==2)
 for(j=1;j<=p;j++)
 { b[i][j]=-1*b[i][j];


   c[i]=1;
   d[i]=-1*d[i];


 }
```

```
}
cout<<"\n\nstandard form is  ::::\n";
if(m==1)
cout<<"MIN   ";
else
cout<<"MAX   ";
for(i=1;i<=p;i++)
cout<<a[i]<<"x"<<i<<" + ";
cout<<"\b\nconstinct equation are\n";
for(i=1;i<=q;i++)
{
  for(j=1;j<=p;j++)
  {
    cout<<b[i][j]<<"x"<<j<<" + ";
    if(j==p)
    { cout<<"\b\b\b";
      if(c[i}==1)
      cout<<" <= "<<d[i];
      else if(c[i])
      cout<<" >= "<<d[i];
    }
  }
  cout<<"\n";
}

//Now finding dual of equations=====================
=========================

for(i=1;i<=p;i++)
for(j=1;j<=q;j++)
b[i][j]=b[j][i];
m=1;
cout<<"\n\ndual is";
if(m==1)
cout<<"MIN   ";
```

```
else
cout<<"MAX    ";
for(i=1;i<=q;i++)
cout<<d[i]<<"x"<<i<<" + ";
cout<<"\b\nconstinct equation are\n";
for(i=1;i<=p;i++)
{
   for(j=1;j<=q;j++)
   {
      cout<<b[i][j]<<"x"<<j<<" + ";
      if(j==q)
      { cout<<"\b\b\b";
        c[i]=2;
        if(c[i]==1)
        cout<<" <= "<<a[i];
        else
        cout<<" >= "<<a[i];
      }
   }
   cout<<"\n";
}
getch();
}
```

## TRANSPORTATION PROBLEM (NORTH-WEST)

```
#include<iostream.h>
#include<conio.h>
#include<math.h>
void main ()
{
clrscr() ;
int sum=0, row=0, col=0, s,d,u[10], v[10],
cost[10][10],c[10][10],i, j,k=0,l=0;
cout<<"\nenter the no. of sources:";
```

```
cin>>s ;
cout<<"\nenter the no. of demand centres:
cin>>d;
cout<<"\nenter  the   cost  units:\n";
for(i=0;i<s;i++)
{
for(j=0;j<d;j++)
{
cin>>cost[i] [j] ;
}
}
cout<<"\nenter  the  supply values : \n";
for(i=0;i<s;i++)
cin>>u[i] ;
cout<<"\nenter the demand values : \n";
for(j=0;j<d;j++)
cin>>v[j];
for(i=0;i<s;i++)
{
cout<<endl; for(j=0;j<d;j++)
{
cout<<cost [i][ j ] <<"\t";
}
cout<<u [i];
}
cout<<endl; for'(j=0; j<d;j++)
{
cout<<v [j ] << " \t";
}
for(i=0;i<s;i++)
k+=u[i];
for(j=0;j<d;j++)
1+=v[j];
cout<<"\nui  = "<<k;
```

```
cout<<"\nvj    = "<<l;
cout<<"\ndifference : ";
cout<< fabs (k-1) ;
if(fabs(k-l)==0)
{
cout<<"\nthe  problem is   balanced\n\n\n";
}
else
     {
     cout<<"\nthe problem is unbalanced";
     if(k>1)
     {
            d=d+1;
            for(i=0;i<s;i++)
            {
            cost[i][d-1]=0;
            }
            v[d-l]=k-l;
            cout<<endl;
            for(i=0;i<s;i++)
            {
             cout<<"\n";
            for(j=0;j<d;j++)
            {
            cout<<cost[i][j]<<"\t";
            }
            cout<<u[i];
            }
            cout<<endl;
            for(j=0;j<d;j++)
            cout<<v[j]<<"\t";
      }
      else
      {
```

```
            s=s+1;
            for(j=0;j<d;j++)
            {
            cost[s-1][j]=0;
             }
            u[s-1]=1-k;
            cout<<endl;
            for(i=0;i<s;i++)
            {
            cout<<"\n";
            for(j=0;j<d;j++)
            {
            cout<<cost[i][j]<<"\t";
            }
            cout<<u[i] ;
            }
            cout<<endl;
            for(j=0;j<d;j++)
            cout<<v[ j ] ;
      }
  }
  cout<<"\n\n\n";
  start:
  if (row<s && col<d)
  {
  if(v[col]<u[row])
  {
      c[row][col]=v[col];
      u[row]=u[row]-v[col];
      v[col]=0;
      for(i=row+1;i<s;i++)
      c[i][col]=0;
      col++;
      goto start;
```

```
}
else if(v[col]>u[row])
{
    c[row][col]=u[row];
    v[col]=v[col]-u[row]
    u[row]=0;
    for(i=col+1;i<d;i++)
    c[row][i]=0;
    row++;
    goto start;
}
else if(u[row]==v[col])
{
    c[row][col]=u[row];
    u[row]=0;
    v[col]=0;
}
}
else
goto end;
end:
for(i=0;i<s;i++)
{
cout<<"\n"; for(j=0;j<d;j++)
  {
cout<<c[i][j]<<"\t";
  }
cout<<u[i];
}
cout<<endl;
for(j=0;j<d;j++)
cout<<v[j]<<"\t";
for(i=0;i<s;i++)
{
```

```
for(j=0;j<d;j++)
{
sum+=c[i][j]*cost[i][j];
}
}
cout<<"\ntotal cost = ";
cout<<sum;
getch();
}
```

## ASSIGNMENT PROBLEM

**Method I**

```
#include<iostream.h>
#include<conio.h>
#jnclude<string.h>
Void display();
Void mini();
lntcheck();
lntw,cst[10][10],cst2[10][10],ass[10][10],l,j,k,min[10],m=0;
Void main()
{ClrscrO;
Cout<<"Enter the number of worker and jobs : \n";
Cin>>w;
Cout<<"Enter the costs of assignment: \n";
For(i=1;i<=w;i++)
{
Cout<<"Now enter costs for job"<<i<<"and\n";
For(j=l;j<=w;j++)
{
Cout<<"for factory "<<j<<" : ";
Cin>>cst[i][j];
Ass[i][j]=0;
}}
Display();
```

```
Mini();
Display();
M= check();
Cout<<"The indicated '1' are the optimal solution :
\n";
Cout<<"WORKERS————JOBS——\n";
For(i=1;i<=w;i++)
{
For(j=1;j<=w;j++)
{
Cout<<"   "<<ass[i][j];
} Cout<<"\n";
}Getch();
}
Void display()
{
Cout<<"WORKERS--JOBS---\n";
For(i=1;i<=w;i++)
{
For(j=1;j<=w;j++)
{cout<<"    "<<cst[i][j];
}
cout<<"\n";
}}
Void mini()
{
For (i=1;i<=w;i++)
{min[i]= 1000;
For(j=1;j<=w;j++)
{
If(cst[i][j]<min[i])
Min[i]=cst[i][j];
}}
 For(i=1;if=w;i++)
```

```
{
For(j=1;j<=w;j++)
{
Cst[i][j]=cst[i][j]-min[i];
}}}
Intcheck()
{
lnt c=0;d=0;
For(i=1;i<=w;i++)
{
For (j=1;j<=w;j++)
  {
lf(cst[i][j]==0)
{for(k=1;k<l;K++)
     {if(ass[k][j]==0)
           Ass[i][j]=1;
           }
lf(i==1)
     Ass[i][j]=1;
Break;
}}}
For(i=0;i<=w;i++)
{For(j=0;j<=w;j++)
{
if(ass[i][j]==1)
  C++;
if(ass[j][i]==1)
  D++;
}}
if((c==(w-1)||(d==(w-1)))
M=1;
Returen(m);}
```

**Method II**

```
#include<iostream.h>
#include<conio.h>
void main()
{
int  t[10][10],i,j,n;
clrscr();
cout<<"how many workers and job do you have :";
cin>>n;
cout<<"\nEnter time of\n\n";
for(i=1;i<=n;i++)
for(j=1;j<=n;j++)
{ cout<<"worker "<<i<<"  job "<<j<<"\t:";
cin>>t[i][j];
}
cout<<"\nDATA YOU ENTERED IS :\n\t\t";
for(i=1;i<=n;i++)
cout<<"job"<<i<<"\t";
for(i=1;i<=n;i++)
{
cout<<"\nworker"<<i<<"\t\t";
for(j=1;j<=n;j++)
{
cout<<t[i][j]<<"\t";
}
}
getch();

int min[10]= {1000,1000,1000,1000, 1000,1000,1000,
1000,1000};
for(i=1;i<=n;i++)
{

for(j=1;j<=n;j++)
```

```
{
if(t[i][j]<=min[i])
  {
min[i]=t[i][j];
  }
}
}
cout<<endl;

for(i=1;i<=n;i++)
{
for(j=1;j<=n;j++)
{
t[i][j]-=min[i];
}
}
cout<<endl;

cout<<"\n\n*************data after row minimum
decrement is**************\n\t\t";

for(i=1;i<=n;i++)
cout<<"job"<<i<<"\t";
for(i=1;i<=n;i++)
{
cout<<"\nworker"<<i<<"\t\t";
for(j=1;j<=n;j++)
{
cout<<t[i][j]<<"\t";
}
}
getch();

int zerr[10]={1000,1000, 1000,1000,1000,1000},
zerc[10]={1000,1000,1000,1000,1000,1000};
```

```
for(i=1;i<=n;i++)
{
for(j=1;j<=n;j++)
{
if(t[i][j]==0)
{zerr[i]=0;
zerc[j]=0;
}
}
}
int f=0,y=0;
for(i=1;i<=n;i++)
{  if(zerr[i]!=0)
   f=1;
}
for(i=1;i<=n;i++)
{  if(zerc[i]!=0)
   y=1;
}
if((f==1)||(y==1))
{
int mn[10]={1000, 1000,1000, 1000,1000, 1000, 1000,
1000,1000};
for(j=1;j<=n;j++)
{

for(i=1;i<=n;i++)
{
if(t[i][j]<=mn[j])
mn[j]=t[i][j];
}
}

for(j=1;j<=n;j++)
```

```
{
for(i=1;i<=n;i++)
{
t[i][j]-=mn[j];
}
}

cout<<"\n\n*************data after column minimum
decrement is**************\n";

for(i=1;i<=n;i++)
cout<<"job"<<i<<"\t";
for(i=1;i<=n;i++)
{
cout<<"\nworker"<<i<<"\t\t";
for(j=1;j<=n;j++)
{
cout<<t[i][j]<<"\t";
}
}
getch();
}

cout<<"\n\n\n************Final job assignment
is*************";

for(i=1;i<=n;i++)
{
for(j=1;j<=n;j++)
{
if(t[i][j]==0)
{cout<<"\nassign job "<<j<<" to worker "<<i;
 for(int z=1;z<=n;z++)
 {  if(z!=i)
```

```
    {   if(t[z][j]==0)
t[z][j]=1000;
    }
 }

 for(int l=1;l<=n;l++)
 {  if(l!=j)
    {   if(t[i][l]==0)
t[i][l]=1000;
    }
 }
}
}
}
cout<<endl;

getch();
}
```

## SEQUENCING PROBLEM

**Method I**

```
#include<iostream.h>
#include<conio.h>
void main()
{
clrscr();
int seq[10],min,n,i,j,k,l,a[10][10],x,y;
x=0;
cout<<"\nenter the number of jobs:";
cin>>n;
y=n-1;
for(i=0;i<2;i++)
{
cout<<"\nfor m/c #"<<i+1<<" enter : \n";
```

```
for(j=0;j<n;j++)
{
cout<<"time for job#"<<j+1<<" : ";
cin>>a[i][j];
cout<<endl;
}
}
cout<<endl;
for(i=0;i<n;i++)
cout<<"J"<<i+1<<"\t";
for(i=0;i<2;i++)
{
cout<<"\n";
for(j=0;j<n;j++)
cout<<a[i][j]<"\t";
}
start:
min=1000;
for(i=0;i<2;i++)
{
for(j=0;j<n;j++)
{
if(a[i][j]<min)
{
min=a[i][j];
k=i;
l=j;
}
}
}
if(k==0)
{
 seq[x]=l+1;
x++;
```

```
a[k][l]=2000;
a[k+1][l]=2000;
}
else
{
seq[y]=l+1;
y--;
a[k][l]=2000;
a[k-1][l]=2000;
}
if(x>y)
goto end;
else
goto start;
end:
cout<<"\n\nsequence is :";
cout<<endl;
for(i=0;i<n;i++)
{
cout<<seq[i]<<"\t";
}
getch();
}
```

**Method II**

```
#include<stdio.h>
#include<conio.h>
void main()
{
clrscr();
int time[10][10],time1[10][10],i,j,job,s[10],
arr[10][10];
printf("N Jobs 2 machines problem\n");
printf("Enter number of jobs :");
```

```
scanf("%d",&job);
printf("\n\nEnter time of each job on each machine:");
for(i=1;i<=2;i++)
{ printf("\nEnter time for machine %d\n",i);
for(j=1;j<=job;j++)
{
scanf("%d",&time[i][j]);
time1[i][j]=time[i][j];
}
}
printf("\ndata you entered is\n\n\n\t");
for(i=1;i<=job;i++)
{
printf("job %d    ",i);
}
for(i=1;i<=2;i++)
{printf("\nM%d",i);
for(j=1;j<=job;j++)
{
printf("\t%d",time[i][j]);
}
}
int p=1,q,min=1000,r,n;
q=job;
for(int z=1;z<=job;z++)
{min=time[1][1];
for(i=1;i<=2;i++)
{
for(j=1;j<=job;j++)
{
    if(time[i][j]<=min)
    {min=time[i][j];
    r=i;
    n=j;
```

```
}}}
if(r==1)
{s[p]=n;
time[r][n]=2000;
time[r+1][n]=2000;
P++;
}
else if(r==2)
{s[q]=n;
time[r-1][n]=2000;
time[r][n]=2000;
q--;
}}
printf("\n\n Sequence of job assignment is\n\n"};
for(z=1;z<=job;z++)
{printf("J%d\t",s[z]);
}
for(i=1;i<=job;i++)
{
for(j=1;j<=6;j++)
{
if((i==1)&&(j==1))
arr[i][j]=0;
if((j==1)&&(i!=1))
arr[i][j]=arr[i-1][j+2];
if((i==1)&&(j==4))
arr[i][j]=arr[i][j-1];
if((j==4)&&(i!=1))
{if(arr[i][j-1]>=arr[i-1][j+2])
arr[i][j]=arr[i][j-1];
else
arr[i][j]=arr[i-1][j+2];
}
if(j==2)
```

```
{p=s[i];
arr[i][j]=time1[1][p];
}
if(j==5)
{P=s[i];
arr[i][j]=time1[2][p];
}
if(j==3)
arr[i][j]=arr[i][j-2]+arr[i][j-1];
if(j==6)
  arr[i][j]=arr[i][j-2]+arr[i][j-1];
}}
printf("\n\n\t\tm1\t\t\tm2\n\tlP\tP.T\tOUT\tlN\tP.T.\tOUT\n");
for(i=1;i<=job;i++)
{ printf("J%d\t"s[i]);
for(j=1;j<=2*3;j++)
{
printf("%d\t",arr[i][j]); }
printf("\n");
}
printf("\n\n TET is %d",arr[job][6]);
getch();}
```

## QUEUING THEORY

```
#include<iostream.h>
#include<conio.h>
#include<stdio.h>
void main()

{

float lam,mu,uf,Lq,Ls,Wq,Ws;
clrscr();
```

```
cout<<"<<<<<<<<<<<<<<<<<<<QUEUING THEORY>>>>>>
>>>>>>>>>>>>>>>>\n\n";

 cout<<"Enter the value Average arrival
rate(lambda).\n";
 cin>>lam;
 cout<<"Enter the Average service rate(mu).\n";
 cin>>mu;
 Lq=(lam*lam)/(mu*(mu-lam));
 Ls=lam/(mu-lam);
 Ws=1/(mu-lam);
 Wq=(lam/mu)*(1/(mu-lam));
 uf=lam/mu;
 cout<<"QUEUE LENGTH                =" <<Lq<<"\n";
 cout<<"SYSTEM LENGTH               =" <<Ls<<"\n";
 cout<<"WAITING TIME IN QUEUE       =" <<Wq<<"\n";
 cout<<"TOTAL TIME IN THE SYSTEM    =" <<Ws<<"\n";
 cout<<"UTILIZATION FACTOR          =" <<uf<<"\n";
 getch();
}
```

## PERT AND CPM (NETWORK ANALYSIS)

```
#include<iostream.h>
#include<conio.h>
#include<stdio.h>
void main()
{
int activity[30][4],time[30][3],i,j,n,node=0;
clrscr();
cout<<"How many data are there    :";
cin>>n;
cout<<"\nEnter ACTIVITY(i-j) and DURATION\n";
for(i=1;i<=n;i++)
{ cout<<endl<<i<<"  from where to where\n";
```

```
for(j=1;j<=3;j++)
{  if(j!=3)
   {cin>>activity[i][j];
   if(activity[i][j]>=node)
     node=activity[i][j];
   }
   if(j==3)
   { cout<<"\t duration      ";
     cin>>activity[i][j];

   }
}
}

clrscr();
/*
cout<<"Data Entered is   : \n \n activity(i-
j)\tduration\n";
for(i=1;i<=n;i++)
{ for(j=1;j<=3;j++)
  { cout<<activity[i][j];
    if(j==1)
    cout<<"-";
    else if(j==2)
    cout<<"\t\t";
  }
  cout<<endl;
}  */

//*******************calculating EST*********
**********

for(i=1;i<=n;i++)
for(j=1;j<=2;j++)
{
```

```
if(j==1)
time[i][j]=0;
else
time[i][j]=1000;
}

int sum,t1,t2;
for(i=1;i<=n;i++)
{ for(j=1;j<=2;j++)
   { if(activity[i][j]==1)
      {}
      else if(j==2)
      { t1=activity[i][j];
         t2=activity[i][j-1];
         sum=time[t2][1]+activity[i][j+1];
         if(sum>=time[t1][1])
         time[t1][1]=sum;
      }
   }
}
int maxest=0;
//cout<<"\n EST is\t";
for(i=1;i<=node;i++)
{// cout<<time[i][1]<<"\t";
if(time[i][1]>=maxest)
maxest=time[i][1];
}

//*******************calculating LFT*************
**********
int dif;
time[node][2]=maxest;
for(i=n;i>=1;i--)
{ for(j=1;j<=2;j++)
```

```
{ if(activity[i][j]==node)
  {}
  else if(j==1)
  { t1=activity[i][j];
    t2=activity[i][j+1];
    dif=time[t2][2]-activity[i][j+2];
    if(dif<=time[t1][2])
    time[t1][2]=dif;
  }
}
}
/*
cout<<"\n LFT is\t";
for(i=1;i<=node;i++)
cout<<time[i][2]<<"\t";    */

//***************finding out Critical Path*******
***************

cout<<"\n\nCritical path is\n";
for(i=1;i<=node;i++)
{if(time[i][1]==time[i][2])
 { cout<<i;
   if(i!=node)
   cout<<" > ";
 }
}

//***************finding EST, LST, EFT, LFT,
TF*********************************

int est[30],lst[30],eft[30],lft[30],tf[30];
for(i=1;i<=n;i++)
{
```

```
        est[i]=time[activity[i][1]][1];
        lft[i]=time[activity[i][2]][2];
        eft[i]=est[i]+activity[i][3];
        lst[i]=lft[i]-activity[i][3];
        tf[i]=lft[i]-eft[i];
    }

    cout<<"\n\nactivity(i-j)\tduration\ tEST\tLST\ tEFT\
tLFT\ttotal  float\n";
    for(i=1;i<=n;i++)
    {
    cout<<activity[i][1]<<"-"<<activity[i][2]<<"\t\t"
<<activity[i][3]<<"\t\t";
       cout<<est[i]<<"\t" <<eft[i]<<"\t" <<lst[i]<<"\t"
<<lft[i]<<"\t"<<tf[i]<<"\n";
    }

    getch();
    }
```

## GAME THEORY

```
    #include<iostream.h>
    #include<conio.h>
    #include<stdio.h>
    void main()
    {
    clrscr();
    int  mx,mn,i,j,a[10][10],min1[10],max1[10],r,c;
    print)"enter  no.  of  rows  and  columns  in  payoff
matrix\n";
    scanf("%d%d",&r,&c);
    //INPUT OF PAYOFF MATRIX
    printf("enter the payoff matrix\n");
    for(i=1;i<=r;i++)
    {
```

```
for(j=1;j<=c;j++)
  {
  scanf("%d",&a[i][j]);
  }
}

//SOLVING GIVEN PROBLEM

for(i=1;i<=r;i++)
{
min1[i]=1000;
for(j=1;j<=c;j++)
  {
  if(a[i][j]<=min1[i])
  min1[i]=a[i][j];
  }
}
for(j=1;j<=c;j++)
{
max1[j]=-1000;
for(i=1;i<=r;i++)
  {
  if(a[i][j]>=max1[j])
  max1[j]=a[i][j];
  }
}

mx=-1000;
for(i=1;i<=r;i++)
{

if(min1[i]>=mx)
{mx=min1[i];  }
```

```
}
mn=1000;
for(i=1;i<=c;i++)
{
if(max1[i]<=mn)
{mn=max1[i];   }
}

for(i=1;i<=r;i++)
{     cout<<i<<"st\t";
for(j=1;j<=c;j++)
{
cout<<a[i][j]<<"\t";
}
cout<<min1[i]<<"\t|";
if(i==1)
{cout<<mx;}
cout<<"\n";
}
cout<<"\t";
for(i=1;i<=c;i++)
cout<<max1[i]<<"\t";
cout<<"\n---------------------------";
cout<<"\n\t"<<mn;

if(mx==mn)
printf("\nsaddle point=%d",mx);
else
printf("\nsaddle point doesn't exist");
getch();
}
```

# APPENDIX B

## Normal Distribution

Table 1: Givesthe standard normal probability distribution between normal variate Z = 0 (the mean) and a positive value of Z. Area for negative value of Z are obtained by symmetry.

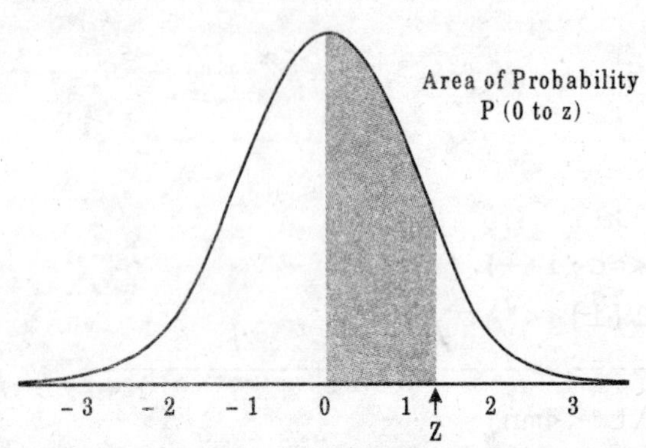


| Z To First Decimal | Second Decimal 0.00 | 0.01 | 0.02 | 0.03 | 0.04 | 0.05 | 0.06 | 0.07 | 0.08 | 0.09 |
|---|---|---|---|---|---|---|---|---|---|---|
| 0.0 | 0.00000 | 0.00399 | 0.00798 | 0.01197 | 0.01595 | 0.01994 | 0.02392 | 0.02790 | 0.03188 | 0.03586 |
| 0.1 | 0.03983 | 0.04380 | 0.04776 | 0.05172 | 0.05567 | 0.05966 | 0.06360 | 0.06749 | 0.07142 | 0.07535 |
| 0.2 | 0.07926 | 0.08317 | 0.08706 | 0.09095 | 0.09483 | 0.09871 | 0.10257 | 0.10642 | 0.11026 | 0.11409 |
| 0.3 | 0.11791 | 0.12172 | 0.12552 | 0.12930 | 0.13307 | 0.13683 | 0.14058 | 0.14431 | 0.14803 | 0.15173 |
| 0.4 | 0.15542 | 0.15910 | 0.16276 | 0.16640 | 0.17003 | 0.17364 | 0.17724 | 0.18082 | 0.18439 | 0.18793 |
| 0.5 | 0.19146 | 0.19497 | 0.19847 | 0.20194 | 0.20540 | 0.20884 | 0.21226 | 0.21566 | 0.21904 | 0.22240 |
| 0.6 | 0.22575 | 0.22907 | 0.23237 | 0.23565 | 0.23891 | 0.24215 | 0.24537 | 0.24857 | 0.25175 | 0.25490 |
| 0.7 | 0.25804 | 0.26115 | 0.26424 | 0.26730 | 0.27035 | 0.27337 | 0.27637 | 0.27935 | 0.28230 | 0.28524 |
| 0.8 | 0.28814 | 0.29103 | 0.29389 | 0.29673 | 0.29955 | 0.30234 | 0.30511 | 0.30785 | 0.31057 | 0.31327 |

| | | | | | | | | | | |
|---|---|---|---|---|---|---|---|---|---|---|
| **0.9** | 0.31594 | 0.31859 | 0.32121 | 0.32381 | 0.32639 | 0.32894 | 0.33147 | 0.33398 | 0.33646 | 0.33891 |
| **1.0** | 0.34134 | 0.34375 | 0.34614 | 0.34849 | 0.35083 | 0.35314 | 0.35543 | 0.35769 | 0.35993 | 0.36214 |
| **1.1** | 0.36433 | 0.36650 | 0.36864 | 0.37076 | 0.37286 | 0.37493 | 0.37698 | 0.37900 | 0.38100 | 0.38298 |
| **1.2** | 0.38493 | 0.38686 | 0.38877 | 0.39065 | 0.39251 | 0.39435 | 0.39617 | 0.39796 | 0.39973 | 0.40147 |
| **1.3** | 0.40320 | 0.40490 | 0.40658 | 0.40824 | 0.40988 | 0.41149 | 0.41308 | 0.41466 | 0.41621 | 0.41774 |
| **1.4** | 0.41924 | 0.42073 | 0.42220 | 0.42364 | 0.42507 | 0.42647 | 0.42785 | 0.42922 | 0.43056 | 0.43189 |
| **1.5** | 0.43319 | 0.43448 | 0.43574 | 0.43699 | 0.43822 | 0.43943 | 0.44062 | 0.44179 | 0.44295 | 0.44408 |
| **1.6** | 0.44520 | 0.44630 | 0.44738 | 0.44845 | 0.44950 | 0.45053 | 0.45154 | 0.45254 | 0.45352 | 0.45449 |
| **1.7** | 0.45543 | 0.45637 | 0.45728 | 0.45818 | 0.45907 | 0.45994 | 0.46080 | 0.46164 | 0.46246 | 0.46327 |
| **1.8** | 0.46407 | 0.46485 | 0.46562 | 0.46638 | 0.46712 | 0.46784 | 0.46856 | 0.46926 | 0.46995 | 0.47062 |
| **1.9** | 0.47128 | 0.47193 | 0.47257 | 0.47320 | 0.47381 | 0.47441 | 0.47500 | 0.47558 | 0.47615 | 0.47670 |
| **2.0** | 0.47725 | 0.47778 | 0.47831 | 0.47882 | 0.47932 | 0.47982 | 0.48030 | 0.48077 | 0.48124 | 0.48169 |
| **2.1** | 0.48214 | 0.48257 | 0.48300 | 0.48341 | 0.48382 | 0.48422 | 0.48461 | 0.48500 | 0.48537 | 0.48574 |
| **2.2** | 0.48610 | 0.48645 | 0.48679 | 0.48713 | 0.48745 | 0.48778 | 0.48809 | 0.48840 | 0.48870 | 0.48899 |
| **2.3** | 0.48928 | 0.48956 | 0.48983 | 0.49010 | 0.49036 | 0.49061 | 0.49086 | 0.49111 | 0.49134 | 0.49158 |
| **2.4** | 0.49180 | 0.49202 | 0.49224 | 0.49245 | 0.49266 | 0.49286 | 0.49305 | 0.49324 | 0.49343 | 0.49361 |
| **2.5** | 0.49379 | 0.49396 | 0.49413 | 0.49430 | 0.49446 | 0.49461 | 0.49477 | 0.49492 | 0.49506 | 0.49520 |
| **2.6** | 0.49534 | 0.49547 | 0.49560 | 0.49573 | 0.49585 | 0.49598 | 0.49609 | 0.49621 | 0.49632 | 0.49643 |
| **2.7** | 0.49653 | 0.49664 | 0.49674 | 0.49683 | 0.49693 | 0.49702 | 0.49711 | 0.49720 | 0.49728 | 0.49736 |
| **2.8** | 0.49744 | 0.49752 | 0.49760 | 0.49767 | 0.49774 | 0.49781 | 0.49788 | 0.49795 | 0.49801 | 0.49807 |
| **2.9** | 0.49813 | 0.49819 | 0.49825 | 0.49831 | 0.49836 | 0.49841 | 0.49846 | 0.49851 | 0.49856 | 0.49861 |
| **3.0** | 0.49865 | 0.49869 | 0.49874 | 0.49878 | 0.49882 | 0.49886 | 0.49889 | 0.49893 | 0.49896 | 0.49900 |

Table 2: Gives a probability that a statistic is less than Z (i.e. between negative infinity and Z).

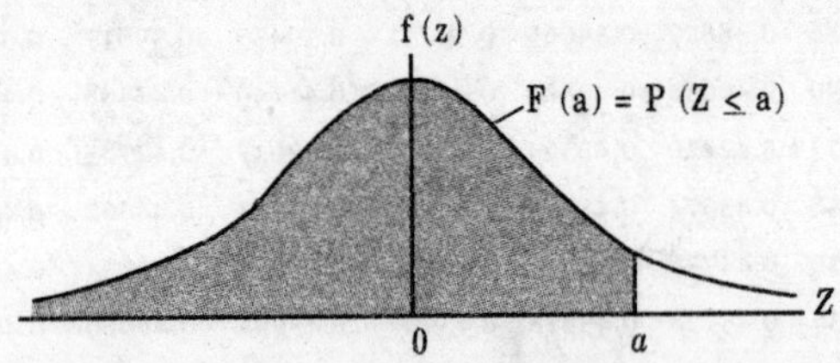


| To First Decimal | *Second Decimal* 0.00 | 0.01 | 0.02 | 0.03 | 0.04 | 0.05 | 0.06 | 0.07 | 0.08 | 0.09 |
|---|---|---|---|---|---|---|---|---|---|---|
| 0.0 | 0.50000 | 0.50399 | 0.50798 | 0.51197 | 0.51595 | 0.51994 | 0.52392 | 0.52790 | 0.53188 | 0.53586 |
| 0.1 | 0.53983 | 0.54380 | 0.54776 | 0.55172 | 0.55567 | 0.55966 | 0.56360 | 0.56749 | 0.57142 | 0.57535 |
| 0.2 | 0.57926 | 0.58317 | 0.58706 | 0.59095 | 0.59483 | 0.59871 | 0.60257 | 0.60642 | 0.61026 | 0.61409 |
| 0.3 | 0.61791 | 0.62172 | 0.62552 | 0.62930 | 0.63307 | 0.63683 | 0.64058 | 0.64431 | 0.64803 | 0.65173 |
| 0.4 | 0.65542 | 0.65910 | 0.66276 | 0.66640 | 0.67003 | 0.67364 | 0.67724 | 0.68082 | 0.68439 | 0.68793 |
| 0.5 | 0.69146 | 0.69497 | 0.69847 | 0.70194 | 0.70540 | 0.70884 | 0.71226 | 0.71566 | 0.71904 | 0.72240 |
| 0.6 | 0.72575 | 0.72907 | 0.73237 | 0.73565 | 0.73891 | 0.74215 | 0.74537 | 0.74857 | 0.75175 | 0.75490 |
| 0.7 | 0.75804 | 0.76115 | 0.76424 | 0.76730 | 0.77035 | 0.77337 | 0.77637 | 0.77935 | 0.78230 | 0.78524 |
| 0.8 | 0.78814 | 0.79103 | 0.79389 | 0.79673 | 0.79955 | 0.80234 | 0.80511 | 0.80785 | 0.81057 | 0.81327 |
| 0.9 | 0.81594 | 0.81859 | 0.82121 | 0.82381 | 0.82639 | 0.82894 | 0.83147 | 0.83398 | 0.83646 | 0.83891 |
| 1.0 | 0.84134 | 0.84375 | 0.84614 | 0.84849 | 0.85083 | 0.85314 | 0.85543 | 0.85769 | 0.85993 | 0.86214 |
| 1.1 | 0.86433 | 0.86650 | 0.86864 | 0.87076 | 0.87286 | 0.87493 | 0.87698 | 0.87900 | 0.88100 | 0.88298 |
| 1.2 | 0.88493 | 0.88686 | 0.88877 | 0.89065 | 0.89251 | 0.89435 | 0.89617 | 0.89796 | 0.89973 | 0.90147 |
| 1.3 | 0.90320 | 0.90490 | 0.90658 | 0.90824 | 0.90988 | 0.91149 | 0.91308 | 0.91466 | 0.91621 | 0.91774 |
| 1.4 | 0.91924 | 0.92073 | 0.92220 | 0.92364 | 0.92507 | 0.92647 | 0.92785 | 0.92922 | 0.93056 | 0.93189 |
| 1.5 | 0.93319 | 0.93448 | 0.93574 | 0.93699 | 0.93822 | 0.93943 | 0.94062 | 0.94179 | 0.94295 | 0.94408 |
| 1.6 | 0.94520 | 0.94630 | 0.94738 | 0.94845 | 0.94950 | 0.95053 | 0.95154 | 0.95254 | 0.95352 | 0.95449 |
| 1.7 | 0.95543 | 0.95637 | 0.95728 | 0.95818 | 0.95907 | 0.95994 | 0.96080 | 0.96164 | 0.96246 | 0.96327 |
| 1.8 | 0.96407 | 0.96485 | 0.96562 | 0.96638 | 0.96712 | 0.96784 | 0.96856 | 0.96926 | 0.96995 | 0.97062 |
| 1.9 | 0.97128 | 0.97193 | 0.97257 | 0.97320 | 0.97381 | 0.97441 | 0.97500 | 0.97558 | 0.97615 | 0.97670 |
| 2.0 | 0.97725 | 0.97778 | 0.97831 | 0.97882 | 0.97932 | 0.97982 | 0.98030 | 0.98077 | 0.98124 | 0.98169 |
| 2.1 | 0.98214 | 0.98257 | 0.98300 | 0.98341 | 0.98382 | 0.98422 | 0.98461 | 0.98500 | 0.98537 | 0.98574 |
| 2.2 | 0.98610 | 0.98645 | 0.98679 | 0.98713 | 0.98745 | 0.98778 | 0.98809 | 0.98840 | 0.98870 | 0.98899 |
| 2.3 | 0.98928 | 0.98956 | 0.98983 | 0.99010 | 0.99036 | 0.99061 | 0.99086 | 0.99111 | 0.99134 | 0.99158 |
| 2.4 | 0.99180 | 0.99202 | 0.99224 | 0.99245 | 0.99266 | 0.99286 | 0.99305 | 0.99324 | 0.99343 | 0.99361 |
| 2.5 | 0.99379 | 0.99396 | 0.99413 | 0.99430 | 0.99446 | 0.99461 | 0.99477 | 0.99492 | 0.99506 | 0.99520 |
| 2.6 | 0.99534 | 0.99547 | 0.99560 | 0.99573 | 0.99585 | 0.99598 | 0.99609 | 0.99621 | 0.99632 | 0.99643 |
| 2.7 | 0.99653 | 0.99664 | 0.99674 | 0.99683 | 0.99693 | 0.99702 | 0.99711 | 0.99720 | 0.99728 | 0.99736 |
| 2.8 | 0.99744 | 0.99752 | 0.99760 | 0.99767 | 0.99774 | 0.99781 | 0.99788 | 0.99795 | 0.99801 | 0.99807 |
| 2.9 | 0.99813 | 0.99819 | 0.99825 | 0.99831 | 0.99836 | 0.99841 | 0.99846 | 0.99851 | 0.99856 | 0.99861 |
| 3.0 | 0.99865 | 0.99869 | 0.99874 | 0.99878 | 0.99882 | 0.99886 | 0.99889 | 0.99893 | 0.99896 | 0.99900 |

# Index


A

Advantages of Duality, 73
Algebraic Method for (3×3)
or higher order matrix game, 275
Assignment Problem,314

B

Basic Feasible Solution, 33
Basic Solution, 33
Big M Method or Charne's
Penalty Method, 42

C

Canonical and Standard Forms of LPP, 35
Characteristics of Operation Research, 3
Classification of queueing models
Distribution of Arrival
and Service Time, 170
Column-Minima Method, 92
Common Errors in Network
Construction, 227
Corner point method, 16
Crashing of Netowrk
(Time-cost Relationship), 245
Critical path method (C.P.M.), 230
Customer Behaviour, 168

D

Degeneracy In Linear Programming
Problems, 49
Degeneracy in Transportation
Problems, 100
Dominance Rules, 264
Dual Simplex Method, 74

E

Economic Interpretation of Duality, 73
Enumeration Method, 127

F

Floats (Slack), 235
Formulation of LPP, 294
Game Theory, 331

G

Games without saddle point 262
Graphical Methods for 2 × n or m × 2
Games, 267
Graphical methods for solving linear
programming problems (2 variables), 14

H

History, 2
Hungarian Method, 127

## I

Important Definitions and Results, 32
Initial Basic Feasible Solution, 86
Input Sources, 166
Introduction,1, 126, 165, 204, 226, 258

## K

Kendall's Notation for
Representing Queueing Models, 169

## L

L.P. method for solving m×n
matrix game, 279
Limitations of Operation Research, 5

## M

Mathematical Formulation of LPP, 7
Matrix Method (Arithmetic method) or
Method of Oddment for (3×3) Game, 276
Matrix-Minima or Least-Cost Method, 90
Maximization Transportation
Problem, 104
Meaning and Definition of O.R., 1
Method of subgames for 2×n or m×2 game
not reducible by dominance rules, 271
Minimax and Maximin Principle, 259
Miscellaneous Assignment Problems, 141
Model II. (M/M/2) : (μ/FCFC) Multiservice
model (C=2), 185
Modi (Modified Distribution)
Method— Optimal Solution, 94
Multiple Optimal
(Alternative Optimal Solution), 39

## N

Non-Degenerate Basic Feasible Solution, 33
North-West Corner Rule, 87
North-West Corner Rule, 87

## P

PERT (Project Evaluation And Review
Technique), 237
PERT and CPM (Network Analysis), 327
Perturbation Technique, 49
Primal Into Dual, 306
Procedure to Convert Primal into
Dual LPP, 65

## Q

Queue Discipline, 167
Queueing cost structure, 165
Queueing Process, 167
Queueing System, 166
Queuing Theory 326

## R

Resolution of Degeneracy
During Initial Stage, 100
Resolution of Degeneracy
During Solution Stage, 102
Results on Duality
(Dual-Primal Relationship), 69
Row-Minima Method, 91
Rules for Network Construction, 227

## S

Scheduling Problem, 145
Scope of Operation Research, 4
Sequencing of 2 jobs on m machines
(Graphical Method), 215
Sequencing of n Jobs on m Machines, 210
Sequencing of n Jobs on two Machines, 205
Sequencing Problem, 321
Service Mechanism, 167
Simplex Algorithm , 42
Simplex Method (Algorithm), 35
Simplex Method, 127

Simplex method, 299
Simplex Table, 296
Slack, Surplus and Artificial Variables, 32
Solution of 2 × 2 Game
  Without Saddle Point, 262
Solution of Assignment Problem, 127
Solution of model–I [ (M/M/1):(μ/FCFS) ]
(Birth and death process), 170
State of the System,168
Steps to Resolve
  Tie in Entering Variables, 49
Steps to Resolve
  Tie in Leaving Variables, 49
Strategy, 258
Summary of Systematic Methods
  for Solving Rectangular Games, s261
Symbols and notations, 169

## T

Transportation Problem
  (North-West), 309
Transprotation Method, 127
Travelling Salesman Problem
  or Routing Problem, 141
Two Person Zero Sum Game, 258
Two-phase Method, 45
Types of Sequencing Problems, 205

## U

Unbalanced Transportation Problem, 102
Unbounded Solution, 41

## V

Vogel Approximation Method (VAM) or
Unit Cost Penalty Method, 92

## W

Why do We Study Operation Research, 6